CREEP

Edited by **Tomasz Tański, Marek Sroka**
and **Adam Zieliński**

Creep
http://dx.doi.org/10.5772/intechopen.68393
Edited by Tomasz Tański, Marek Sroka and Adam Zieliński

Contributors

Han-Yong Jeon, Dan Liu, Dirk John Pons, Zak Abdallah, Robert Joseph Lancaster, Spencer Jeffs, Tao Xu, Guanglei Zhou, Mark Whittaker, Veronica Gray, William Harrison, Vahid Monfared, Mieczyslaw Rękas, Krystyna Schneider, Rolf Sandström, Grzegorz Golański, Joanna Jasak, Cezary Kolan, Tomasz Arkadiusz Tański, Marek Sroka, Adam Zieliński

Published by InTech
Janeza Trdine 9, 51000 Rijeka, Croatia

Publishing Process Manager Martina Usljebrka
Technical Editor SPi Global
Cover InTech Design team

First published January, 2018
Printed in Croatia
Legal deposit, Croatia: National and University Library in Zagreb

Additional hard copies can be obtained from orders@intechopen.com

Creep, Edited by Tomasz Tański, Marek Sroka and Adam Zieliński
p. cm.
Print ISBN 978-953-51-3724-5
Online ISBN 978-953-51-3725-2

For EU product safety concerns:
IN TECH d.o.o., Prolaz Marije Krucifikse Kozulić 3, 51000 Rijeka, Croatia,
info@intechopen.com or visit our website at intechopen.com.

Meet the editors

Editor, Prof. Tomasz Tański is the head of the Institute of Engineering Materials and Biomaterials Silesian University of Technology; is a member of the Metallurgy Section of Committee of Metallurgy of the Polish Academy of Sciences; is a specialist in materials such as nonferrous alloys, composite and nanostructured materials, and manufacturing and surface engineering; and studied the properties and structures of engineering materials. He authored or coauthored more than 300 scientific publications worldwide including 12 monographs and books and more than 70 publications in the Philadelphia list; won 18 awards and honors nationally and internationally; and is and/or was a supervisor or contractor of more than 15 research and didactic projects in Poland and abroad and reviewer and promoter of numerous scientific papers, including 4 doctoral research in the field of nanotechnology and materials.

Co-Editor, Dr. Marek Sroka, PhD and MSc in Engineering, is an assistant professor in the Institute of Engineering Materials and Biomaterials at the Silesian University of Technology in Gliwice, Poland. During his scientific activity, he participated and organized many scientific international conferences. His scientific interests include materials science, materials for service at elevated temperatures, high-temperature creep resistance, the creep tests, and computer aid in material engineering. He is an author and coauthor of ca. 70 scientific publications worldwide including more than 15 publications in the Philadelphia list; he won 10 awards and honors, national and international; he is and/or was a contractor of more than 5 research and didactic projects in Poland and abroad; and a reviewer of numerous scientific publications.

Co-Editor, DSc. Adam Zieliński is a researcher at the Institute for Ferrous Metallurgy. He is an expert in the area of materials engineering. He is well versed in materials for service at elevated temperatures, in high-temperature creep resistance, in the creep tests, and in diagnostics of the high-pressure power, chemical, and petrochemical installations. He collaborates closely with the power boiler manufacturers, as well as the power industry repair plants in the area of the power installation diagnostics, residual life, and damage processes. He is an author and coauthor of ca. 135 scientific publications worldwide including more than 30 publications in the Philadelphia list; he won 10 awards. He is an author and coauthor of more than 400 researches and is an expert in the direct instructions of the energy industry and petrochemical industry.

Contents

Preface

This book contains 12 chapters with original and innovative research studies in the issues related to the broadly defined creep effect, which concerns not only the area of construction materials but also natural phenomena.

The technical progress observed for several years and a variety of interesting results of research, both with regard to the insight into the creep effect and methods for its analysis, have been enhancing the increase in requirements for the strength of new materials and development of new research methods. In particular, it concerns the construction elements and natural phenomena where creep effect that is generally understood as a process of deformation at a specific temperature takes place. Thus, creep is particularly important in many fields of science, such as power engineering, aviation, chemical industry, optical industry, earth sciences, etc., and the issues presented in the study are of great practical, and often utilitarian, significance.

The research carried out for the range of materials have revealed that creep has a significant impact in the event of an increase in the temperature of widely understood materials subjected to analysis. In steel elements, creep intensifies even at approx. 400 °C, while for materials with low melting temperature, the creep effect can already be observed at room temperature. Therefore, it can be concluded that the need to consider the creep effect in calculations depends on the material, temperature, time, and existing load.

This book aims to provide the readers, including, but not limited to, students and doctoral students and also the research personnel and engineers involved in the operation of equipment and structural components as well as specialists in high-temperature creep-resisting materials, with a comprehensive review of new trends in the field of creep-exposed materials and their research methodology. The chapters of this book were developed by respected and well-known researchers from different countries. We hope that after studying this book, you will have objective knowledge about new aspects in topic concerning creep phenomena.

Editor: Prof. Tomasz Tański
Silesian University of Technology, Poland

Co-editor: Dr. Marek Sroka
Silesian University of Technology, Poland

Co-editor: DSc. Adam Zieliński
Institute for Ferrous Metallurgy, Poland

Chapter 1

Introductory Chapter: Why Creep is Continuously Interesting for Science

Tomasz Tański, Marek Sroka and Adam Zieliński

Additional information is available at the end of the chapter

http://dx.doi.org/10.5772/intechopen.72495

1. Introduction

The following study presents issues related to the broadly defined creep effect, which concerns not only the area of construction materials but also natural phenomena.

The advantage of the book is a very extended set of both theoretical and practical examples of creep testing for various categories of analysis of this process, carried out by scientists of renown in this field from a number of scientific and research centres all over the world. As presented in the book, the emphasis on the discussion of a new trend of experimental creep testing, which binds the classic creep methods to seek the correlation of parameters obtained in tests, deserves particular attention.

The technical progress observed for several years and a variety of interesting results of research, both with regard to the insight into the creep effect and methods for its analysis, have been enhancing the increase in requirements for the strength of new materials and development of new research methods. In particular, it concerns the construction elements and natural phenomena where creep effect that is generally understood as a process of deformation at a specific temperature takes place. Thus, creep is particularly important in many fields of science, such as power engineering, aviation, chemical industry, optical industry, earth sciences, etc., and the issues presented in the study are of great practical, and often utilitarian, significance.

The basic concept of creep defines a process that takes place under constant load, which ensures that stresses are within the Hooke's law range, and at elevated temperature. Therefore, the applied load causes stresses and strains whose values vary not only as a result of overload but also when loading remains unchanged, resulting in stresses lower than the yield strength.

The research carried out for the range of materials has revealed that creep has a significant impact in the event of an increase in the temperature of widely understood materials subjected to analysis. In steel elements creep intensifies even at approx. 400°C, while for materials with low melting temperature, the creep effect can already be observed at room temperature. Therefore, it can be concluded that the need to consider the creep effect in calculations depends on the material, temperature, time and existing load.

To sum-up the following study, it should be stated that the creep tests presented in individual chapters make a significant contribution to solving the issue of developing destructive and non-destructive methods. This concerns not only the creep methods but also the tools to provide information on a failure mode at a specific point of creep progress. Undoubtedly, this brings us closer to development of a set of research methods to ensure that failures and catastrophes in different industries can be avoided.

2. Outline of this book

To outline details of each chapter of the book, the concise summary of each of them is given below.

Whittaker et al. presented a modern philosophy for creep lifing in engineering alloys. The traditional view of creep in materials science is based on the derivation of a power law relationship between stress and the creep rate in the so-called secondary phase. A transition in the value of the exponent of the relationship, n, is often assumed to be representative of a change from dislocation to diffusional creep processes as applied stress levels decrease. However, in operational environments for high-performance alloys such as power generation plant and gas turbine engines, there is little evidence of significant contribution to the overall creep rate from diffusional creep. Furthermore, power law-based approaches have been shown to be flawed in many engineering alloys due to high and unrealistic exponent and activation energy values along with the breakdown of the power law at high stresses. However, modern philosophy, known as the Wilshire equations, is built around the dominance of dislocation activity in creep strain accumulation and has shown significant potential.

Golański et al. presented a study on degradation of the microstructure and mechanical properties of high-chromium steels used in the power industry. High-chromium martensitic steels are one of the basic creep-resisting construction materials used for the modernisation of the old and construction of the new power units. During the service under creep conditions, the metastable microstructure of martensitic steels undergoes gradual degradation. The rate of degradation mostly depends on the temperature of work, but it is also affected by the stresses. The changes in the microstructure of martensitic steels have an influence on the decrease in their mechanical properties, including creep resistance. The knowledge and description of the changes in the microstructure of steels working in creep conditions allow extending the time of safe operation of the elements of power systems. The paper presents and describes the main mechanisms of degradation of martensitic steels of the 9–12%Cr type on the basis of the independent studies and literature data.

Rękas et al. presented an article on the subject of high-temperature creep of metal oxides. This chapter presents a comprehensive review of the creep technique used for the study of defect structure and diffusion in metal oxides, both single crystals and ceramics. At high temperatures, the creep rate is proportional to the diffusion coefficient of the slowest species in solid compounds, whatever deformation mechanisms are present (Nabarro viscous creep, recovery creep or pure climb creep). The creep rate dependence on deviation from stoichiometry can be determined from this diffusion. In the case of metal oxides, the departure from stoichiometry is controlled by the oxygen activity which usually is identified with oxygen partial pressure (pO_2). The pO_2 dependence of the creep rate provides direct information about the nature of minority point defects. On the other hand, studies of the temperature dependency of the creep rate inform us about the activation energy of the diffusion coefficient.

Han-Yong et al. presented the review of long-term durable creep performance of geosynthetics by constitutive equations of reduction factors. The exact nature of the time dependence of the mechanical properties of a polymer sample depends upon the type of stress or straining cycle employed. During creep loading, a constant stress is applied to the specimen at t=0, and the strain increases rapidly at first, slowing down over longer time periods. In an elastic solid, the strain stays constant with time. In this case, the strain is held constant, and the stress decays slowly with time. The increase in strain is not linear, and the curve becomes steeper with time and also as the stress rate is increased. If different constant strain rates are used, the variation of stress with time is not linear. The slope of the curve tends to decrease with time, but it is steeper for higher strain rates. The variation of both strain and stress with time is linear for constant stress and strain rate tests upon elastic materials.

Liu et al. presented a unified creep-fatigue equation with application to engineering design. The chapter reviews the existing creep-fatigue equations. It then describes the principles of the unified creep-fatigue equation. This equation is then applied to a variety of materials: SS316, SS304, Inconel 718, GP91 and AL2024-T3 at multiple temperatures and cyclic times. Next, we assess the economy of this theory by evaluating the life prediction error with overall experimental cost. This leads to the presentation of a simplified form for the unified formulation, which has minimal testing requirements for material characterisation, hence good usefulness to design. Finally, a case study for a gas turbine disk is given, showing how the equation may be applied to material properties in finite element analysis (FEA).

Abdallah presented creep lifing models and predictions. The chapter will help researchers to employ such models on the various materials and alloys under varying conditions of stresses and temperatures under creep. This chapter presents novel methods that are able to model, simulate and reconstruct full creep curves based on short-term measurements. The chapter also shows to explore the creep behaviour of materials that are being used in the gas turbine aero engine. This work will also help researchers to understand the theory behind the long-term creep behaviour and predictions. The models will provide a tool through which physically meaningful interpolation and extrapolation of the creep data can be obtained.

Lancaster et al. presented the creep method—small punch creep (SPC). A thorough characterisation of the creep properties of any modern alloy designed for a structural application can be an expensive and timely process. As such, significant effort is now being placed in identifying suitable alternative characterisation techniques. The small punch creep (SPC) test is now widely regarded as an effective tool for ranking and establishing the creep properties of a number of critical structural materials from numerous industrial sectors.

Over recent years, the SP creep test has become an attractive miniaturised mechanical test method ideally suited for situations where only a limited quantity of material is available for qualification testing. Typically, the method requires only a modest amount of material and can provide key mechanical property information for highly localised regions of critical components. As such, SP creep testing offers a feasible option for determining the creep properties of novel alloy variants still at the experimental stage and the residual life of service-exposed material.

Xu et al. in the chapter, "Thermo-Mechanical Time-Dependent Deformation and Fracturing of Brittle Rocks," proposed a thermo-mechanical numerical model to describe the time-dependent brittle deformation of brittle rocks under different constant temperatures and confining pressures. The mesoscale model accounts for material heterogeneity through a stochastic local failure stress field and local material degradation using an exponential material softening law.

Importantly, the model introduces the concept of a mesoscopic renormalisation to capture the cooperative interaction between microcracks in the transition from distributed to localised damage.

Monfared showed the review on creep analysis and solved the problem. Creep in solids subjected to high stress, and temperature is one of the important topics in the scientific societies: therefore, the creep analysis becomes more significant in various industries. So, the creep analysis is vital and important for applications connecting high temperature and high stress. Therefore, a thorough knowledge of creep characteristics and deformation mechanisms of reinforced and non-reinforced materials is required to utilise the materials in high-stress and high-temperature applications.

Harrison et al. presented an article on advanced methods for creep in engineering design. Novel creep deformation and damage methods are considered for creep-resistant alloys and compared to existing methods. Focusing on the three candidate models, the theta-projection technique, a true-stress model and a new method based on the Wilshire equations, the merits of alternative approaches to full creep curve representation have been considered. The evaluation of parameters for these models has been investigated with respect to the micromechanical phenomenon. These models have been implemented in the commercially available finite element analysis software, Abaqus, and in doing so, an evaluation of creep hardening models has been made. An alternative approach to creep hardening is presented which relates creep rate to material state variables instead of the traditional approach of time or strain hardening. The ability of these models to represent transient creep has been assessed by comparing predictions to experimental test results at nonconstant creep conditions.

Sandström showed the fundamental models for the creep of metals. Analysis of creep properties has traditionally been made with empirical methods involving a number of adjustable parameters. This makes it quite difficult to make predictions outside the range of the original data. In recent years, the author has formulated basic models for prediction of creep properties, covering dislocation, particle and solid solution hardening. These models do not use adjustable parameters. In the present chapter, these models are further developed and utilised.

The dislocation mobilities play an important role. The high-temperature climb mobility is extended to low temperatures by taking vacancies generated by plastic deformation into account. This new expression verifies the validity of the combined climb and glides mobility that has been used so far. By assuming that the glide rate is controlled by the climb of the jogs, a dislocation glide mobility is formulated.

The role of the mobilities is analysed, and various creep properties are derived. For example, secondary creep rates and strain versus time curves are computed and show good agreement with experimental data.

Author details

Tomasz Tański[1*], Marek Sroka[1] and Adam Zieliński[2]

*Address all correspondence to: tomasz.tanski@polsl.pl

1 Silesian University of Technology, Gliwice, Poland

2 Institute for Ferrous Metallurgy, Gliwice, Poland

Chapter 2

A Modern Philosophy for Creep Lifing in Engineering Alloys

Mark Whittaker, Veronica Gray and
William Harrison

Additional information is available at the end of the chapter

http://dx.doi.org/10.5772/intechopen.71829

Abstract

Lifing of components which are likely to be subject to high temperature creep deformation is a critical area to a range of industries, particularly power generation and aerospace. In particular, extrapolation of short term data to predict long-term allowable creep stresses is an area of significant importance, since no appropriate method of accelerating tests has been discovered. Traditional methods for extrapolation are mainly based around power law type equations that have historically formed the basis of creep mechanism understanding. The current chapter however, seeks to offer alternative approaches in the field, particularly emphasising the need to link lifing approaches to observable micro-mechanical behaviour.

Keywords: creep, Wilshire equations, dislocations, region splitting

1. Introduction

The following Chapter is dedicated to the late Brian 'George' Wilshire, a large, unforgettable and missed presence in the field of creep.

Creep is a time dependant process of deformation that affects materials operating under stress and temperature. This elongation of material which can occur at in-service stresses and temperatures means that design engineers needs to consider this dynamic process, otherwise risk catastrophic and possibly fatal failure. A specific case which requires consideration is the power generation industry which requires long design lives that are often extended to over 30 years or over 250,000 h [1]. This chapter discusses issues with reference to specific materials used in power generation and extends the approaches to the aerospace nickel superalloy Waspaloy. It should be noted the issues discussed here are relevant to all engineering materials that experience creep.

Initially defined as a field of study in the late 1940s via publication of the first test standards [2–5], creep and the methods used to predict it are a relatively new field of study. As such, forming the backbone of almost all methods to predict creep are power-law derived relationships that use the dependency of $\acute{\varepsilon}_m$ and t_f on σ, with equations of the form:

$$M/_{t_f} = \acute{\varepsilon}_m = A\sigma^n \exp\left(-Q_c/_{RT}\right) \tag{1}$$

Using this relationship the values of A, stress exponent n, and activation energy Q_c vary with test conditions. Examining this relationship it is comprised of three empirical components:

$$M = \acute{\varepsilon}_m \, t_f \tag{2}$$

$$\acute{\varepsilon}_m \, \alpha \, \sigma^n \tag{3}$$

$$\acute{\varepsilon}_m \, \alpha \, \exp\left(-Q_c/_{RT}\right) \tag{4}$$

These components are Eq. 2, which is the Monkman-Grant relationship, stating that the product of the minimum creep rate and time to failure equals a constant, M [6]. Eq. 3 is Norton's law which relates minimum creep rate to stress through the exponent n [7]. Lastly, Eq. 4 is the Arrhenius relationship which is an empirical description of reaction rates, particularly for diffusion in chemistry [8]. From these relationships we see M, n, and Q_c are defined as constants over a stress and temperature range associated with a creep mechanism. Changes in these values should reflect a change in creep mechanism and signal a new creep regime ultimately affecting the prediction made using Eq. 1. Furthermore, within this equation there is the ability to infer the creep mechanism. The activation energy, Q_c, described by the Arrhenius component provides a measure of energy within the system that can be linked with activation energies for mechanisms such as diffusion [8]. The power-law also has traditionally linked the value of n to mechanisms through some first principles derivation, but mostly is reliant on empirical evidence [7].

1.1. Power-law creep mechanisms

The power-law equations were derived to describe creep deformation behaviour based upon secondary, or steady-state creep being present in the creep curve. However, detailed inspection of normal creep strain/time curves show that a minimum creep rate, $\acute{\varepsilon}_m$, not a steady-state value is usually reached where the decaying primary rate is offset by the tertiary acceleration as seen in **Figure 1**. Rather than seeking to identify *'steady-state'* mechanisms, emphasis should therefore be directed to the deformation processes governing strain accumulation and the various damage mechanisms causing the creep rate to accelerate, leading to eventual fracture [9].

Power-law approaches are based on having n and Q_c characterise the creep mechanism which raises issues when applied to high performance materials such as those used in gas turbines and power generation. To begin with, it is necessary to consider the power-law in Eq. 1, where n and Q_c are derived from different empirical relationships to represent the same physical

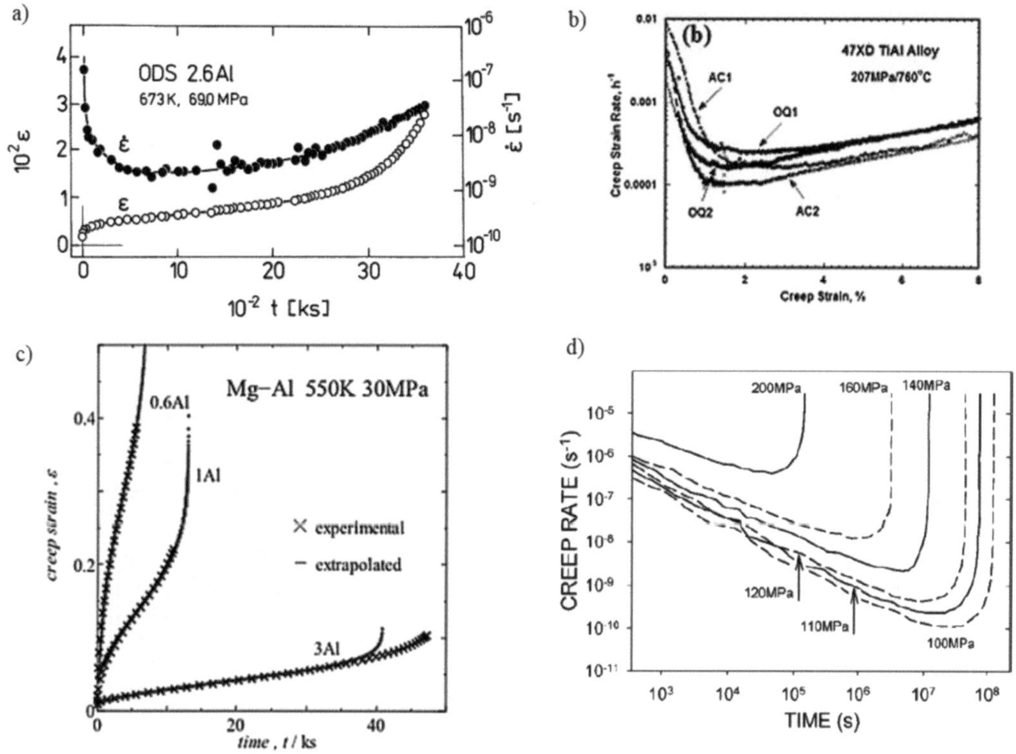

Figure 1. Examples of minimum creep rate, rather than secondary or steady-state creep for (a) aluminium alloy [10], (b) TiAl alloy [11], (c) MG-Al alloy [12], and (d) grade 91 steel [13].

process i.e. a creep mechanism. Activation energy or Q_c is a numerical value obtained from the application of the Arrhenius equation where the gradient of $\dot{\varepsilon}_m$ vs. $1/T$ for a given stress should yield a constant for the same creep mechanism. The value of Q_c is then also comparable to physically derived known quantities such as the activation energies of specific diffusion processes. The changing value of Q_c with regards to Eq. 1 should produce *'regions'* where expected creep life behaviour changes with creep mechanism.

Considering the stress exponent, *n*, its definition and application is even less physically reliable. The stress exponent *n* can be derived from first principles for very limited and specific cases such as those for ice derived in the 1950s [14–18]. From specific derivation a value of $n = 1$ is associated with diffusion whilst $n = 4–7$ is considered to indicate dislocation creep mechanisms [19]. The actual meaning of *n* is material dependant and often requires significant microstructural investigation and understanding. Beyond this, if n is an indicator of creep mechanism then it should act similarly to Q_c. In other words, *n* should remain constant for the stress-temperature range where a specific creep mechanism is dominant then transition to a different appropriate constant value when the mechanism changes. Indeed, *n* is only seen to remain constant over limited stress-temperature ranges then become non-constant, known as Power-Law Breakdown [19].

Values derived from Eq.1 should allow inference of the underlying creep mechanism. It is useful to compare studies of similar material datasets in order to consider the effectiveness of the technique and the consistency of its application:

- For Grade 122 steel, using the data provided by NIMS [20] several analyses have been conducted [21]. One analysis saw a decrease of $n \approx 16$ to $n \approx 5$ when the temperature of the tests increased from 823 to 1023°K with Q_c varying from 680 to 500 kJ/mol [22]. Another analysis [20, 23] for the same steel saw a high stress regime with $n \approx 60$ and $Q_c = 1045$ kJ/mol, and a low stress regime at $n \approx 5$ and $Q_c = 640$ kJ/mol.

- For Grade 91 steel, the stress exponent ranges from $n > 16$ to $n \approx 4$ with $Q_c = 400$–780 kJ/mol [24].

- Seen also in Grade 92, $n \approx 18$ drops to $n \approx 3$ and $Q_c = 540$–760 kJ/mol over the test range [25].

- In Grade 23 $n \approx 13$ at high stress and drops to $n \approx 6$ at low stresses whilst $Q_c = 300$–535 kJ/mol [26]

When considering these results, the combination of n and Q_c values do not produce consistency and indeed show themselves to highly dependent on their application, with the value of Q_c particularly showing a significant stress dependency. It is also extremely difficult to align the values gained out of the analyses of these typical engineering materials with the values proposed by traditional theory.

1.2. Power-law constants

One of the backbones of creating reliable modelling is reproducibility. In terms of n and Q_c it has already been shown that for the case of Grade 122, interpretation by those conducting the analysis can produce widely differing results. Considering the power-law based equations used to predict creep properties the use of a Time–Temperature Parameter (TTP) is often employed. The TTP is derived from $log(t_f)$ vs. $1/T$ (or T in some cases) where the gradient, and/or intercept are used to define the relationship.

For the Larson-Miller approach [27], a graph of $log(t_f)$ vs. $1/T$ is made where the gradient is the Larson-Miller Parameter, P_{LM}, and the intercept at $1/T$~0 is the Larson-Miller Constant, C_{LM}. These TTPs are then related to stress via:

$$P_{LM} = f(\sigma) = T\left(C_{LM} + log\left(t_f\right)\right) \quad (5)$$

This means to use this equation, values for P_{LM} are mapped against stress and a function chosen. A similar process is conducted for a variety of methods including Orr et al. [28], Manson and Succop [29], Manson and Haferd [30], and Goldhoff-Sherby [31] which each offer differences in curve shape. Additionally, these methods offer different approaches to activation energy:

- Larson-Miller: P_{LM} is equivalent to Q_c determined by mapping *log(t_f)* vs. *1/T* at constant stress. From Eq. 5, P_{LM} and therefore Q_c is expected to vary with stress in a continuous manner rather than step change with creep mechanism.
- Goldhoff-Sherby: PGS is also equivalent to Qc but requires the linear fit of *log(tf)* vs. *1/T* at constant stress to converge ensuring that Qc cannot be constant.
- Orr-Sherby-Dorn: COSD is equivalent Qc but is constrained to be constant.

A description of these methods, their implementation and efficacy can be found in [32, 33].

In considering these power-law derived methods that implement a TTP, there are issues ensuring reproducibility and physical accuracy. During the implementation of the aforementioned creep models, a parameter is mapped in terms of stress e.g. P_{LM} vs. σ. Consequently a function is chosen to describe this relationship. The definition of this function is not in a fixed form and therefore each implementation of the power-law based relationships is subject to *'user'* preference/choice. This is especially problematic when extrapolating as the functions chosen to represent stress dependency can produce unphysical results and therefore be extremely subjective and unreliable when extrapolating beyond the sampled test data.

1.3. Power-law summary

Power law based equations form the basis of almost all creep modelling approaches used today. Its underlying principles are a unification of the Monkman-Grant, Norton, and Arrhenius relationships into an empirical equation. As creep has advanced as a field and included more extreme conditions and high performance materials, the power-law has struggled to keep up with the array of mechanisms. From the power-law's application to steel, its inability to deal with these mechanisms through the combination of the activation and stress exponent terms, Q_c and n, is clear. Furthermore the application of power-law based models often requires users to choose the TTP-stress relationship introducing 'bespoke' modelling unique to the persons/group that have performed it. In identifying the weaknesses of the power-law, this chapter looks at the work of Wilshire and colleagues in their attempt to address some of these issues.

2. The Wilshire approach

The last 70 years of creep has been reliant on using power-law based analysis and stems from the widely held assumption that dislocation creep mechanisms are dominant at high stresses, with transition to diffusion based creep mechanisms at low temperatures. For pure aluminium, pure copper and the precipitate hardened Al7010, Wilshire and Whittaker [9], examined the creep curve beyond the *'steady-state'* to identify the dominant creep mechanisms.

For the precipitate hardened alloy Al7010, the role of diffusional creep in low stress regimes assumed by the power-law was challenged. For this material, precipitate-free zones sometimes found in the vicinity of grain boundaries in the low stress regime increased in width with increasing test duration, offering preferred locations for dislocation movement. This suggests the low stress behaviour of Al7010, and indeed other precipitate hardened materials, is not solely diffusional creep as assumed previously, thus leading to the work by Wilshire to develop a new approach to creep lifing. These mechanisms should be therefore supported and reflected in the value of the activation energy which can be derived either from the time to failure, t_f, or the minimum creep rate, $\acute{\varepsilon}_m$ with respect to *1/T*.

Considering the power-law approach to creep modelling a number of issues have been highlighted which need to be addressed. Seeing the issues facing the field of creep and its inability to unite results with microstructural processes, Wilshire[1] and colleagues proposed an alternative to the power-law.

2.1. An alternative to power-law based equations

Utilising the power-law and its Arrhenius basis, the activation energy is calculated at constant stress. For a creep mechanism to occur it requires a certain amount of energy known as activation energy. For a creep test, kinetic energy is created by applying a stress and thermal energy by applying temperature. This energy is used by the system to activate creep mechanisms. From the Arrhenius or power-law approach it assumes that the dominant energy process or mechanism is determined by the absolute kinetic energy and therefore measures activation energy at constant stress.

Unlike a chemical or gaseous system, a solid has limitations where the maximum thermal energy is $0\text{-}T_m$°K and the maximum kinetic energy of the system is $0\text{-}\sigma_{UTS}$ MPa.[2] Using this idea, the temperature dependant UTS provides the upper limit of a solid system in terms of both kinetic and thermal energy. To implement this idea the stress is normalised such that:

$$M/t_f = \acute{\varepsilon}_m = A^* \left(\frac{\sigma}{\sigma_{UTS}} \right)^n exp\left(-Q_c^*/RT\right) \tag{6}$$

It should be noted in Eq. 6 that $A \neq A^*$, and $Q_c \neq Q_c^*$. In chemistry terms, both Q_c and Q_c^* are apparent activation energies rather than true activation energies. By definition true activation energy is a chemical or thermal reaction rate derived from first principles for a mechanism [34]. Apparent activation energy is the reaction rate that is dependent on a time to failure or a cumulative reaction process such as $\acute{\varepsilon}_m$. The minimum creep rate, $\acute{\varepsilon}_m$, is often used instead of t_f due to it being less sensitive to failure mechanisms and anomalies, but it is still is a time dependant cumulative value and as such produces an apparent activation energy.

---

[1]Wilshire is used as an abbreviation for Wilshire and colleagues. We recognise research is collaborative, but for simplicity we use Wilshire to represent all those involved in developing this work. Collaborators can be found in the references at the end of the chapter.

[2]When referred to in this work, the Ultimate Tensile Stress, UTS or σ_{UTS} is always temperature dependant.

The existing literature often uses minimum creep rate, $\dot{\varepsilon}_m$, and secondary creep rate, $\dot{\varepsilon}_s$, interchangeably, however as previously shown in **Figure 1** a secondary phase does not exist for many materials under a range of conditions and is not reliably observed in test data [35]. Therefore rather than seeking to identify *'steady-state'* creep mechanisms, the focus of the Wilshire approach is directed at the deformation processes that define strain accumulation, and the damage mechanisms that accumulate and lead to eventual failure [9]. This allows a minimum creep rate to be defined as a decaying primary rate is offset by tertiary acceleration.

From Eq. 6, the normalisation of the stress means that the normalised activation energy, Q_c*, is evaluated at on a relative scale. This reflects the physical limitation of the system such that the material at its UTS will undergo failure no matter the specific amount of stress or temperature applied. The approach promotes the idea that a solid system has a maximum energy state, and that systems in the same relative energy should be experiencing the same mechanism. To evaluate Q_c*, the gradient of $log(t_f)$ vs. $1/T$ is evaluated at constant normalised stress i.e. constant σ/σ_{UTS}, or, using regression analysis fitting Q_c* from $ln(-ln(\sigma/\sigma_{UTS}))$ vs. $ln(t_f\ exp.(-Q_c*/RT))$. It should be noted that some initial investigations using this method normalised the test stress by the temperature dependant yield stress rather than the UTS. This normalisation focused on the elastic limit of the material as it is a defined point of mechanism change. If σ_Y/σ_{UTS} remains fairly constant with temperature then activation energy values will be equivalent to conducting normalisation by the UTS.

The goal of creep modelling is to create a relationship where the apparent activation energy is equivalent to the true activation energy of the dominant creep mechanism. From the previous section it can be seen that for the power-law approach, the apparent activation energy ranging up to 1045 kJ/mol does not correlate with the true activation energy of any known creep mechanism. Using a normalised stress approach to calculate Q_c*, Wilshire found for Grade 91, 92 and 122 that for all these steels $Q_c* \approx 300$ kJ/mol, equivalent to lattice diffusion for the material [35].

2.2. The Wilshire equations

Although normalising creep test stress has shown to provide more reasonable Q_c* values, as well as a more consistent relationship between stress and time to failure, it still contains the problematic stress exponent, n. Indeed, Eq. 6 still suffers from the vulnerabilities outlined in §1.1 and 1.2. To deal with n and power-law relationships, Wilshire developed a series of equations:

$$\sigma/\sigma_{UTS} = exp\left\{-k_u\left[t_f\cdot exp\left(-Q_c^*/RT\right)\right]^u\right\} \tag{7}$$

$$\sigma/\sigma_{UTS} = exp\left\{-k_v\left[\dot{\varepsilon}_m\cdot exp\left(-Q_c^*/RT\right)\right]^v\right\} \tag{8}$$

$$\sigma/\sigma_{UTS} = exp\left\{-k_w\left[t_\varepsilon\cdot exp\left(-Q_c^*/RT\right)\right]^w\right\} \tag{9}$$

These equations use the apparent activation energy Q_c*, evaluated at constant σ/σ_{UTS}. The constants k_u, u, k_v, and v are evaluated at $ln(-ln(\sigma/\sigma_{UTS}))$ vs. $ln(t_f.exp.(-Q_c*/RT))$, and $ln(-ln(\sigma/\sigma UTS))$ vs. $ln(\dot{\varepsilon}_m.exp.(-Q_c*/RT))$ respectively. Unlike the power-law, constants in Eqs. 7 and 8 are indeed constant over each creep region and therefore do not require the user to choose relationships between variables. In contrast, Eq. 9 is designed to predict tims to specific strains and can potentially be utilised to model the whole creep curve where k_w and w are functions of strains. This however is not discussed further here as full creep curve modelling is discussed in a separate chapter.

Another important aspect of the Wilshire Equations is incorporation of basic physical limits. As $\acute{\varepsilon}_m \to 0$, then $t_f \to \infty$ when $\sigma/\sigma_{UTS} \to 0$. Also, as $\acute{\varepsilon}_m \to \infty$, then $t_f \to 0$ when $\sigma/\sigma_{UTS} \to 1$. This minimum physical limitation is inherent in the Wilshire equations but not so in the power-law meaning it is possible for the power-law to produce unphysical results.

3. Application of the Wilshire equations

3.1. Yield behaviour of polycrystalline copper

To demonstrate the application of the Wilshire Equations with respect to the power-law, the example of polycrystalline pure copper is first considered [9, 36].

Samples were machined from cold-drawn 10 mm rods of oxygen-free high-conductivity copper i.e. 99.95% wt. Cu. Test pieces had a gauge length of 25 mm and diameter of 4 mm. In order to obtain a grain size of ~40 μm, the test pieces were annealed under a vacuum of 10^{-4} Pa for

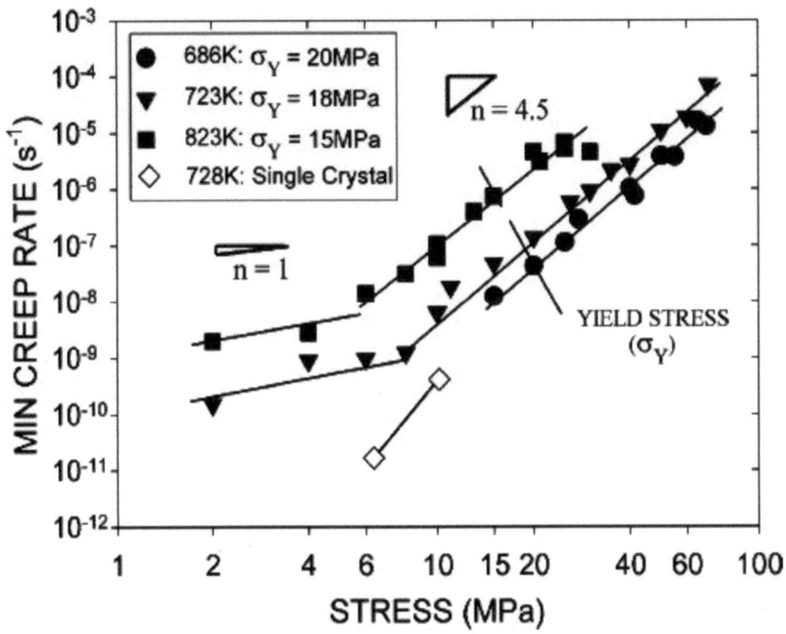

Figure 2. Power-law analysis of polycrystalline copper with n = 1, 4.5 [36].

http://dx.doi.org/10.5772/intechopen.71829

3600s at 973°K. To ensure no inhomogeneous grain growth, tests were conducted at 686, 723 and 823°K, markedly below the annealing temperature. Upon investigation, no dynamic recrystallization or grain growth was observed under the test stress-temperature combinations used. Tests were conducted on 10:1 constant stress Andrade-Chalmers cam creep machines complying with BS EN 10291:2000.

Conducting power law analysis using the minimum creep rate, $\dot{\varepsilon}_m$, two distinct creep regimes can be observed in **Figure 2** with $n = 1$ and $n = 4.5$ and $Q_c = 110$ kJ/mol. From a classical power-law interpretation, the change in n is consistent with a change from diffusion to dislocation based creep at ≤0.61T_m and aligns with understood microstructural evolution [37–39]. Immediately from this analysis it was noted that the value of the constant M from Eq. 2 with longer tests and higher temperatures reduces from ~0.07 to ~0.02.

If the normalised power-law in Eq. 6 is applied using an activation energy of 110 kJ/mol, then the relationship between normalised stress, σ/σ_Y, and $t_f.exp(Q_c/RT)$ is approximately linear as seen in **Figure 3**.

Classically the yield strength denotes the transition from elastic to plastic behaviour and as such a change in material behaviour either side would be expected. From power-law analysis there is no indication that creep behaviour changes in the higher stress regime. Looking at the test data in **Figure 4**, the shape of the creep curve changes as test conditions go from above to below yield, σ_Y. With higher temperatures and lower stresses, less primary creep was seen, whilst more tertiary creep is observed, also noting the lack of steady-state creep. Also, with

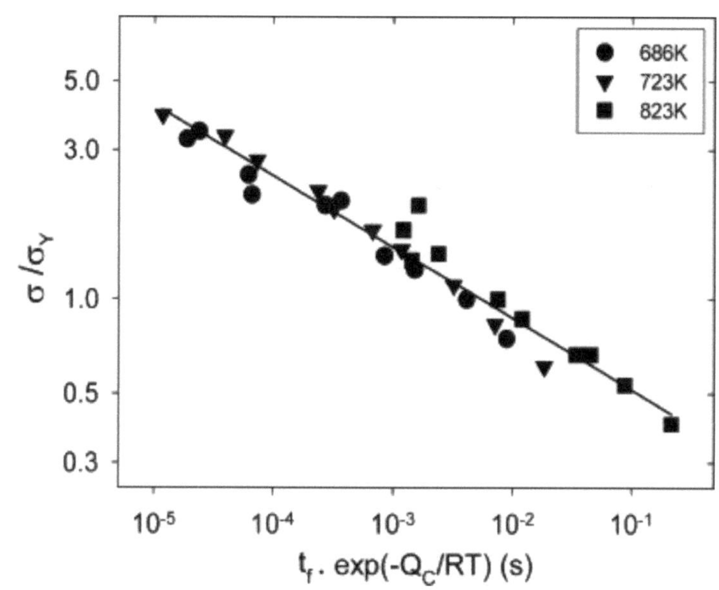

Figure 3. Normalised power-law analysis of polycrystalline copper showing linear trend between normalised stress and time to failure with $Qc = 110$ kJ/mol [36].

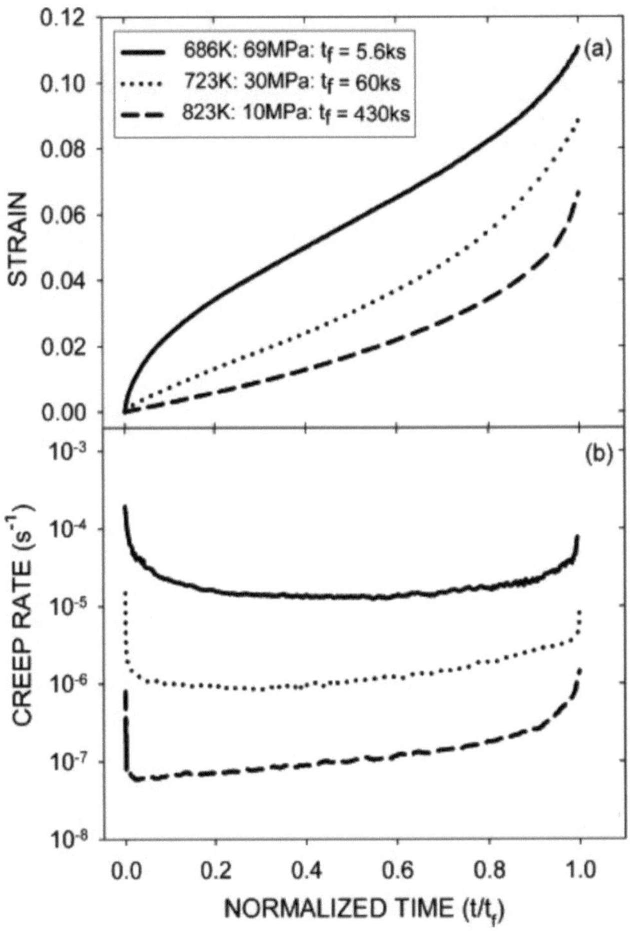

Figure 4. Changing creep curve shape of polycrystalline copper (a) strain vs. normalised time, and (b) creep rate vs. normalised time [36].

increasing test stress the point at which the minimum creep rate reached occurs later in the test. This supports the Wilshire approach of considering creep in terms of primary and tertiary creep rather than steady-state as polycrystalline creep curves transition from primary-dominated to tertiary-dominated curves with increasing temperature and test duration. Furthermore, it suggests in conjunction with other evidence [36], that power-law changes in n are not related to creep mechanism transitions, but rather the complex relationship between changes in $\dot{\varepsilon}_m$ with changing test conditions.

Knowing there is a change in the shape of creep curves as test conditions pass over the yield stress, the equations used to model the data should reflect this transition. From the power-law and normalised power-law we see no indication of a different creep regime around the yield stress. However, by use of the Wilshire approach, the normalised activation energy

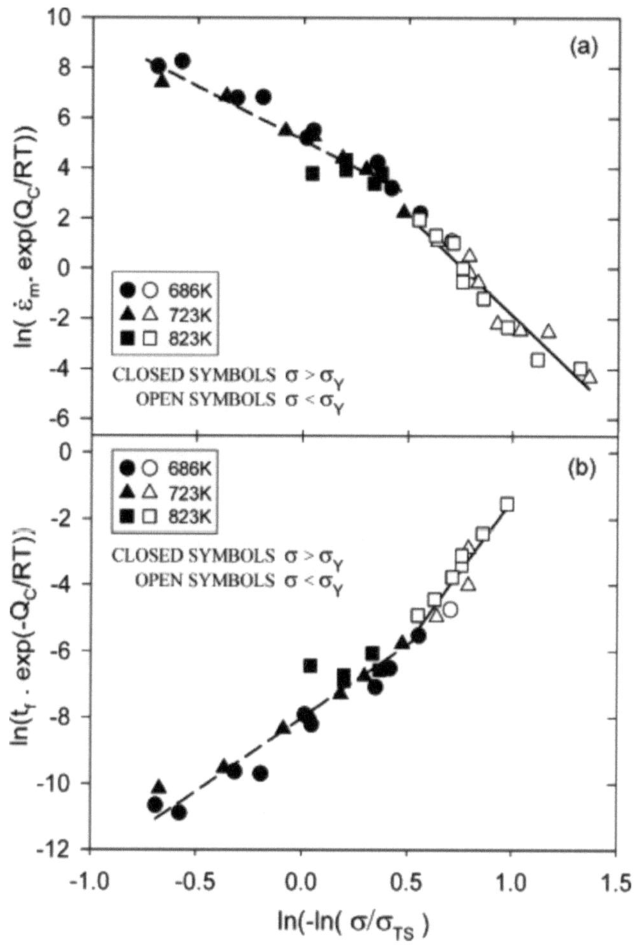

Figure 5. Application of the Wilshire equations to polycrystalline copper demonstrating two regions, above and below yield [36].

was found to be $Q_c^* \approx 110$ kJ/mol which coincides with the associated value for grain boundary diffusion. This mechanism is confirmed by microstructural observations made in Ref. [36]. Implementing Eq. 8, the Wilshire approach shows two linear regions in **Figure 5** when determining k_v and v. Notably results from above yield tests lie in one regime and below yield in another.

In identifying two regimes, the Wilshire Equations are implemented in two regimes, above and below the yield stress. The results are observed in **Figure 6** where the Wilshire approach in comparison to the power-law produces a predictive method that reflects changes in creep mechanism at the yield stress. The power-law on the other hand suffers from changing n and inevitably power-law breakdown without capturing the change in behaviour that occurs at yield.

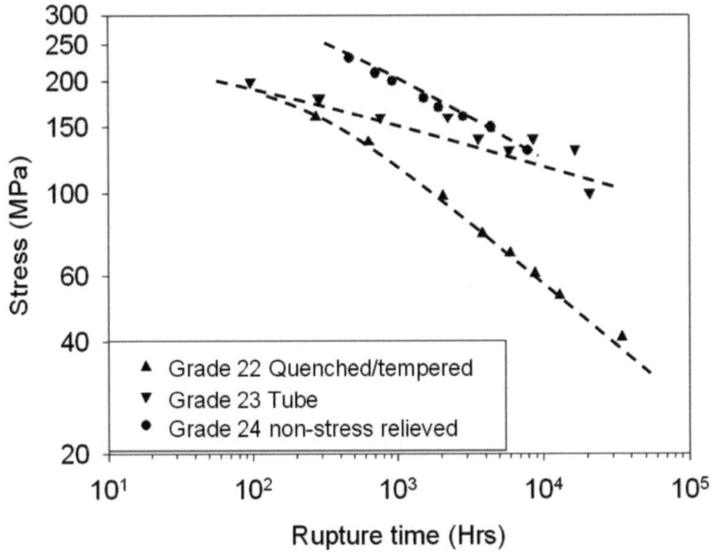

Figure 6. The stress dependence of the creep life at 873 K (600°C) for Gr. 22, 23, and 24 steels [47].

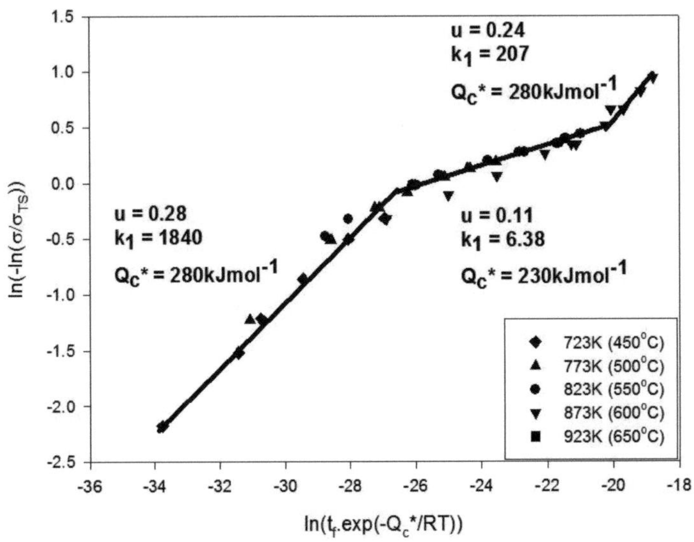

Figure 7. Adopting Eq. 7, k_u, u and Q_c* values are determined by plotting of $ln[t_f.exp.(-Q_c^*/RT)]$ vs. $ln[-ln(\sigma/\sigma_{UTS})]$, for Gr.22 tube [47].

3.2. Long-term prediction of 9–12% steels

Creep is a significant factor in modern engineering, especially for the power generation industry. Having a design life of over 25 years, power generation relies on extrapolating creep test data rather than performing costly long-term tests. The need to be able to predict long-term creep behaviour from short term tests is therefore crucial for any creep modelling method to be successful [40].

To design for such a long working life under often non-steady stress and temperature, the benchmark used is nominally the 'allowable tensile creep strengths' [41]. These strengths are 0.67 or 0.8 of the minimum stress causing rupture in 100,000 h or, the stress that results in a creep rate of 0.01% per 1000 h i.e. ~3×10^{-11} s^{-1}. To conduct tests to 100,000 h or ~11 years is unfeasible and therefore short term test data is extrapolated to these values. One group of materials particularly used in this industry is 9–12% Cr steels or Grade 91, 92 and 122. Having undergone these long-term tests, this dataset provided by NIMS [42–44] allows the long-term extrapolation capability of a creep model to be evaluated.

Using the normalised power-law approach of Eq. 6, predicting long-term behaviour of 9–12% Cr steels is problematic. Overall for Grade 91, 92 and 122 the constant M, increases with longer tests. Specifically for Grade 122, Q_c goes from ~500 to 680 kJ/mol with decreasing stress and increasing temperature. Over the same range n falls from ~16 to ~5. These ranges of values are common for high stress tests on alloys strengthened by fine precipitates or insoluble particles [45]. Due to this inconsistency, predicting long-term creep behaviour using the normalised power-law method was abandoned for a parametric power-law approach such as those detailed in §1.2. Due to the problems with these methods discussed in §1.2, these methods are unable to extrapolate long-term data with sufficient accuracy to 100,000 h from 30,000 h tests [46].

It is therefore appropriate to apply the Wilshire Equations to evaluate its ability to predict long-term creep behaviour from short term creep tests. Normalising by σ_{UTS} the activation energy for Grade 91, 92 and 122 was found to be 300 kJ/mol for these martensitic steels. This activation energy is close to that of lattice diffusion in the alloy steel [13]. To test the long-term extrapolation capability of the Wilshire Equations, data from <5000 h and <30,000 h was used. Using Q_c^* = 300 kJ/mol, it was observed that for Grade 92 and 122 there were 2 regions similar to polycrystalline copper, i.e. two linear regimes with different values of k_u and u. Only one region was observed for Grade 91. This can be corroborated with micro-structural behaviour. When examining the Reduction of Area (RoA) with regards to $\dot{\varepsilon}_m$, it was noticed that Grade 92 and 122 displays behaviour consistent with longer duration and higher temperature tests having increasing boundary cavities and thus experiencing a

Temperature	Grade 91	Grade 92	Grade 122
823°K (550°C)	154 (149)	182 (178)	196 (199)
848°K (575°C)	117 (112)	140 (137)	151 (155)
873°K (600°C)	87 (82)	104 (103)	113 (116)
898°K (625°C)	62 (59)	76 (75)	81 (84)
923°K (650°C)	43 (39)	53 (53)	53 (58)

Table 1. Creep rupture strengths (100,000 h) predictions for Grade 91, 92 and 122 from t_f < 30,000 h, and t_f < 5000h (in brackets), MPa [47].

Analysis method	823°K (550°C)	873°K (600°C)	923°K (650°C)
ECCC	166	94	49
Wilshire t_f<30,000 h	154	87	43
Wilshire t_f<5000 h	149	82	39
Tenaris	153	86	46
ECCC-MRM	152	86	44
ECCC-MC	149	84	44
ECCC-LM	157	94	51
ECCC-LM (ed)	150	98	42

Details can be found in [47].

Table 2. Creep rupture strength predictions for a number of methods for comparison.

transition from ductile transgranular to brittle intergranular failure. This was not observed for Grade 91.

The values of k_u and u showed a small variation between the values from tests <30,000 h and those <5000 h. In applying this 2 region approach, it yields creep rupture strengths listed in **Table 1** where it can be seen the extrapolation of creep data from <5000 h tests using the Wilshire approach is almost equivalent to the extrapolations from data <30,000 h.

From **Table 2**, in comparison to other employed extrapolation methods by NIMS, ECCC etc. that use datasets up to <70,000 h, the Wilshire Equations show a remarkable improvement in extrapolation capability. Furthermore, the Wilshire Equations, unlike the other employed methods, reflects the observed change in material ductility thus demonstrating its capability to reflect microstructural and mechanism changes.

Through wider application, the method of implementing the Wilshire equations has evolved especially as demonstrated when applied to the 2.25Cr series of alloys, Grade 22, 23 and 24 [47]. These steels provide a test to the Wilshire equations as Grade 22 (2.25Cr-1Mo) steel has long been utilised and has a wealth of results thanks to NIMS providing both short and long-term data. Furthermore, the development of Grade 23 (2.25Cr-1.6 W) and particularly Grade 24 (2.25Cr-1Mo-0.3 V) steel offers the potential to implement an improved material, if confidence can be provided in predictive methods such as the Wilshire equations.

Initial application of the Wilshire Equations proved difficult, since rather than the two distinct regions seen in previous alloys described above, a third region was observed for longer duration/high temperature tests [47]. Evaluation of test pieces revealed that this behavioural change was most likely due to the degradation of the bainitic microstructure in the material, with the high temperature and long exposure leading to an overaged ferritic microstructure with coarse molybdenum carbide particles also present. In order to provide optimised fits to each of the three materials, the value of the activation energy was allowed to vary in each of the three regions seen in **Figure 6**. With surprising consistency the Q_c* value was found to be approximately 280 kJ/mol for tests conducted above the initial yield stress of the material, 230 kJ/mol for the intermediate region, and, 280 kJ/mol for low stress/higher temperature tests [47].

The three regions of seemingly complex behaviour can be rationalised in a manner consistent with the observed microstructural behaviour, as well as elongation and reduction in area measurements. Above the yield stress, creep takes place through the generation and movement of new dislocations, formed at appropriate sources, due to the fact that the yield stress of the material is exceeded. When the applied stress falls below the yield stress, the creep lives become longer than that expected from direct extrapolation of results when $\sigma > \sigma_Y$. This is because grain deformation is restricted, thus creep occurs only by grain boundary zone deformation when $\sigma < \sigma_Y$. A second break occurs in **Figure 6** which is a consequence of the bainite regions in the initial ferrite/bainite microstructure degrading to ferrite and molybdenum carbide particles in long-term tests with very coarse carbides along the grain boundaries. As a result, the creep rates are faster, and the creep lives are significantly shorter than expected by extrapolation of the intermediate stress data when the bainitic microstructures are present [48].

Subsequent application of this varying activation energy was found to improve fits in a range of other materials [49, 50], whilst remaining consistent with the observed micromechanical behaviour.

3.3. Mechanisms of Waspaloy

In an initial investigation of Al7010, Wilshire observed that power-law equations struggle to deal with precipitation strengthened materials [9] due to their inherit assumption of diffusional creep. As such a study was carried out on the nickel superalloy Waspaloy. Details of the material and creep experiments performed can be found in Whittaker et al. [50]. Constant stress creep tests were performed from 550 to 800°C and 140–1150 MPa. From power-law analysis the activation energy at high stresses is ~350 kJ/mol but increases to 700 kJ/mol at low stress with the threshold between high and low stress occurring at ~600 MPa.

Applying the Wilshire Equations, two values of activation energy were observed. For high stress tests, $Q_c^* = 400$ kJ/mol with $k_v = 434$ and $v = -0.24$. For low stress tests $Q_c^* = 340$ kJ/mol with $k_v = 72$ and $v = -0.14$, noting the threshold between the regimes determined from fitting approximately coincides with the yield stress. As with the previous examples, when the creep curve is examined above and below this threshold (yield), the shape of the creep curve goes from primary-dominated to tertiary-dominated (see **Figure 8**).

Rather than relying solely upon creep curve shape, further investigation was conducted in order to verify the creep mechanisms. To do so, a series of creep tests was undertaken at graded levels of stress both above and below the yield stress at a temperature of 700°C. From the starting material, transmission electron microscopy (TEM) in **Figure 9** shows a bimodal distribution of γ' precipitates, with the secondary precipitates approximately 200 nm in

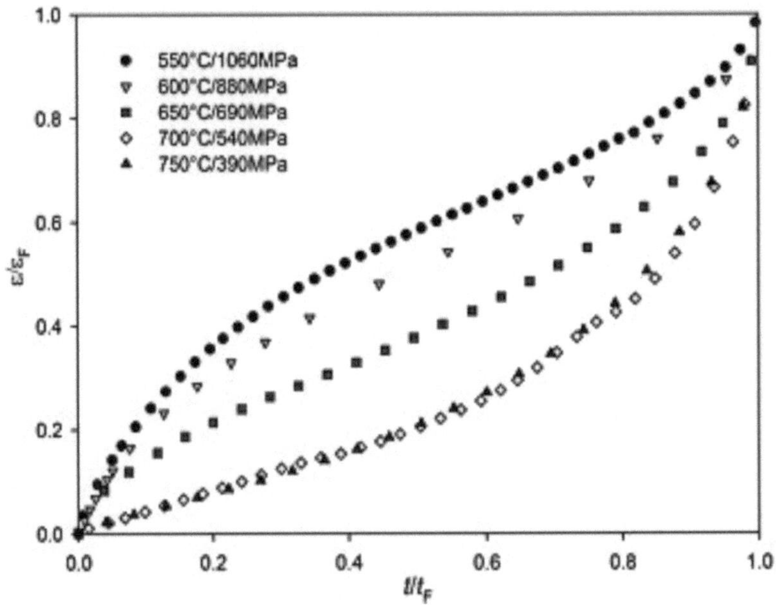

Figure 8. Changing creep curve shape of Waspaloy plotted on a normalised scale [50].

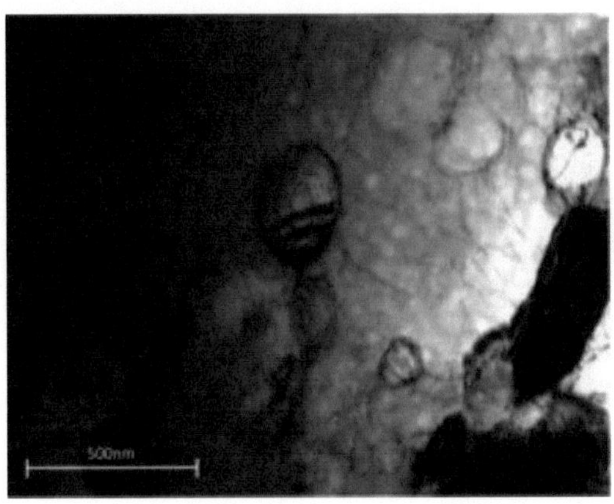

Figure 9. TEM of Waspaloy microstructure in original, un-tested state [50].

diameter and the tertiary of the order of 50 nm diameter. In the γ matrix, dislocations are present in moderate numbers in loose networks that avoid passing through the secondary precipitates. On the grain boundaries some carbides were visible and additionally there was evidence of strain recovery in a series of sub-grains with low angle boundaries.

In order to separate the mechanisms of plasticity from creep, tensile tests to ~0.2% proof stress (740 MPa) at 700°C were examined by TEM, **Figure 10**. It was found that the dislocation

Figure 10. TEM of Waspaloy microstructure after tensile testing [50].

density had increased with most single dislocations cutting through the tertiary γ', and other dislocations looping around the secondary precipitates. The low angle sub-grains in the tensile sample were the same as those observed in the virgin material.

To investigate creep mechanisms, creep tests were performed at 500, 600, 700 and 800 MPa at 700°C and interrupted once the minimum creep rate was reached. The test pieces were air cooled under reduced load and then prepared for TEM. In **Figure 11a**, at 500 MPa there is a low dislocation density where the dislocations Orowan loop around the secondary γ', but also the tertiary γ' which was observed to have the occasional stacking fault. The sub-grains at the boundaries also exhibited this dislocation behaviour.

For specimens crept at 600 MPa and 700 MPa there is a change in behaviour to that seen at 500 MPa. In **Figure 10b** for 600 MPa, noting a reduction in Orowan looping, the dislocation density has increased and stacking faults in the secondary and tertiary γ' are evident as precipitates are cut by superlattice partial dislocations. From these stacking faults, slip is occurring on more than one slip plane. Although the dislocation spacing between precipitates is increased compared to 500 MPa, it is still below the tertiary precipitate spacing thus indicating the main impediment to slip is the tertiary precipitates. Looking at the grain boundaries, low angle sub-grain are still present but now have a sub-structure which is a combination of dislocation recovery and strain hardening processes.

At above yield conditions of 800 MPa the dislocation density of the material is extremely high to the point where individual dislocations are difficult to identify. This is seen in **Figure 11c**, where the dislocations are extremely dense but mostly confined to the γ and the tertiary γ' with occasional stacking faults in the secondary γ'. The larger precipitates even in above yield conditions, continue to resist cutting by lattice dislocations.

From this investigation it can be seen that there is a change in dislocation behaviour above and below yield. Below the yield stress, dislocation densities are relatively low and increase

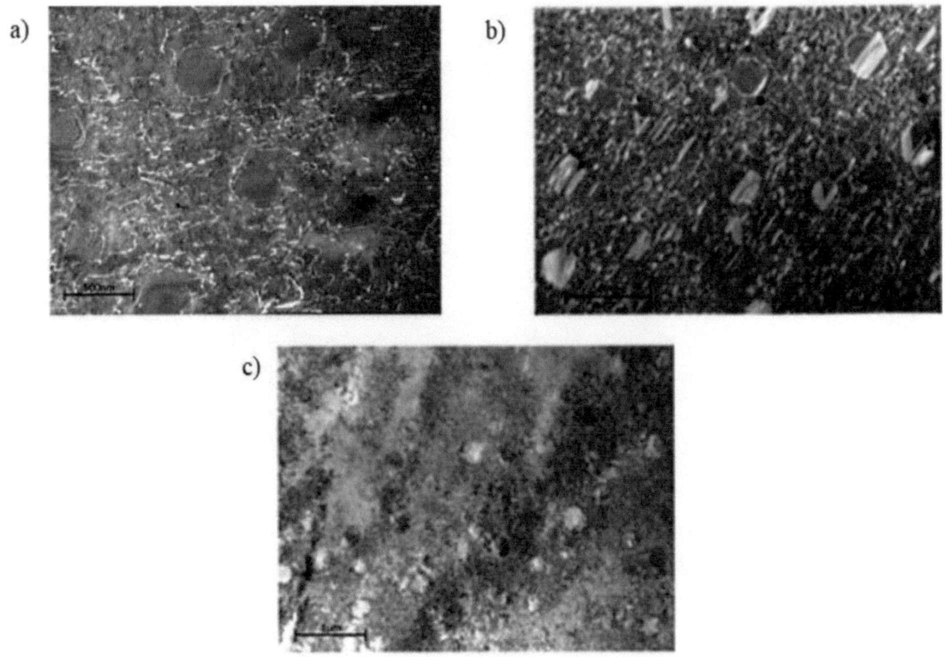

Figure 11. TEM of Waspaloy microstructure after creep at 700°C at (a) 500 MPa, (b) 600 MPa, and (c) 800 MPa [50].

slightly with stress with the sub-grains at the boundaries fairly unaffected. Above the yield point, dislocation density increases to the point where the dislocation spacing is closer than that of the precipitates and the sub-grain structure around the grain boundaries is obliterated by the increase in dislocations throughout. This is interpreted as below yield creep being controlled by dislocation climb around tertiary γ' precipitates. Above yield creep is controlled by the climb and recovery of dislocation tangles. When considered in light of the Wilshire Equations, the change in activation energy at approximately yield from 340 to 400 kJ/mol reflects this change in mechanism.

4. Region splitting

Having a method of analysing creep data that reflects actual microstructural or mechanism change is crucial to advance the field of creep. Region splitting has long been considered but only recently become wider spread in creep modelling [51]:

> *'According to the concept of the deformation and fracture mechanisms map we should accept the assumption that creep rupture data attained within the domain where creep or creep fracture is governed by one dominant mechanism cannot be used for prediction to another domain, where creep and fracture are controlled by different dominant mechanisms. As soon as*

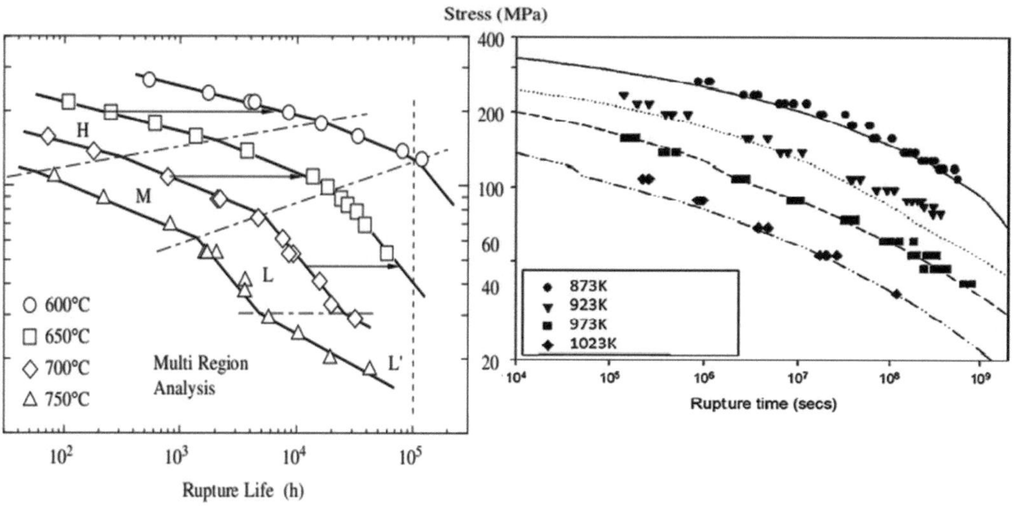

Figure 12. Region splitting by Maruyama et al. for 316 stainless steel [53] (left), and, Wilshire et al. for 316H stainless steel [49] (right).

the condition is changed and the boundary is crossed into another domain, the prediction becomes unreliable' [52].

Therefore region splitting is the segmenting of creep data according to dominant creep mechanism.

For power-law models of creep, region splitting has two major implementations. The first method of implementing region splitting is to fit a model and use linear regression to define regions. For 316 steel shown in **Figure 12**, it can be seen that this approach implemented by Maruyama et al. [53] results in multiple regions not associated with specific microstructural mechanism changes. The alternate power-law approach used by Kimura et al. [54] is to region split at ~$0.5\sigma_{0.2\%PS}$ which for most steels coincides with the elastic limit of the material. Kimura et al. use the same activation energy across both regions and using the stress exponent, n, to reflect the mechanism change. Both these power-law approaches to region splitting are problematic. The first essentially reduces the dataset to obtain the best model fit regardless of creep mechanism and would prove difficult to independently reproduce. The second method uses a known material property limit to split data but does not have the flexibility to apply to different materials such as superalloys, and furthermore to multiple regions as seen in Grade 22, 23, and 24 [47].

When used in conjunction with the Wilshire equations region splitting becomes significantly more consistent and reproducible. Throughout the examples in this chapter, region splitting for the Wilshire equations has been implemented through the change constants k_u, u, k_v and v, as well as, the change in activation energies Q_c^*. In each case, these changes in constants have coincided with changes in creep mechanism, with microstructural evidence provided. Furthermore, these regions have not been determined subjectively but rather as a result of the implementation of the Wilshire equations.

Furthering research into modelling creep over wider ranges of stress and temperature, region splitting is becoming more important in characterising the behaviour seen. Power-law approaches region split either by regression or through selecting a material property limit associated with a change in creep mechanism. This is problematic as it is not linked to the underlying mechanisms, or, is predefined and unable to model multiple regions. The Wilshire equations throughout this chapter have implemented region splitting producing better long-term predictions and regions that reflect changes in actual creep mechanisms.

5. Conclusion

The power-law governing creep behaviour is an empirical relationship that inherently assumes a transition from diffusion to dislocation driven creep. This relationship uses a combination of stress exponent, n, and activation energy, Q_c, to determine the underlying creep mechanism which often do not agree, and do not describe the actual dominant creep mechanism. As an alternate approach the Wilshire Equations were developed.

In a fair comparison to the power-law for polycrystalline copper, the Wilshire Equations reflected the change in creep curve shape seen at the yield stress. The power-law did not indicate any change in creep behaviour then proceeded to 'power-law breakdown'.

In developing a creep modelling technique, confidence in extrapolation is critical. For 9–12% Cr Steels Grade 91, 92 and 122, it was shown that data from tests <5000 h in duration was able to predict 100,000 h creep strengths. Furthermore, the Wilshire equations again reflected a change in creep behaviour with Grade 92 and 122 having 2 regions with Grade 91 only having 1. This mirrored changes in the Reduction of Area and ductility of the tests with Grade 91 behaving differently to Grades 92 and 122. Further development of the equations in the 2.25Cr series of alloys revealed the requirement of a non-constant activation energy, which it is argued more closely represents the observed micromechanical behaviour observed in a range of materials.

Support for this argument was achieved through the application of the Wilshire approach to Waspaloy where a change in creep mechanism around the yield stress was clearly observed. Through TEM observations, below yield dislocation interactions with γ' particles were identified as the primary hardening mechanism, whereas above yield forest hardening was identified leading to the distance between dislocations becoming the critical factor.

The evidence therefore supports the Wilshire Equations satisfying 3 critical criteria, firstly to reflect changing creep behaviour, secondly to be able to extrapolate to long lives, and, finally to reflect the underlying physical process of the creep mechanism. In achieving this it is recognised that the Wilshire Equations provide a significant step forward in creating accurate methods of analysis and long-term prediction based on the underlying physics and creep mechanisms.

Author details

Mark Whittaker*, Veronica Gray and William Harrison

*Address all correspondence to: m.t.whittaker@swansea.ac.uk

Institute of Structural Materials, College of Engineering, Swansea University, UK

References

[1] Boyce MP. Gas Turbine Engineering Handbook. 2nd ed. Woburn, MA: Gulf Professional Publishing; 2002. 816 p

[2] British Standard A23(5); 1948

[3] British Standard 1686(6); 1950

[4] British Standard 1687(7); 1950

[5] British Standard 1688(8); 1950

[6] Monkman FC, Grant NJ. An empirical relationship between rupture life and minimum creep rate. In: Grant NJ, Mullendore AW, editors. Deformation and Fracture at Elevated Temperatures. Boston: MIT Press; 1965

[7] Norton FH. The Creep of Steels at High Temperatures. New York: McGraw-Hill; 1929. 112 p

[8] Arrhenius SA. Über die Dissociationswärme und den Einfluß der Temperatur auf den Dissociationsgrad der Elektrolyte. Zeitschrift für Physikalische Chemie. 1889;**4**:96-116

[9] Wilshire B, Whittaker MT. The role of grain boundaries in creep strain accumulation. Acta Materialia. 2009;**57**(14):4115-4124

[10] Cadek J, Zhu SJ, Milicka K. Threshold creep behaviour of aluminium dispersion strengthened by fine alumina particles. Materials Science and Engineering A. 1998;**252**:1-5

[11] Zhu H, Seo DY, Maruyama K, Au P. Effect of microstructural stability on creep behaviour of 47XD TiAl alloys with fine-grained fully lamellar structure. Scripta Materialia. 2005; **52**:45-50

[12] Sato H, Fujita K. Quantification of creep rate changes and life prediction by means of strain-accelerated-parameter. Keikinzoku. 2010;**60**(7):353-357

[13] Wilshire B, Sharning PJ. A new methodology for analysis of creep and creep fracture data for 9-12% chromium steels. International Materials Reviews. 2008;**53**(2):91-104

[14] Weertman J. Theory of steady-state creep based on dislocation climb. Journal of Applied Physics. 1955;**26**:1213

[15] Weertman J. Steady-state creep through dislocation climb. Journal of Applied Physics. 1957;**28**:362

[16] Glen JW. The creep of polycrystalline ice. Proceedings of the Royal Society A. 1955;**228**(1175): 519-538

[17] Glen JW. Rate flow of polycrystalline ice. Letters to Nature. 1953;**172**:721-722

[18] Nye JF. The flow of ice from measurements in glacier tunnels, laboratory experiments and the Jungfraufirn borehole experiment. Proceedings of the Royal Society A. 1139;**219**:1953

[19] Kassner WE, Pérez-Prado M-T. Five-power-law creep in single phase metals and alloys. Progress in Materials Science. 2000;**45**(1):1-102

[20] NIMS Japan. Creep Data Sheet No: 51; 2006

[21] Wilshire B, Sharning PJ. Long-term creep life prediction for a high chromium steel. Scripta Materialia. 2007;**56**(8):701-704

[22] Kimura K, Shibli IA, Holdsworth SR, Merckling G. Creep and Creep Fracture in High Temperature Components-Design and Life Assessment Issues. DEStech: Lancaster, PA; 2005

[23] Maruyama K, Lee JS, Shibli IA, Holdsworth SR, Merckling G. Design and life issues. In: Creep and Creep Fracture in High Temperature Components. Lancaster, PA: DEStech; 2005. pp. 372-379

[24] Wilshire B, Sharning PJ. Rationalization and extrapolation of creep and creep fracture data from grade 91 steel. Materials at High Temperatures. 2008;**25**(2):55-65

[25] Wilshire B, Sharning PJ. A new approach to creep data assessment. Materials Science and Engineering A. 2009;**510**:3-6

[26] Whittaker MT, Wilshire B. Creep and creep fracture of 2.25Cr-1W steels (grade 23). Materials Science and Engineering A. 2010;**527**(18):4932-4938

[27] Larson FR, Miller J. A time-temperature relationship for rupture and creep stresses. Transactions ASM. 1954;**74**:765-775

[28] Orr R, Sherby O, Dorn J. Correlation of rupture data for metals at elevated temperatures. Transaction ASM. 1954;**46**:113-118

[29] Manson S, Succop G. Stress-rupture properties of inconnel 700 and correlation on the basis of several time-temperature parameters. ASTM Special Publication. 1956;**174**:40

[30] Manson S, Haferd A. A linear time-temperature relation for extrapolation of creep and stress-rupture data. NASA Technical Note 2890. 1953

[31] Goldhoff R, Hahn G. Correlation and extrapolation of creep-rupture data of several steels and superalloys using time-temperature parameters. ASM Publication. 1968;**D-8-100**:199-247

[32] Abdullah Z, Gray V, Whittaker MT, Perkins KM. A critical analysis of the conventionally employed creep lifing methods. Materials. 2014;**7**(5):3371-3398

[33] Gray V, Abdullah Z, Whittaker MT. Determining the Parameters and Constants of Failure. Swansea University, Berlin, Germany: ResearchGate; 2015. DOI: 10.13140/RG.2.1.3639.9520

[34] Gibbs GB. The thermodynamics of creep deformation. Physica Status Solidi B. 1964; **5**(3):693-696

[35] Wilshire B, Sharning PJ. A new methodology for analysis of creep and creep fracture data for 9-12% chromium steels. International Materials Review. 2008;**53**(2):91-104

[36] Wilshire B, Battenbough AJ. Creep and fracture of polycrystalline copper. Materials Science and Engineering A. 2007;**443**:156-166

[37] Burton B, Greenwood GW. The limit of the linear relation between stress and strain rate in the creep of copper and copper-zinc alloys. Acta Metallurgica. 1970;**18**(12): 1237-1242

[38] Feltham P, Meakin JD. Creep in face-centred cubic metals with special reference to copper. Acta Metallurgica. 1959;**7**(9):614-627

[39] Burton B, Greenwood GW. The contribution of grain-boundary diffusion to creep at low stresses. Metal Science Journal. 1970;**4**(1):215-218

[40] Evans RW, Wilshire B. Creep of Metals and Alloys. London, UK: Institute of Metals; 1985. p. 314

[41] Holdsworth SR, Bullough CK, Orr J. BS PD 6605 creep rupture data assessment procedure. ECCC Recommendations. 2001

[42] NIMS Creep Data Sheet No. 43; 1996

[43] NIMS Creep Data Sheet No. 48; 2002

[44] NIMS Creep Data Sheet No. 51; 2006

[45] Arzt E. Creep of dispersion strengthened materials: A critical assessment. Mechanics Research. 1991;**31**:399-453

[46] Kimura K. Present status and future prospect on NIMS creep data sheet. In: Mishra RS, Earthman JC, Raj SV, editors. Creep Deformation and Fracture, Design and Life Extension. Pittsburgh: MS&T; 2005. pp. 97-106

[47] Whittaker MT, Wilshire B. Advanced procedures for long-term creep data prediction for 2.25 chromium steels. Metallurigcal and Materials Transactions A. 2013;**44**:136-153

[48] Whittaker MT, Harrison WJ. Evolution of the Wilshire equations for creep life prediction. Materials at High Temperatures. 2014;**31**(3):233-238

[49] Evans M, Whittaker MT, Wilshire B. Long-term creep data prediction for type 316H stainless steel. Materials Science and Engineering A. 2012;**552**:145-150

[50] Whittaker MT, Harrison WJ, Deen C, Rae C, Williams S. Creep deformation by dislocation movement in Waspaloy. Materials. 2017;**10**:61

[51] Gray V, Whittaker MT. The changing constant of creep: A letter on region splitting in creep lifing. Materials Science and Engineering A. 2015:**632**:96-102

[52] Foldyna V, Kubon Z, Jakobova A, Vodarek V. Development of advanced high chromium Ferritic steels. In: Microstructural Development and Stability in High Chromium Ferritic Power Plant Steels. London, UK: The Institute of Materials; 1997. pp. 73-92

[53] Maruyama K, Ghassemi Armaki H, Yoshimi K. Multiregion analysis of creep rupture data of 316 stainless steel. International Journal of Pressure Vessels and Piping. 2007;**84**(3):717

[54] Kimura K. Creep rupture strength evaluation with region splitting by half yield. Proceedings of the ASME 2013 Pressure Vessels & Piping Division Conference. 2013

Chapter 3

Degradation of the Microstructure and Mechanical Properties of High-Chromium Steels Used in the Power Industry

Grzegorz Golański, Cezary Kolan and Joanna Jasak

Additional information is available at the end of the chapter

http://dx.doi.org/10.5772/intechopen.70552

Abstract

High-chromium martensitic steels are one of the basic creep-resisting construction materials used for the modernization of old and the construction of new power units. During the service under creep conditions, the metastable microstructure of martensitic steels undergoes gradual degradation. The rate of degradation mostly depends on the operating temperature, but it is also affected by stresses. The changes in the microstructure of martensitic steels have an influence on the decrease in their mechanical properties, including creep resistance. The knowledge and description of the changes in the microstructure of steels working under creep conditions allow extending the time of safe operation of the elements of power systems. The paper presents and describes the main mechanisms of degradation of 9–12%Cr martensitic steels on the basis of the independent studies and literature data.

Keywords: creep-resistant steel, 9–12%Cr steel, microstructure degradation, precipitates, mechanical properties

1. Introduction

The need to reduce emissions of pollutants (mainly CO_2) to the atmosphere enforced by increasingly stringent EU directives has contributed to the development of conventional energy. Restrictions on emissions to the atmosphere caused by the combustion of fossil fuels have forced the power industry to increase the thermal efficiency of power units (from 33–35 to 40%, and ultimately up to 50%). On the other hand, the need to increase the thermal efficiency of power units involves a significant increase in steam parameters (pressure, temperature). This requires the construction of new and the modernization of existing power units to allow them to operate at the so-called supercritical or ultra-supercritical steam parameters. The increase in steam parameters in new and existing power units was possible due to the materials revolution in the power industry and was associated with the introduction of new grades of steels and cast steels with higher resistance to creep and oxidation than that of the materials used so far. The implementation of new materials in the power industry took place primarily through modifications and optimizations of steels that were already being used in the power sector. It has contributed not only to the increase in steam parameters, but also to the reduction in overall dimensions of boiler components, and thus in their weight, which has also a significant impact on the reduction in the energy production costs [1, 2].

One of the new steel groups introduced to the power industry was high-chromium martensitic steels containing 9–12%Cr. By the optimization of carbon content and the introduction of additions and micro-additions such as W, Co, V, Nb, N, B, and Cu to these steels, the construction materials characterized by high mechanical properties were obtained. For example, their creep strength is higher by approx. 20–25% than that of the steels used so far [2, 3]. The expected high reliability and long life of up to a minimum of 200,000 h of pressure parts made from, among others, 9–12%Cr steels require understanding and describing the effects and microstructure degradation processes for these materials. Based on many years of Authors' own research and literature data, the main steel/martensitic cast steel microstructure degradation mechanisms and their impact on mechanical properties were described and characterized in this paper.

2. Microstructure degradation and properties of 9–12%Cr steels

The basic requirement for creep-resistant steels used in the broadly understood power industry is to maintain—for a relatively long time of operation (at present, 200–250,000 h)—the assumed mechanical properties at the operating temperature of power equipment components. The maintenance of the required mechanical properties of creep-resistant steels during long-term service depends on the stability of their microstructure. The structural components of the power equipment are influenced by the continuous destruction process, which has a significant impact on the life and time of safe service of a specific component. Therefore, the time of safe service for devices used in the power industry is one of the most important parameters related to their life, and it determines their applicability in this sector [1, 4, 5].

During their long-term service, progressive changes in microstructure of creep-resistant steels take place—the process of degradation of their microstructure occurs. For 9–12%Cr martensitic steels, the main microstructure degradation mechanisms include [6–8]:

- matrix recovery and polygonization processes,
- coagulation of $M_{23}C_6$ carbides,
- precipitation of secondary phases: Laves phase and Z-phase, and
- depletion of alloying elements in matrix.

Martensitic steels in the hardened condition are characterized by high dislocation density within the martensite laths (10^{16}–10^{18} m^{-2}). Due to high-temperature tempering of hardened steel, a more thermodynamically stable microstructure with still high dislocation density at 10^{12}–10^{14} m^2 (including free dislocations) within the subgrains formed during the tempering is obtained. The dislocation substructure in 9–12%Cr steels is characterized by small elongated subgrain (of 200–400 nm in width) and low-angle boundaries [6, 7, 9, 10]. High dislocation density and microstructure refinement with dislocation boundaries has a very intensive impact on the 9–12%Cr steel hardening with the dislocation hardening mechanism and the grain boundary hardening mechanism, respectively. The calculations in [11] showed that the gain in yield strength in martensitic steels for the above-mentioned mechanisms is 18 and 33%. In addition to this hardening, the following further mechanisms are additionally used to form the structure and mechanical properties of 9–12%Cr steels: solution hardening with interstitial and substitution elements and precipitation hardening [9, 11, 12]. In high-chromium martensitic steels, the precipitation hardening mechanism is mainly performed by the secondary particles precipitated when tempering $M_{23}C_6$ carbides and MX precipitates. In 9–12%Cr creep-resistant steels, three types of MX precipitates can occur [3, 13, 14]:

- Primary niobium-rich spheroidal NbC carbides (carbonitrides)
- Secondary lamellar VN (VX) nitrides (carbonitrides), which are precipitated within the martensite laths during high-temperature tempering
- Precipitate complexes consisting of the spherical NbX precipitate in which the VN precipitate nucleates, referred to as the "V-wings"

The degree of hardening with secondary phase precipitates depends mainly on the amount and size of precipitates and their distribution within the matrix.

High-chromium martensitic steels in the as-received condition (i.e., after quenching and tempering) have a metastable microstructure, which will be affected by gradual evolution as a result of long-term service (**Figure 1**).

Figure 1. The microstructure of GP91 cast steel in the as-received condition [8].

Long-term effect of the temperature and time, in the case of creep as well as stress, leads to a decrease in strengthening with the dislocation mechanisms and the grain boundaries. A decrease in free dislocation density within grains and increase in size of the subgrains take place. The decrease in dislocation density with the time of service/aging is associated with the progressing process of their regrouping, arranging, annihilation, and entangling the dislocations in grain boundaries as well as the formation of cellular dislocation and subgrain microstructure (**Figure 2**). The matrix recovery and polygonization process results in the disappearance of martensite lath microstructure and the formation of polygonised ferrite microstructure **Figures 3** and **4**.

In the microstructure of martensitic steels, the formation of polygonal structure during service takes place due to the progressive increase in the size of subgrains. This process is slow because of low mobility of these subgrains. The stability of subgrain size has a positive impact on the maintenance of high mechanical properties, including creep resistance [16]. The increase in the size of subgrains occurs due to the migration or coalescence of the sub-boundaries.

The increase in the size of sub-boundaries usually takes place with the "Y" mechanism [15, 17, 18]. The migration with this mechanism is based on the movements of "Y" nodes that are the place where three sub-boundaries meet, which allows the coalescence of two low-angle boundaries. The increase in the size of subgrains with the "Y" node movement mechanism is shown in **Figure 5**.

The matrix recovery and polygonization process takes place in the presence of secondary dispersion phases, which act as a stabilizing agent. The lath microstructure stability depends on the stability of $M_{23}C_6$ carbides precipitated on these phases (**Figure 6**). $M_{23}C_6$ carbides precipitate on the tempered martensite lath boundaries and on the subgrain boundaries preventing

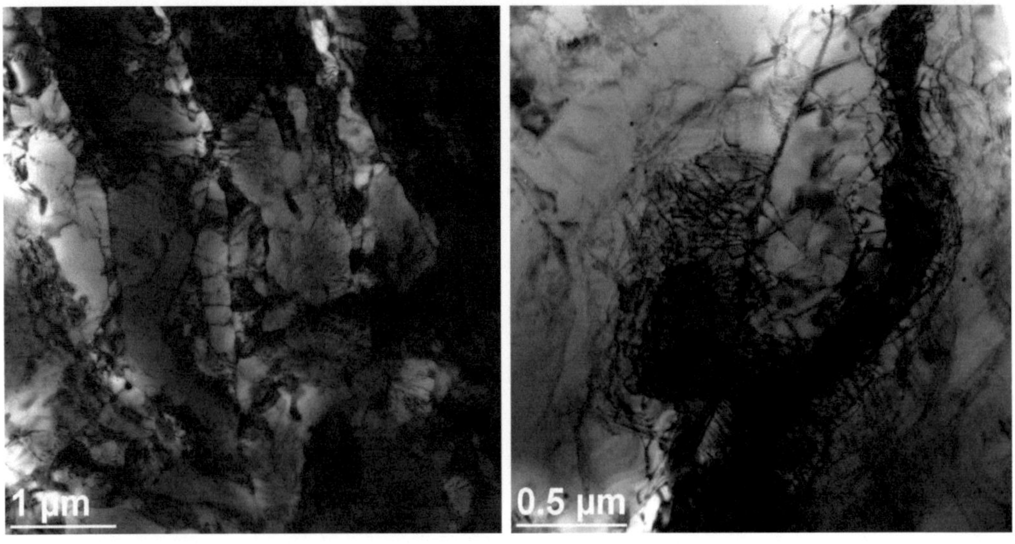

Figure 2. The interaction of dislocations with lath/subgrain boundaries, PB2 steel, TEM.

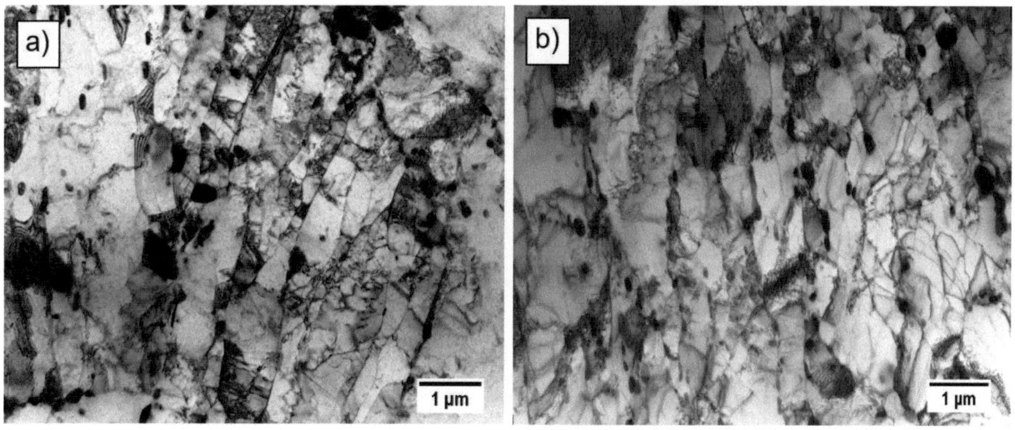

Figure 3. The microstructure of: (a) T91 steel after long-term service [12], (b) GP91 cast steel after 70,000 h aging at 600°C.

their growth due to the matrix polygonization, repolygonization, and recrystallization processes [17, 19]. In addition, the elongated shape of $M_{23}C_6$ carbides precipitated on the subboundaries (**Figures 1**, **3**, and **5**) has a positive impact on anchoring the boundaries by them as their contact surface with the boundary on the same volume fraction is bigger than for spherical particles [18].

The thermodynamic thermal stability of $M_{23}C_6$ carbides is not too high—the $Cr_{23}C_6$ carbide formation enthalpy is: –20 kJ/mol [20]. In the as-received condition, the size of $M_{23}C_6$ carbides

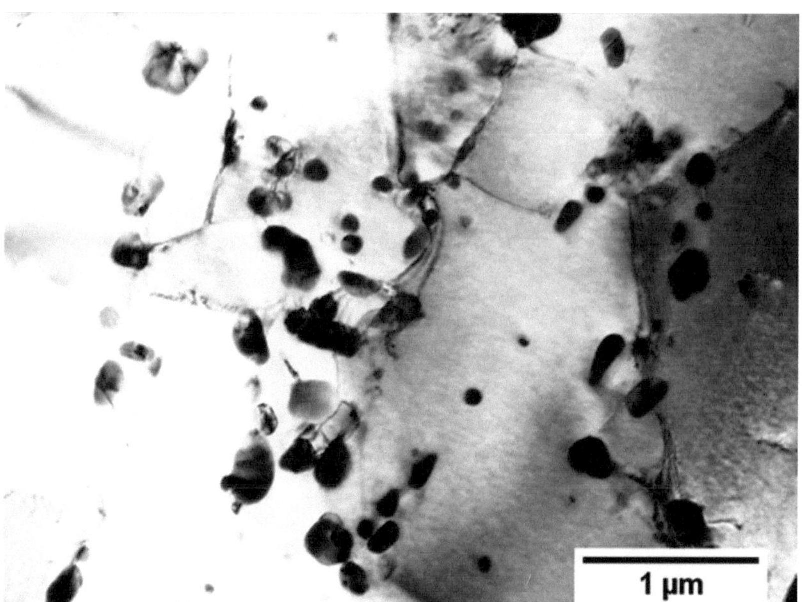

Figure 4. The microstructure of polygonised ferrite with numerous precipitates in GP91 cast steel after low-cycle fatigue at 600°C with strain amplitude of 0.60%.

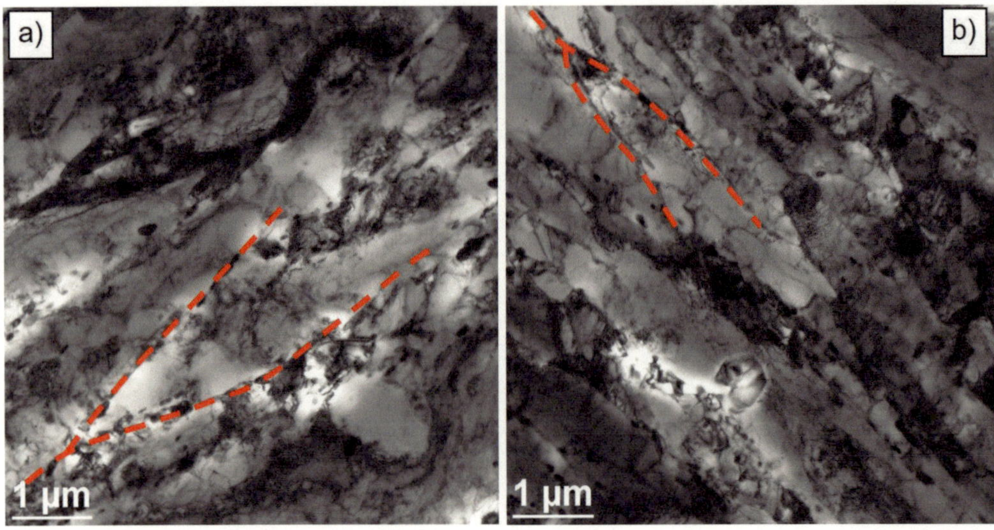

Figure 5. The morphology of "Y" nodes in: (a) P91, (b) PB2, TEM [15].

in martensitic steels is 50–150 nm [8, 21]. The long-term service/aging contributes to changing the morphology of $M_{23}C_6$ carbides. These precipitates show a high tendency to coagulation. The process of coagulation of precipitates is determined by two basic factors: the thermodynamic and the kinetic one. The thermodynamic factor results from a large value of the surface energy of the interphase boundaries. As a result of coagulation, the surface energy decreases and aims at reaching the energy equilibrium. The kinetic factor of coagulation, on the other hand, is connected with the diffusion and reactions occurring on the boundary surface. They run at different rates, and the slowest one determines the rate of particle growth in the system, and thereby determines the kinetics of coagulation. The constant of the rate of growth of particles K_p in the matrix of martensitic steels is presented in **Table 1**.

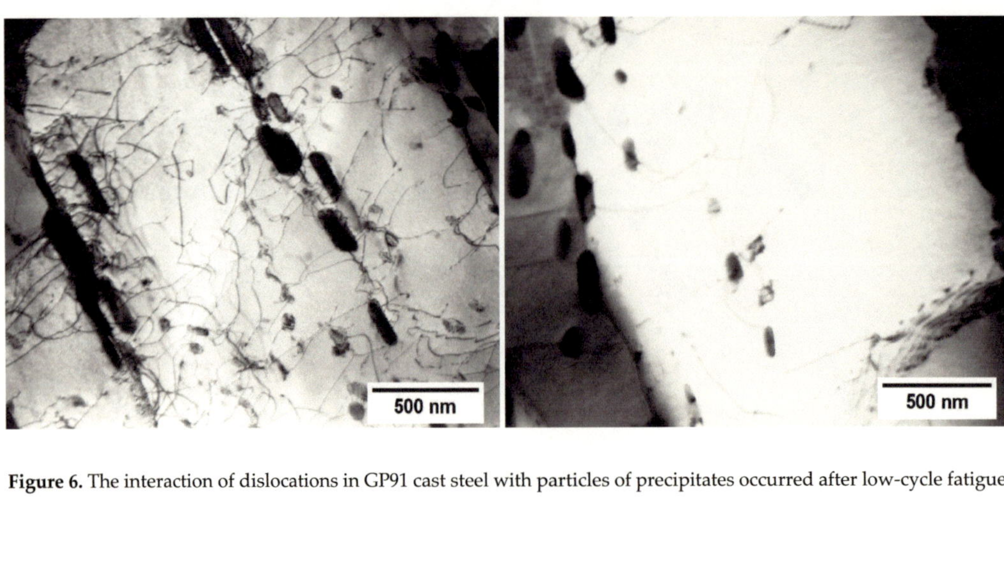

Figure 6. The interaction of dislocations in GP91 cast steel with particles of precipitates occurred after low-cycle fatigue.

Precipitate	Steel	Temperature, °C	Based on solubilities at tempering temperature		Based on solubilities at exposure temperature	
			K_p, $m^3\ s^{-1}$	γ, J/m^2	K_p, $m^3\ s^{-1}$	γ, J/m^2
MX	P92	600	1.17×10^{-32}	0.5	8.58×10^{-33}	0.5
		650	9.5×10^{-32}	0.5	65.5×10^{-33}	0.5
Laves phase	P92	600	–	–	2.91×10^{-31}	1.0
		650			41.6×10^{-31}	1.0
$M_{23}C_6$	P92	600	0.12×10^{-29}	0.1	1.88×10^{-30}	0.1
		650	1.37×10^{-29}	0.1	4.78×10^{-30}	0.1
	P91	600	2.88×10^{-29}	1.0	7.67×10^{-30}	0.5
		650	25.3×10^{-29}	0.8	59.8×10^{-30}	0.3

Table 1. Calculated coarsening rate constants K_p of MX, Laves phase, and $M_{23}C_6$ precipitated in P91 and P92 steel based on the shown fit values for the interfacial energy γ [22].

The coagulation of $M_{23}C_6$ carbides reduces their amount with almost the same volume fraction and results in the increase in distance between these precipitates. Also, nonuniform distribution of these precipitates within matrix makes them become a less effective factor controlling the increase in the size of subgrains (**Figure 7**). The literature data [23] show that the subgrain boundaries with mutual disorientation angle of less than 20° are not the points of preferential carbide precipitation. According to the research in [24], only about 8% of $M_{23}C_6$ carbides was precipitated at the low-angle boundaries with mutual disorientation angle of 8–15°. The low-angle boundaries represent at least 60% of the total amount of boundaries in

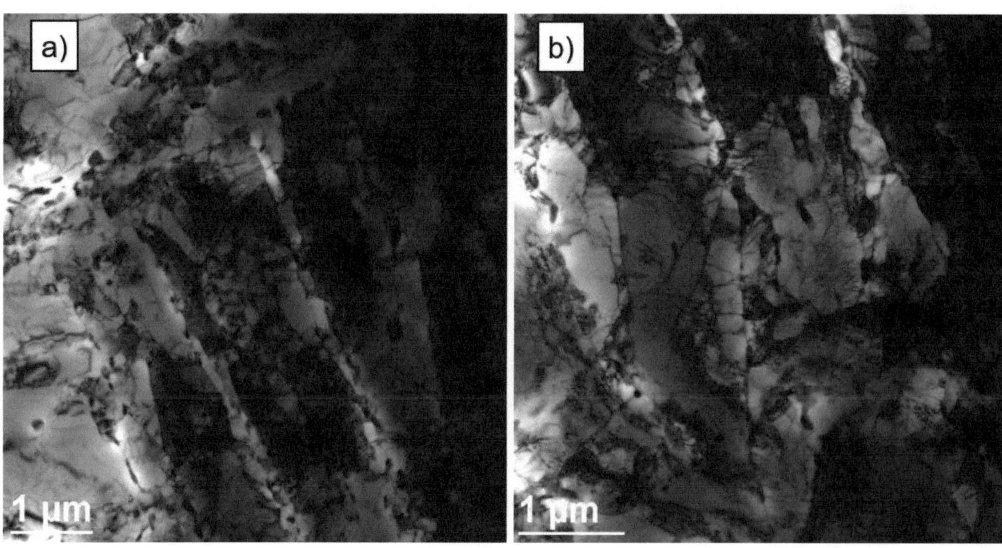

Figure 7. The microstructure of martensitic steels with both wide and narrow martensite laths visible: (a) P91, (b) PB2, TEM.

tempered martensite. This limits their role and reduces their effectiveness as a substructure stabilizer, which results in the reduction in creep strength. The precipitate-free boundaries show higher mobility, which results in an increase in their width. The increase in the size of $M_{23}C_6$ carbides precipitated at the boundaries is also conducive to the reduction in ductility of 9–12%Cr steels [6, 8, 22–26].

In steel with micro-addition of boron, the carbon atoms in $M_{23}C_6$ carbides are partially replaced by boron during the tempering, which results in the formation of $M_{23}(C, B)_6$ carboborides. Like $M_{23}C_6$ carbides, these precipitates occur at the grain boundaries and at the martensite lath boundaries. However, these precipitates are more finely dispersed and characterized by higher thermodynamic stability compared to $M_{23}C_6$ carbides [9, 14, 27]. This results in a slower increase in the size of these precipitates, which has a positive effect on the stability of tempered martensite lath microstructure and results in a higher creep resistance. Also according to [24], vanadium plays a significant role as a factor controlling the process of coagulation of $M_{23}C_6$ carbides. Vanadium dissolved in the matrix is conducive to decreasing of the coefficient of chromium diffusion in ferrite. Similar influence is also observed in the case of tantalum [28]. The temperature of work has a considerable effect on the rate of coagulation of $M_{23}C_6$ carbides. Elevating the temperature of service by 50°C can cause a growth of the rate of coagulation of these precipitates even by an order of magnitude.

The martensitic steels gain high creep resistance mainly due to the precipitation hardening provided by: MX nitrides, carbonitrides (where: M = V, Nb; X = C, N). MX precipitates are characterized by nanometric dimensions of about 10–50 nm, and in spite of their low volume fraction of 0.020–0.025, they ensure very strong hardening of creep-resistant steels (**Figure 8**).

The hardening with these precipitates is ensured by anchoring and hindering the motion of dislocations [6, 7, 9, 14, 15, 18, 29]. The calculations made for the P91 steel showed that the stress required for dislocation to "bypass" the carbide and nitride particles with the Orowan mechanism is as follows: for $M_{23}C_6$—39 MPa, for NbC—15 MPa, and for VN—106 MPa [30]. The MX precipitates in 9%Cr steels have a very high thermal stability (**Table 1**). The approximate formation enthalpy for these precipitates is as follows: for VC and NbC carbides, 55 and 70 kJ/mol, respectively, and for VN nitride, 125 kJ/mol [19]. High stability of MX precipitates and their coherent (semi-coherent) interphase boundaries cause that after approx. 100,000 h creep at 600°C, their size is similar to that as in the as-received condition [24, 30, 31].

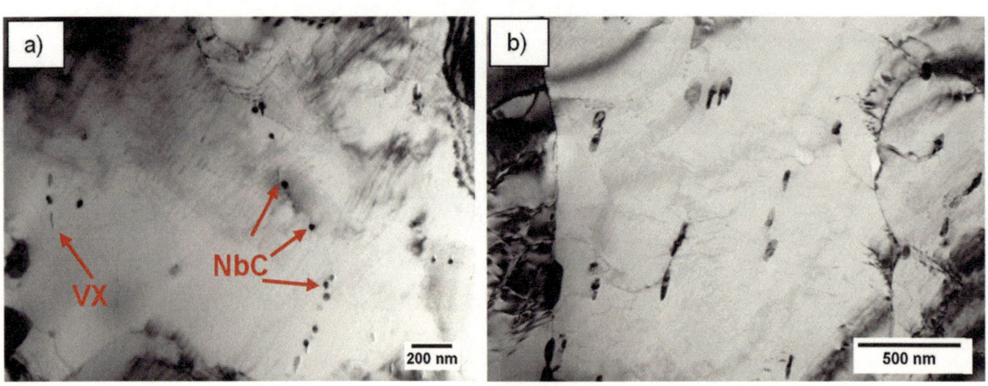

Figure 8. The MX precipitates in T91 steel after service: (a) NbC and VX, (b) V-wings [12].

In martensitic steels used at above 600°C and containing at least 10%Cr, MX precipitates represent a metastable phase and undergo a transition into a more thermodynamically stable Z-phase (**Figure 9a**) [30–34]. The disappearance of finely dispersed MX precipitates in the microstructure of these steels due to the MX carbide → Z-phase transition (complex Cr(V, Nb) N nitride) results in a very fast decrease in creep resistance [31, 32, 34, 35].

According to [32, 35], one large Z-phase precipitate is formed at the expense of dissolving approx. 1000–1500 finely dispersed MX precipitates in the matrix. The disappearance of MX precipitates in martensitic steels during service in favor of Z-phase eliminates the effect of precipitation hardening with these particles. Nevertheless, as shown in [12], the interaction of MX precipitates with dislocations (**Figure 9b**) is still observed in the microstructure of P91 steel after service, and single Z-phase precipitates do not adversely affect its properties, and thus the creep strength (**Figure 10**).

Unlike the MX precipitates, both the chemical composition and the size of Z-phase depend on chemical composition of the steel it precipitates in and on creep duration. The Z-phase in 9%Cr steels is approx. 80–100 nm, whereas in steels with 11–12%Cr it is much larger and amounts to approx. 0.5–2 μm. Consequently, in 9%Cr steels the Z-phase precipitation is accompanied by a slight reduction in the volume fraction of MX precipitates, whereas in 12%Cr steels MX precipitates are virtually completely transformed into this complex nitride [31, 32, 36]. In addition, in 9%Cr steels the Z-phase precipitates after approx. 10^5 h at the earliest, while in 12%Cr steels the precipitation of this phase can be observed as early as after 10^3 h. Hence, the effect of Z-phase precipitates on creep strength is slight in steels with 9%Cr, whereas in 12%Cr steels it is significant [25, 32, 35, 36].

In high-chromium martensitic steels, the Z-phase precipitation may proceed with two mechanisms [32, 33]. The schematic transition of MX precipitates into Z-phase in high-chromium martensitic steels is shown in **Figure 11**.

On the other hand, dissolving NbX precipitates in the matrix "releases" carbon atoms, which results in the precipitation of chromium-rich $M_{23}C_6$ carbides, frequently nearby the Z-phase particles. The precipitation of Z phase is preferential near the grain boundaries of prior austenite, and in the steels containing delta ferrite additionally also near the interphase boundary mar-

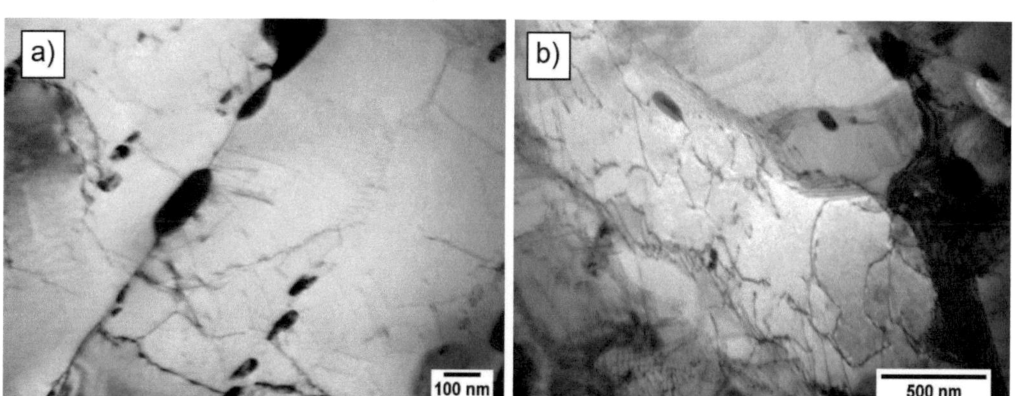

Figure 9. Z-phase precipitate in T91 steel after service (a), interaction of dislocations inside the subgrain with MX precipitates (b).

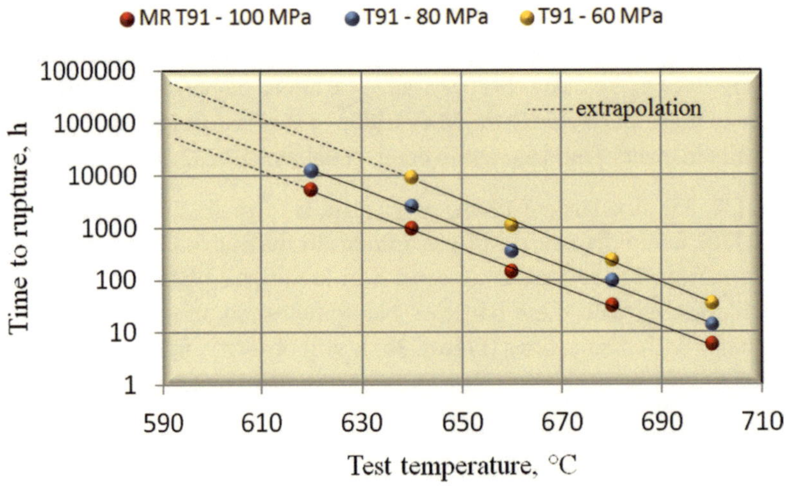

Figure 10. Results of short-term creep tests of T91 steel after service [12].

tensite/delta ferrite [14, 25, 33, 35, 36]. This is due to faster diffusion of substitution elements nearby these defects. It results in the formation of near-boundary areas free from MX precipitates, which leads to the accelerated matrix recrystallization and reduction in strength properties in these areas. Such changes lead to the formation of creep grain, which is unequal in volume, and consequently to a faster destruction of steel during service [32, 35, 36]. The disappearance of

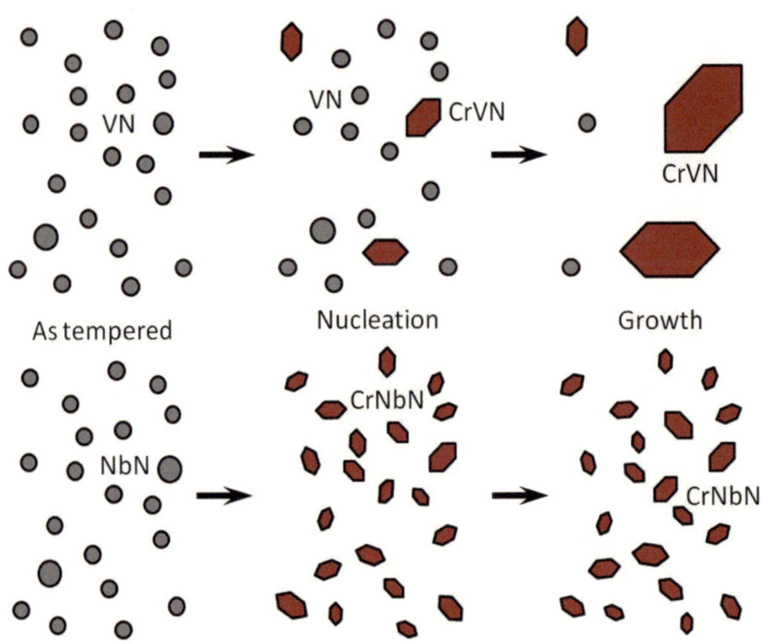

Figure 11. Schematic transition of MX precipitates into Z-phase [32].

finely dispersed MX precipitates also results in disproportionately high reduction in hardness in relation to other mechanical properties [21].

In martensitic steels containing approx. 9%Cr, a more important problem is the increase in the size of $M_{23}C_6$ carbides as well as the precipitation and growth of intermetallic Fe_2Mo Laves phase [6, 7, 25, 37]. In 9–12%Cr steels in the as-received condition, the Laves phase does not occur. The precipitation of this phase takes place during service/aging mainly at the grain/lath boundaries, frequently nearby $M_{23}C_6$ carbides **Figure 12** [37–39]. In the case when the total content of W + Mo in the steel amounts to at least 4.53, the particles of Laves phase precipitate heterogeneously at grain boundaries as well as homogeneously within grains, forming the precipitation free zone on both sides of the grain boundary [40].

It is assumed that due to high dispersion in the initial period of the precipitation the Laves phase has a positive effect on properties of these steels by increasing the precipitation hardening. However, low stability of the Laves phase results in its high coagulability, which results in a very fast increase in its size [15, 25, 37, 39].

The Laves phase precipitating in 9–12%Cr creep-resistant steels makes the matrix deplete of substitution elements (tungsten, molybdenum, chromium), which increases the tendency of these steels to the recovery and polygonization process and reduces their resistance to oxidation. On the other hand, the matrix depletion of substitution elements (Cr, Mo, W), which are also components of $M_{23}C_6$ carbides, has a positive effect on the inhibition of coagulation of these precipitates [12, 38]. According to [41, 42], the nucleation and growth of Laves phase requires the enrichment of micrograin boundaries not only in Mo and Si, but also in phosphorus.

The Laves phase precipitates with average diameter above 130 nm also contribute to the change in cracking mechanism from ductile to brittle (transcrystalline, cleavable fracture) and are the main reason for sudden reduction in creep strength of [40, 41, 43, 44]. The Laves phase and $M_{23}C_6$ carbide precipitates occurring during long-term service form the so-called continuous grid of precipitates at the grain boundaries (**Figure 13**), which contributes to a decrease in ductility of 9% Cr steel [8, 12, 43–49].

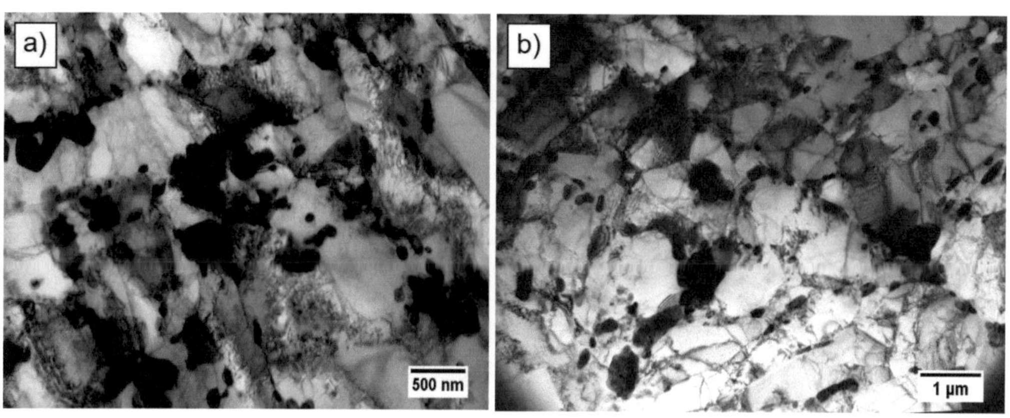

Figure 12. Precipitation of Laves phase at the grain boundary nearby $M_{23}C_6$ carbides: (a) T91 steel, (b) GP91 cast steel [38].

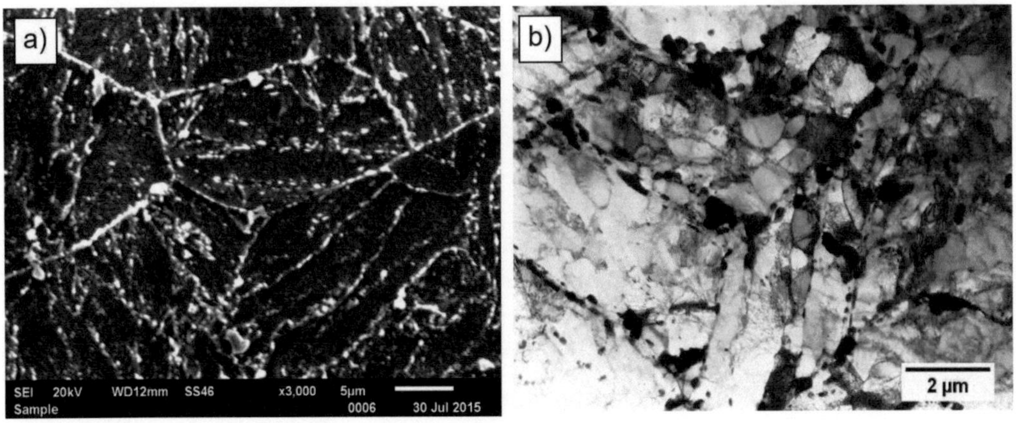

Figure 13. Continuous grid of precipitates at the former austenite grain boundaries (a) P91 steel (SEM), (b) GP91 cast steel (TEM) [49].

The increase in stability of the Laves phase precipitates in these steels can be achieved by the addition of boron and/or tungsten [50, 51]. On the contrary, phosphorus, silicon, and cobalt have an adverse effect as they stimulate the precipitation of the Laves phase [52, 53].

In the tempered structure of martensite in steel of the 9–12%Cr type, in the case of the typical volume fractions of particular precipitates and spacing between them, the Orowan stress can be estimated as shown in **Table 2**.

The precipitation processes as well as the growth of carbides and secondary phases, which occurs during the operation of creep-resistant steels, make the matrix deplete of substitution elements as a result of their diffusion into these precipitates.

The matrix depletion of the above-mentioned elements facilitates the self-diffusion processes, speeds up the matrix recovery and polygonization processes, and reduces the oxidation resistance, thus contributing to the reduction in high-temperature creep resistance and life of these steels [11, 54].

An important factor affecting the basic property of 9–12%Cr steels, i.e., creep strength, is aluminum content in the steel. The addition of aluminum to steel is to deoxide it in the metallurgical process, hence part of aluminum will remain as Al_2O_3 in the steel, while the other part in the atomic form is dissolved in solid solution. The aluminum content in 9–12%Cr steels for

Particle	Volume fraction, %	Diameter, nm	Spacing, nm	Orowan stress, MPa
Fe_2M	1.5	70	410	95
$M_{23}C_6$	2	50	260	150
MX	0.2	20	320	120

Table 2. Volume fraction, diameter, and spacing of each kind of precipitates in high-chromium martensitic steel, together with Orowan stress from the values of interparticle spacing [14].

the power industry should not exceed 0.04%. Due to its greater affinity to nitrogen, aluminum forms large lamellar AlN nitrides of about 0.5–1.0 μm to prevent the formation of finely dispersed vanadium-rich MX precipitates. This results in not only a reduction in the volume fraction of vanadium-rich MX precipitates, but also in a change in their chemical composition. There occurs a decrease in the amounts of vanadium and nitrogen with simultaneous increase in the amounts of niobium and carbon in these precipitates, which makes niobium-rich carbides (NbC) become the main precipitate in martensitic steel. This results in a decrease in creep strength of martensitic steels: for P91—by about 10% at 600°C, whereas for P92 the time to rupture of test specimen was shorter by 7–30 times, depending on test temperature and stress [30, 55]. Higher than permissible aluminum content in 9–12%Cr steels has a positive

Features of microstructure		**As-received condition**	**Creep/aging**
Matrix	Dislocation density	High	Low/very low
	Size, width of subgrains/ martensite laths	High-temperature tempered martensite microstructure with small width of laths	Recovery and polygonization process – transformation of the lath martensite microstructure into the polygonised ferrite grain microstructure
Precipitates	MX	Finely dispersed (~20–50 nm), precipitated inside laths at dislocations, limit grain growth, make steel precipitation-hardened	Finely dispersed (~20–50 nm), precipitated inside laths at dislocations, limit grain growth, make steel precipitation-hardened, change into Z-phase
	$M_{23}C_6$	50–150 nm, precipitated at the martensite lath boundaries and at the former austenite grain boundaries, stabilize substructure	≥220 nm, partially precipitated at the subgrain boundaries, coagulate during creep/ aging resulting in a reduction in creep strength and increase in embrittlement
	Z-phase	Nonexistent	Formed at the expense of finely dispersed MX precipitates, causes sudden decrease in creep strength
	Laves phase	nonexistent	Medium and large size of precipitates (≥0.5–1 μm), precipitated at the grain and subgrain boundaries nearby $M_{23}C_6$, decreases creep strength and reduces ductility

Table 3. Basic features of microstructure in 9–12%Cr steels in the as-received condition and after creep.

effect on the increase in impact strength and reduction in brittle fracture appearance transition temperature, which is the result of positive effect of AlN on the austenite grain refinement. The increase in aluminum content from 0.03 to 0.094% in P92 steel makes the average diameter of austenite grain decrease from 50 μm (which corresponds to the grain grade of 5.5) to 10 μm (grain grade 10) [30, 56]. The negative impact of aluminum on creep resistance requires control of chemical composition of 9–12%Cr steel as early as at its production stage, in particular with regard to elements with high affinity to nitrogen, such as Al and Ti, so as to prevent from the formation of unfavorable AlN or TiN nitrides at the expense of finely dispersed VN precipitates.

The basic features of the microstructure of 9–12%Cr steel in the as-received condition and after long-term service/aging are summarized in **Table 3**.

Author details

Grzegorz Golański*, Cezary Kolan and Joanna Jasak

*Address all correspondence to: grisza@wip.pcz.pl

Institute of Materials Engineering, Czestochowa University of Technology, Czestochowa, Poland

References

[1] Viswanathan R, Bakker W. Materials for ultrasupercritical coal power plants—Boiler materials—Part 1. Journal of Materials Engineering and Performance. 2001;**10**:81-95. DOI:10.1361/105994901770345394

[2] Masuyama F. History of power plants and progress in heat resistant steels. ISIJ International. 2001;**41**:612-625. DOI: 10.2355/isijinternational

[3] Bhadeshia HKDH. Design of ferritic creep-resistant steels. ISIJ International. 2001;**41**:626-640. DOI: 10.2355/isijinternational

[4] Golański G, Stachura S. Characterization of new-low alloy steels for power plant. Hutnik–Wiadomości Hutnicze. 2009;**9**:679-683 (in Polish)

[5] Abe F. New martensitic steels. In: Di Gianfrancesco A, editor Materials for Ultra-Supercritical and Advanced Ultra-Supercritical Power Plants. Woodhead Publishing; 2017. pp. 323-374

[6] Panait CG, Bendick W, Fuchsmann A, Gourgues-Lorenzon AF, Besson J. Study of the microstructure of the grade 91 steel after more than 100,000 h of creep exposure at 600°C. International Journal of Pressure Vessels and Piping. 2010;**87**:326-335. DOI: 10.1016/j.ijpvp.2010.03.017

[7] Panait CG, Zielińska-Lipiec A, Kozieł T, Czyrska-Filemonowicz A, Gourgues-Lorenzon AF, Bendick W. Evolution of dislocation density, size of subgrains and MX-type precipitates in a P91 steel during creep and during thermal ageing at 600°C for more than 100,000 h. Materials Science and Engineering. 2010;**527A**:4062-4069. DOI: 10.1016/j.msea.2010.03.010

[8] Golański G, Mroziński S. Low cycle fatigue and cyclic softening behaviour of martensitic cast steel. Engineering Failure Analysis. 2013;**35**:692-702. DOI: 10.1016/j.engfailanal.2013.06.019

[9] Dudova N, Mishnev R, Kaibyshev R. Effect of tempering on microstructure and mechanical properties of boron containing 10% Cr steel. ISIJ International. 2011;**51**:1912-1918. DOI: 10.2355/isijinternational

[10] Baek JH, Kim SH, Lee CB, Hahn DH. Mechanical properties and microstructural evolution of modified 9Cr-1Mo steel after long-term aging for 50,000 h. Metals and Materials International. 2009;**15**:565-573. DOI: 10.1007/s12540-009-0565-y

[11] Qiang L. Modeling the microstructure–mechanical property relationship for a 12Cr–2W–V–Mo–Ni power plant steel. Materials Science and Engineering. 2003;**361A**:385-391. DOI: 10.1016/S0921-5093(03)00565-3

[12] Golański G, Zielińska-Lipiec A, Zieliński A, Sroka M. Effect of long-term on microstructure and mechanical properties of martensitic 9%Cr steel. Journal of Materials Engineering and Performance. 2017;**26**:1101-1107

[13] Golański G. Evolution of secondary phases in GX12CrMoVNbN9-1 cast steel after heat treatment. Archives of Materials Science and Engineering. 2011;**48**:12-18

[14] Abe F. Precipitate design for creep strengthening of 9% Cr tempered martensitic steel for ultra-supercritical power plants. Science and Technology of Advanced Materials. 2008;**9**:1-15. DOI: 10.1088/1468-6996/9/1/013002

[15] Golański G, Jasak J, Zieliński A, Wieczorek P. Quantitative analysis of stability of 9%Cr steel microstructure after long-term ageing. Archives of Metallurgy and Materials. 2017;**62**:273-281. DOI: 10.1515/AMM-2017-0040

[16] Ghassemi-Armaki H, Chen RP, Maruyama K, Yoshizawa M, Igarashi M. Static recovery of tempered lath martensite microstructures during long-term aging in 9-12%Cr heat resistant steels. Materials Letters. 2009;**63**:2423-2425. DOI: 10.1016/j.matlet.2009.08.024

[17] Abe F. Coarsening behavior of lath and its effect on creep rates in martensitic 9Cr–W steels. Materials Science and Engineering. 2004;**387-389A**:565-569. DOI: 10.1016/j.msea.2004.01.057

[18] Xu LQ, Zhang DT, Liu YC, Ning BQ, Qiao ZX, Yan ZS, Li HJ. Precipitation behavior and martensite lath coarsening during tempering of T/P92ferritic heat-resistant steel. International Journal of Minerals, Metallurgy, and Materials. 2014;**21**:438-447. DOI: 10.1007/s12613-014-0927-4

[19] Nes E, Ryum N, Hunderi O. On the Zener drag. Acta Metallurgica. 1985;**33**:11-22. DOI: 10.1016/0001-6160(85)90214-7

[20] Pickering FB. Historical development and microstructure of high chromium ferritic steels for high temperature applications. In: Strang A, Gooch DJ, editors. Microstructural Development and Stability in High Chromium Ferritic Power Plant Steels. London: The Institute of Materials Cambridge; 1997. pp. 1-29

[21] Kadoya Y, Dyson BF, McLean M. Microstructural stability during creep of Mo- or W-bearing 12Cr steels. Metallurgical and Materials Transactions. 2002;**33A**:2549-2557. DOI: 10.1007/s11661-002-0375-z

[22] Hald J, Korcakova L. Precipitate stability in creep resistant ferritic steels—Experimental investigation and modelling. ISIJ International. 2003;**43**:420-427. DOI: 10.2355/isijinternational

[23] Laws MS, Goodhew PJ. Grain boundary structure and chromium segregation in a 316 stainless steel. Acta Metallurgica. 1991;**39**:1525-1533. DOI: 10.1016/0956-7151(91)90238-V

[24] Xu Y, Zhang X, Tian Y, Chen C, Nan Y, He H. Wang M. Study on the nucleation and growth of $M_{23}C_6$ carbides in a 10% Cr martensite ferritic steel after long-term aging. Materials Characterization. 2016;**111**:122-127. DOI: 10.1016/j.matchar.2015.11.023

[25] Golański G, Zieliński A, Słania J, Jasak J. Mechanical properties of VM12 steel after 30 000 hrs of ageing at 600°C temperature. Archives of Metallurgy and Materials. 2014;**59**:1351-1354. DOI: 10.2478/amm-2014-0230

[26] Yan W, Wang W, Shan YY, Yang K. Microstructural stability of 9-12%Cr ferrite/martensite heat-resistant steels. Frontiers of Materials Science. 2013;**7**:1-27. DOI: 10.1007/s11706-013-0189-5

[27] Abe F, Horiuchi T, Taneike M, Sawada K. Stabilization of martensitic microstructure in advanced 9Cr steel during creep at high temperature. Materials and Science Engineering. 2004;**378A**:299-303. DOI: 10.1016/j.msea.2003.11.073

[28] Xiao X, Liu G, Hu B, Wang J, Ma W. Coarsening behawior for M23C6 carbide in 12%Cr-reduced activation ferrite/martensite steel: Experimental study combined with DICTRA simulation Journal of Materials Science. 2013;**48**:5410-5419. DOI: 10.1007/s10853-013-7334-5

[29] Abe F, Taneike M, Sawada K. Alloy design of creep resistant 9Cr steel using a dispersion of nano-sized carbonitrides. International Journal of Pressure Vessels and Piping. 2007;**84**:3-12. DOI: 10.1016/j.ijpvp.2006.09.003

[30] Magnusson H, Sandström R. Influence of aluminium on creep strength of 9-12% Cr steel. Materials Science and Engineering. 2009;**527A**:118-125. DOI: 10.1016/j.msea.2009.07.060

[31] Danielsen HK, Hald J. Behaviour of Z phase in 9-12%Cr steels. Energy Materials. 2006;**1**:49-57. DOI: 10.1179/174892306X99732

[32] Danielsen HK. Precipitation process of Z-phase in 9-12%Cr steels, In: Gandy A, Shingledecker J, editors. Advances in Materials Technology for Fossil Power Plants. 7th International Conference, Waikoloa, Hawaii, USA. 2013: pp. 1104-1115.

[33] Golpayegani A, Andrén HO, Danielsen HK, Hald J. A study on Z–phase nucleation in martensitic chromium steels. Materials Science and Engineering. 2008;**489A**:310-318. DOI: 10.1016/j.msea.2007.12.022

[34] Sawada K, Kushima H, Kimura K, Tabuchi M. TTP diagrams of Z phase in 9-12%Cr heat-resistant steels. ISIJ International. 2007;**47**:733-739. DOI: 10.2355/isijinternational

[35] Sawada K, Kushima H, Kimura K, Tabuchi M. Z-phase formation and its effect on long-term creep strength in 9-12%Cr creep resistant steels. Transactions of the Indian Institute of Metals. 2010;**63**:117-122. DOI: 10.1007/s12666-010-0016-y

[36] Yoshizawa M, Igarashi M, Moriguchi K, Iseda A, Armaki HG, Maruyama K. Effect of precipitates on long-term creep deformation properties of P92 and P122 type advanced ferritic steels for USC power plants. Materials Science and Engineering. 2009;**510-511A**: 162-168. DOI: 10.1016/j.msea.2008.05.055

[37] Li Q. Precipitation of Fe_2W laves-phase and modeling of its direct influence on the strength of a 12Cr-2W steel. Metallurgical and Materials Transactions. 2006;**37A**:89-97. DOI: 10.1007/s11661-006-0155-2

[38] Golański G, Zielińska-Lipiec A, Mroziński S, Kolan C. Microstructural evolution of aged heat resistant cast steel following strain controlled fatigue. Materials Science and Engineering. 2015;**627A**:106-110. DOI: 10.1016/j.msea.2014.12.120

[39] Aghajani A, Richter F, Somsen C, Fries S, Steinbach I, Eggeler G. On the formation and growth of Mo–Rich laves phase particles during long-term creep of a 12% chromium tempered martensite ferritic steels. Scripta Materialia. 2009;**61**:1068-1071. DOI: 10.1016/j.scriptamat.2009.08.031

[40] Zhu S, Yang M, Song XL, Zhang Z, Wang LB, Tang S, Xiang ZD. A few observations on laves phase precipitation in relation to its effects on creep rupture strength of ferritic steels based on Fe-9Cr(wt%) alloys at 650°C. Materials Science and Engineering. 2014;**619A**:47-56. DOI: 10.1016/j.msea.2014.09.059

[41] Isik MI, Kostka A, Eggeler G. On the nucleation of laves phase particles during high-temperature exposure and creep of tempered martensite ferritic steels. Acta Materialia. 2014;**81**:230-240. DOI: 10.1016/j.actamat.2014.08.008

[42] Isik MI, Kostka A, Yardley VA, Pradeep KG, Duarte MJ, Choi PP, Raabe D, Eggeler G. The nucleation of Mo-rich laves phase particles adjacent to $M_{23}C_6$ micrograin boundary carbides in 12%Cr tempered martensite ferritic steels. Acta Materialia. 2015;**90**:94-104. DOI: 10.1016/j.actamat.2015.01.027

[43] Zieliński A, Golański G, Sroka M, Tański T. Influence of long-term service on microstructure, mechanical properties, and service life of HCM12A steel. Materials at High Temperatures. 2016;**33**:24-32. DOI: 10.1179/1878641315Y.0000000015

[44] Lee JS, Armaki HG, Maruyama K, Muraki T, Asahi H. Causes of breakdown of creep strength in 9Cr-1.8W-0.5Mo-VNb steel. Materials Science and Engineering. 2006;**428A**:270-275. DOI: 10.1016/j.msea.2006.05.010

[45] Zieliński A, Golański G, Sroka M. Assessment of microstructure stability and mechanical properties of X10CrWMoVNb9-2 (P92) steel after long-term thermal ageing for high-temperature applications. Kovove Materialy. 2016;**56**:61-70. DOI: 10.4149/km-2016-1-61

[46] Zieliński A, Dobrzański J, Purzyńska H, Golański G. Changes in properties and microstructure of high-chromium 9-12%Cr steels due to long–term exposure at elevated temperature. Archives of Metallurgy and Materials. 2016;**61**:957-964. DOI: 10.1515/amm-2016-0163

[47] Paul VT, Soraja S, Vijayalakshmi M. Microstructural stability of modified 9Cr–1Mo steel during long term exposures at elevated temperatures. Journal of Nuclear Materials. 2008;**378**:273-281. DOI: 10.1016/j.jnucmat.2008.06.033

[48] Yan P, Liu Z. Toughness evolution of 9Cr-3W-3Co martensitic heat resistant steel during long time aging. Materials Science and Engineering. 2016;**650A**:290-294. DOI: 10.1016/j.msea.2015.09.115

[49] Golański G, Kępa J. The influence of ageing temperatures on microstructure and mechanical properties of GX12CrMoVNb9-1 (GP91) cast steel. Archives of Metallurgy and Materials. 2012;**57**:757-582. DOI: 10.2478/v10172-012-0061-0

[50] Park JS, Kim SJ, Lee CS. Effect of W addition on low cycle fatigue behavior of high Cr ferritic steels. Materials Science and Engineering. 2001;**298A**:127-136. DOI: 10.1016/S0921-5093(00)01291-0

[51] Artinger I, Elarbi Y. Effect of aging time and boron addition on the properties of 9-12%Cr power plant steels—Outcomes from different experimental investigations. Periodica polytechnical/Mechanical Engineering. 2006;**50**:3-10

[52] Yamada K, Igarashi M, Muneki S, Abe F. Effect of co addition on microstructure in high Cr ferritic steels. ISIJ International. 2003;**43**:1438-1443. DOI: 10.2355/isijinternational.43.1438

[53] Helis L, Toda Y, Hara T, Miyazaki H, Abe F. Effect of cobalt on the microstructure of tempered martensitic 9Cr steel for ultra-supercritical power plants. Materials Science and Engineering. 2009;**510-511A**:88-94. DOI: 10.1016/j.msea.2008.04.131

[54] Maruyama K, Sawada K, Koike J. Strengthening mechanisms of creep resistant tempered martensitic steel. ISIJ International. 2001;**41**:641-653. DOI: 10.2355/isijinternational

[55] Naoi H, Ohgami M, Liu X, Fujita T. Effects of aluminium content on the mechanical properties of a 9Cr–0.5Mo–1.8W steel. Metallurgical and Materials Transactions. 1997;**28A**:1195-1203. DOI: 10.1007/s11661-997-0284-2

[56] Foldyna V, Jakobová A, Riman R, Gemperle A. Effects of structural factors on the creep properties of modified chromium steels. Steel Research International. 1991;**60**:453-458. DOI: 10.1002/srin.199100429

Chapter 4

High Temperature Creep of Metal Oxides

Krystyna Schneider and Mieczyslaw Rekas

Additional information is available at the end of the chapter

http://dx.doi.org/10.5772/intechopen.70876

Abstract

This chapter presents a comprehensive review of the creep technique used for the study of defect structure and diffusion in metal oxides, both single crystals and ceramics. At high temperatures, the creep rate is proportional to the diffusion coefficient of the slowest species in solid compounds, whatever deformation mechanisms are present (Nabarro viscous creep, recovery creep or pure climb creep). The creep rate dependence on deviation from stoichiometry can be determined from this diffusion. In the case of metal oxides, the departure from stoichiometry is controlled by the oxygen activity which usually is identified with oxygen partial pressure, p_{O_2}. The p_{O_2} dependence of the creep rate provides direct information about the nature of minority point defects. On the other hand, studies of the temperature dependency of the creep rate inform us about the activation energy of the diffusion coefficient.This review focuses primarily on the creep behavior of transition metal oxides such as $Ni_{1-y}O$, $Co_{1-y}O$, $Fe_{1-y}O$ exhibiting disorder in metal sublattice, as well as ZrO_{2-x} with majority defects in oxygen sublattice. The advantage of these studies is determination of both defect structure and diffusion coefficients of minority defects namely in oxygen sublattice in iron-triad oxides and in zirconium ZrO_2 sublattice.

Keywords: high temperature creep of metal oxides, defect structure, nonstoichiometry, diffusion

1. Introduction

This chapter presents a review of the creep technique used for the study of defect structure and diffusion in metal oxides. Transition metals of the iron-row triad (Fe, Co, Ni) monoxides were chosen. These oxides at higher temperatures exhibit electronic conductivity. Zirconia stabilized with yttria, termed as YSZ, was also the subject of interest. This material is treated as model system of ionic (super ionic) conductor. Taking into account that creep behavior of the oxides is generally dependent of the minority defects properties (types and concentrations) which are especially strictly related with majority point defect, a short description of the point defect structure of each analyzed oxide proceeds the appropriate chapter.

2. Definition of terms

Creep, or more generally high-temperature plastic deformation of metal oxides, is a suitable method for measuring diffusion coefficients and to assess the nature of point defects responsible for diffusion processes. Above a so called Tamman's temperature and under constant applied stress, a steady state is reached where the rate of deformation remains constant. Tamman's temperature of metal oxides assumes value of about $0.45 \times T_{melt}$ [K] [1] creep rates, $\dot{\varepsilon}$, is commonly expressed by the equation [2]

$$\dot{\varepsilon} = \frac{d\varepsilon}{dt} = \frac{A\sigma^m}{d^b} exp\left[-\frac{Q}{kT}\right] \tag{1}$$

where ε is the creep strain, A is constant dependent of the material structure and the creep mechanism, σ is the applied stress, m and b are exponents dependent on the creep mechanism, d is the grain diameter, k is Boltzmann's constant, T is the absolute temperature and Q is an activation energy of the creep.

At high temperatures, the creep rate is proportional to some diffusion coefficient, whatever the exact nature of deformation mechanism: Nabarro-Hearing creep, grain boundary sliding and dislocation movements are valid [3]. This diffusion coefficient, D, is an effective diffusivity involving both the lattice and the grain boundaries diffusivities. In most cases, it reduces to the diffusivity of the slowest species. In case of transition metal oxides, the creep rate dependence on deviation from stoichiometry is carried out through diffusivity. In binary oxides, at high temperatures the deviation from stoichiometry and related majority point defects are controlled by the oxygen partial pressure, p_{O_2}. In most cases, these defects are only located either in metal or oxygen sublattice. Defects on other sublattice are present in a much lower concentration (named as 'minority defects'), and the diffusivity of the corresponding species is very low. Their diffusion coefficient is proportional to the concentration of minority defects, which also varies as some characteristic power law with p_{O_2}, so that creep rate has related to this variable. Accordingly, the p_{O_2} dependence of the creep rate leads to a direct approach to assess the nature of the minority point defects [2]

$$\dot{\varepsilon} = \frac{d\varepsilon}{dt} = \frac{C\sigma^m}{d^b} p_{O_2}^{\frac{1}{n}} exp\left[-\frac{E_{act}}{kT}\right] \tag{2}$$

where $C = Ap_{O_2}^{\frac{-1}{n}}$, $E_{act} = Q$ is the energy activation of diffusion and n is the parameter dependent of the point defect structure.

3. Creep of transition metal monoxides

The transition metal monoxides such as iron-triad group (Fe, Co, Ni) show only metal deficiency: $Fe_{1-y}O$, $Co_{1-y}O$ and $Ni_{1-y}O$. The nonstoichiometry and related point defect structure discussed below for these oxides are based on both thermogravimetric equilibration and electrochemical measurements. The nonstoichiometry, y, is most pronounced with $Fe_{1-y}O$ (y = 0.043–0.167) [4, 5], whereas in $Co_{1-y}O$ [6] and $Ni_{1-y}O$ [7] y is about 0.000–0.012 and 0.0000–0.0010, respectively.

Nonstoichiometry y in $M_{1-y}O$ (M = Fe, Co, Ni) can be modified by changes of oxygen activity, p_{O_2}, in thermodynamic equilibrium with the oxide crystal (Kröger and Vink [8] notation of defects is used throughout this paper):

$$\frac{1}{2}O_2 \Leftrightarrow V_M^{i-} + ih^\bullet + O_O \quad (3)$$

where i = 0, 1, 2 denotes the degree ionization. The nonstoichiometry of the $M_{1-y}O$ is then given by:

$$y = [V_M^x] + [V_M'] + [V_M''] \quad (4)$$

Applying the mass action law to reaction (3) and assuming that interaction among defects may be neglected, as well as supposing appropriate electroneutrality conditions, one arrives at the following relationship between nonstoichiometry and equilibrium oxygen pressure and temperature:

$$y = y_o p_{O_2}^{1/n_y} \exp\left(-\frac{E_y}{RT}\right) \quad (5)$$

where $1/n_y$, or simply n_y, is the parameter depending on the ionization degree of cation vacancies, assuming 2, 4 and 6 values for neutral, single- and double-ionized vacancies, respectively; E_y denotes the temperature coefficient directly related to the enthalpy of defect formation.

$M_{1-y}O$ (M = Fe, Co, Ni) are p-type semiconductors, which electrical conductivity is realized by electron (or electron hole) hopping mechanism between M^{3+} and M^{2+} ions. The concentration of quasi-free electron holes $[h^\bullet]$ is given by:

$$[h^\bullet] = [V_M'] + 2[V_M''] \quad (6)$$

Accordingly, the following relationship between electrical conductivity, σ and equilibrium oxygen pressure and temperature is given by:

$$\sigma = e\mu_h[h^\bullet] = \sigma^o\, p_{O_2}^{1/n_\sigma} \exp\left(-\frac{E_\sigma}{RT}\right) \quad (7a)$$

or

$$[h^{\bullet}] = h^{o}\left(p_{O_2}\right)^{\frac{1}{n_{\sigma}}} \exp\left(-\frac{E_{\sigma}}{RT}\right) \tag{7b}$$

where e is the elementary charge, μ is mobility of the holes, σ^{o} and h^{o} are constants involving mobility of electron holes and entropy terms of the equilibrium constant of the reaction (3), n_{σ} is the parameter depending on the ionization degree of cation vacancies, assuming ∝, 4 and 6 values for neutral, single- and double-ionized vacancies, respectively; E_{σ} denotes the activation energy of the electrical conductivity.

Several minority defects can be proposed in $M_{1-y}O$. Considering only simple defects oxygen vacancies V_O^{j+} (j = 1 or 2) or oxygen interstitials O_i^{k-} (k = 1 or 2) the formation of these defects is described by the following:

$$O_O + j\,h^{\bullet} \Leftrightarrow V_O^{j+} + \frac{1}{2}O_2 \tag{8}$$

and

$$\frac{1}{2}O_2 \Leftrightarrow O_i^{k-} + k\,h^{\bullet} \tag{9}$$

Applying the mass action law to these equilibriums and taking into account the Eq. (7b), we can write the following relationships:

$$\left[V_O^{j+}\right] = K_j^{o}(h^{o})^{j}\left(p_{O_2}\right)^{\left(\frac{j}{n_{\sigma}}-\frac{1}{2}\right)} \exp\left[\frac{-jE_{\sigma} + \Delta H_j)}{RT}\right] \tag{10}$$

and

$$\left[O_i^{k-}\right] = K_i^{o}(h^{o})^{-k}\left(p_{O_2}\right)^{\left(\frac{1}{2}-\frac{k}{n_{\sigma}}\right)} \exp\left[\frac{-\Delta H_k - kE_{\sigma})}{RT}\right] \tag{11}$$

From Eqs. (2) and (10),

$$\frac{1}{n} = \frac{j}{n_{\sigma}} - \frac{1}{2} \tag{12}$$

and

$$E_{act} = jE_{\sigma} + \Delta H_j \tag{13}$$

Similarly, from Eqs. (2) and (11),

$$\frac{1}{n} = \frac{1}{2} - \frac{k}{n_{\sigma}} \tag{14}$$

and

$$E_{act} = \Delta H_k - kE_\sigma \tag{15}$$

So, using Eqs. (12) and (13) or Eqs. (14) and (15) carrying out the creep rates measurements as a function of oxygen partial pressure and temperature, we can determine both parameters n and $E_{act.}$ Then, using the values of electrical properties such as electrical conductivity (or Seebeck coefficient) which are related to the concentration of the electron holes, we can identify the type of minority defect responsible for the creep diffusion process and can determine the activation energy of the diffusion.

3.1. Creep of NiO

Nickel oxide is a nonstoichiometric compound with a deficit of metal. Its chemical formula may be written as $Ni_{1-y}O$. The resulting majority point defects are nickel vacancies formula may be written as $Ni_{1-y}O$. The resulting majority point defects are nickel vacancies: V_{Ni}', V_{Ni}'' and electron holes (**Figure 1**) [7]. The dependence of electrical conductivity, σ, as a function of oxygen partial pressure, p_{O_2} and temperature is shown in **Figure 2** [9, 10]. The reciprocal of oxygen exponent $n_\sigma = n_h$ varies values between 4 and 6.

The results of the creep studies of NiO single crystals [11, 12] and NiO polycrystals [3, 13, 14] are summarized in **Table 1**. The comparison of the activation energies for creep rate with oxygen self-diffusion obtained by Dubois et al. [15, 16] agrees only with a rather large uncertainty [17–19]. Controversy, **Figure 3** presents the diffusion coefficients determined by tracer of nickel [17, 18] and by the tracer self-diffusion of oxygen [19] on one hand, and the diffusion coefficient determined using the creep technique, on the other. The excellent agreement is observed between oxygen diffusion coefficient and the diffusion coefficient from the creep.

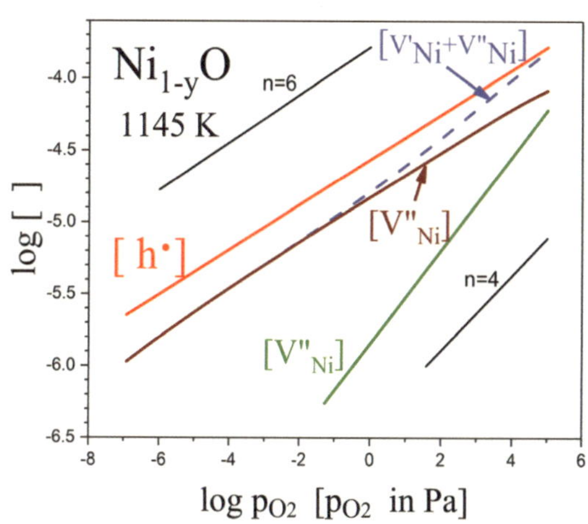

Figure 1. Majority defect concentration in $Ni_{1-y}O$. Defect concentrations ([ ]) are unitless values. They are expressed as Ni-site ratio.

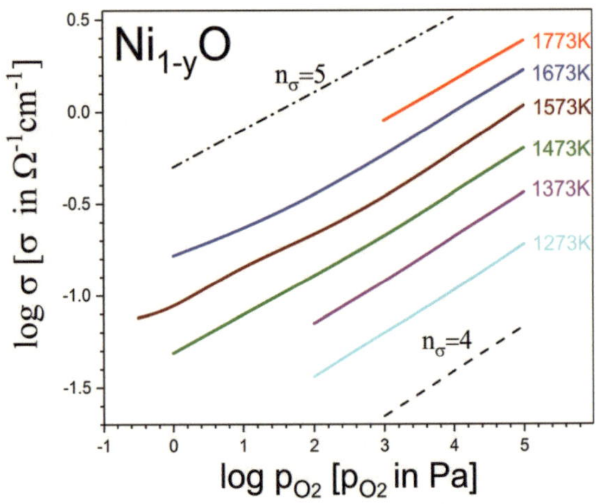

Figure 2. Electrical conductivity of at 1145 K $Ni_{1-y}O$ [9, 10].

Material	T/ T_m**	σ (MPa)	p_{O_2} (Pa)	m	E_{act} (eV)	n	Minority defect	Mechanism of creep	Ref.
NiO s.crystal	0.09–0.66	20–100	0.1–2 × 10^4	NA	NA	NA	NA	NA	[11]
NiO s.crystal	0.6–0.8	15–120	1–2 × 10^4	12 7	5.4–8.5	−0.06 to +0.11	$V_O^{\bullet\bullet}$	Thermally activated glide recovery creep	[12]
NiO polycrytals	0.6–0.8	6–20 20–90 90	1–2 × 10^4	1.5 7.9	3.8	0–0.03	$V_O^{\bullet\bullet}$	Nabarro creep	[13]
NiO polycrytals	0.56–0.61	34.5–79.8	2 × 10^4	3.25 ± 0.18	2.46 ± 0.31	NA	NA	Diffusion controlled climb glide	[25]
CoO s.crystal	0.49–0.66	6.9–31	2.1 × 10^4 1.1 × 10^{-4}	4.6 3.3	2.9 1.77	NA	NA	Diffusion	[24]
CoO s.crystal	0.61–0.75	8.3–14.5	NA	5	2.2	NA	NA	Diffusion controlled dislocation motion	[25]
CoO s.crystal	0.61–0.80	2–20	10^{-6} 1 × 10^5	4.4–5.6	3.1–4.7	2	$V_O^{\bullet\bullet}$ \ O_i^x, O_i^- O_i^{2-}	Oxygen diffusion	[26]
CoO s.crystal	0.6 0.8	5–25	1 2 × 10^4	8.5 6.5	5 2.5–5	0.5 0.1	O_i ” $V_O^{\bullet\bullet}$	Oxygen diffusion	[27]
CoiO polycrytals	0.58–0.62	34.5–79.8	2 × 10^4	3	3.12	NA	Na	Dislocation motion	[14]
FeO polycrystals	0.73–0.95	2.5–15	10^{-10}–3 × 10^{-8}	4.15 ± 0.10	2.8	NA	$V_O^{\bullet\bullet}$	Climbing dislocations	[37]
FeO s.crystal	0.67 0.85	3.2–5.3	3 × 10^{-13}–2 × 10^{-6}	5.3 4.2	3.0 ± 0.4	−0.015 +0.11	O_i'' O_i'	Diffusion controlled recovery	[38]

*Denotation of symbols, see Eq. (2).
**T_m – Melting temperature (K).

Table 1. Creep results of NiO, CoO and FeO*.

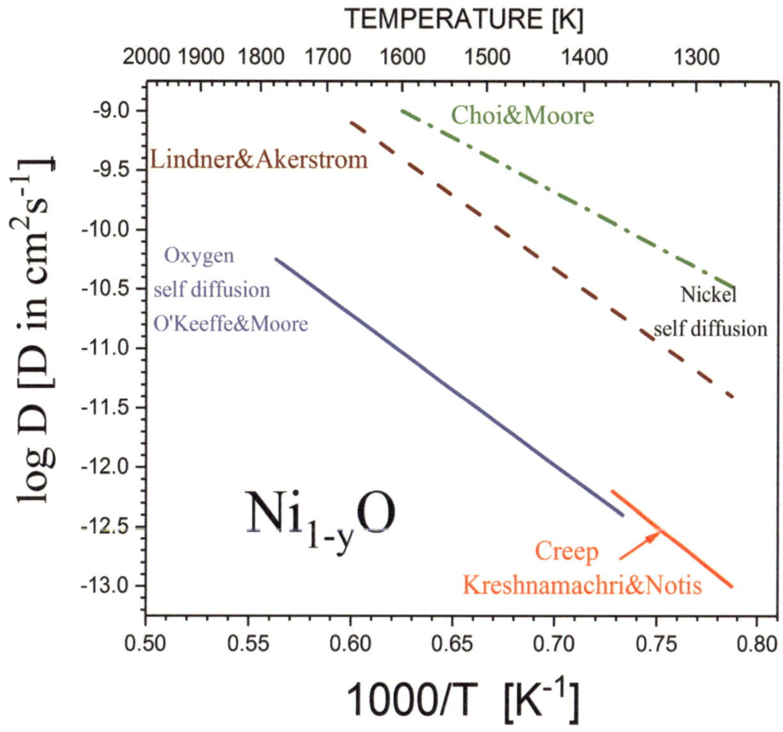

Figure 3. Temperature dependence of diffusion coefficients [14, 17–19].

3.2. Creep of CoO

Taking into account both crystal structure and nonstoichiometry departure, CoO shows similarities to NiO. However, there is a major difference at low temperature (below 1170 K) and high oxygen pressure CoO is unstable and transforms to the spinel-type Co_3O_4 (**Figure 4**). Also the departure of stoichiometry, y, in $Co_{1-y}O$ is about 10 times higher than that of $Ni_{1-y}O$. Traditional analysis of point defect structure in $Co_{1-y}O$ based on the ideal defect model with the assumption that point defects in $Co_{1-y}O$ are randomly distributed and do not interact with each other, leads to the conclusion that at high p_{O_2} (close to CoO/Co_3O_4 border) the predominant ionic point defects are $V_{Co}^{\times}$ and V_{Co}'; at the intermediate region of p_{O_2} both V_{Co}' and V_{Co}'' coexist together as the predominant ionic point defects; and finally near the border Co/CoO the doubly ionized cobalt vacancies V_{Co}'' are predominant [20]. However, the assumption of validity of ideal point defects in CoO is questionable. The ideal point defect model gives satisfactory results when nonstoichiometry departure and related point defect concentration is below 0.1 at%, for higher concentrations interaction between defects must be taken into account [21].

The interaction between defects in $Co_{1-y}O$ can be well described by the Debye-Hückel approach [22]. In addition, the model proposed is based on the assumption that the only type of predominant ionic defects present in cobalt monoxide at elevated temperatures are double-ionized cation vacancies, V_{Co}'' instead of two or even more types of ionic defects considered before. Despite the simplicity of the model, the agreement with the experiment is very good within the entire range of CoO stability.

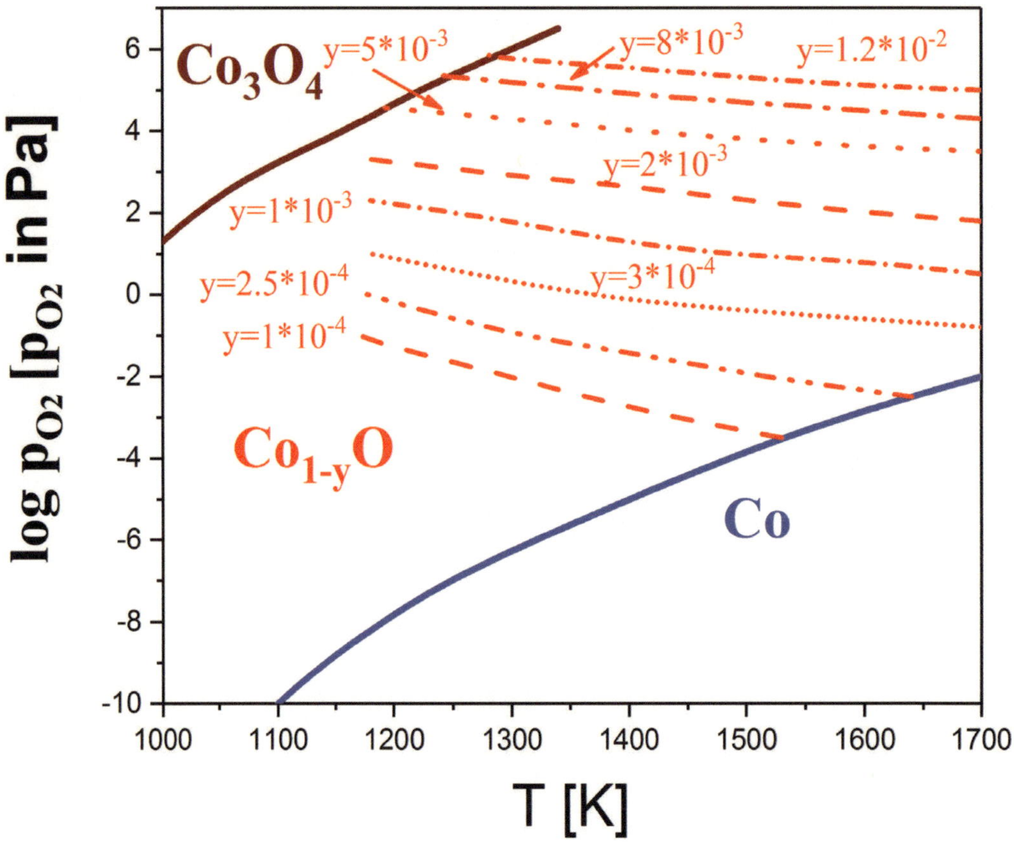

Figure 4. Stability range of $Co_{1-y}O$ [6, 20].

Apart from the Debye-Hückel model, there is also another model making allowance for interaction between the defects which explains well the experimental data of CoO, in particular at high values of deviation from stoichiometric composition (i.e. close to the $Co_{1-y}O/Co_3O_4$. This so called cluster model involves the formation of 4:1 clusters consisting of four double-ionized octahedral cobalt vacancies and one trivalent Co ion in tetrahedral interstitial position: $[(V_{Co})_4Co_i]^{3-}$ [23]. **Figure 5** [20] shows the plots of electrical conductivity, σ, as a function of oxygen partial pressure and temperature. The dependences shown can be used to determine the reciprocal of oxygen exponent $n_\sigma = n_h$ defined in Eq. (7a). The exact value of n_h is needed in the interpretation of the creep rate versus p_{O_2}.

The results of the creep studies of CoO single crystals [24–28] and CoO polycrystals [14] are summarized in **Table 1**. A comparison of the activation energies for creep rate with oxygen self-diffusion obtained by Dubois et al. [15, 16] agrees only with a rather large uncertainty.

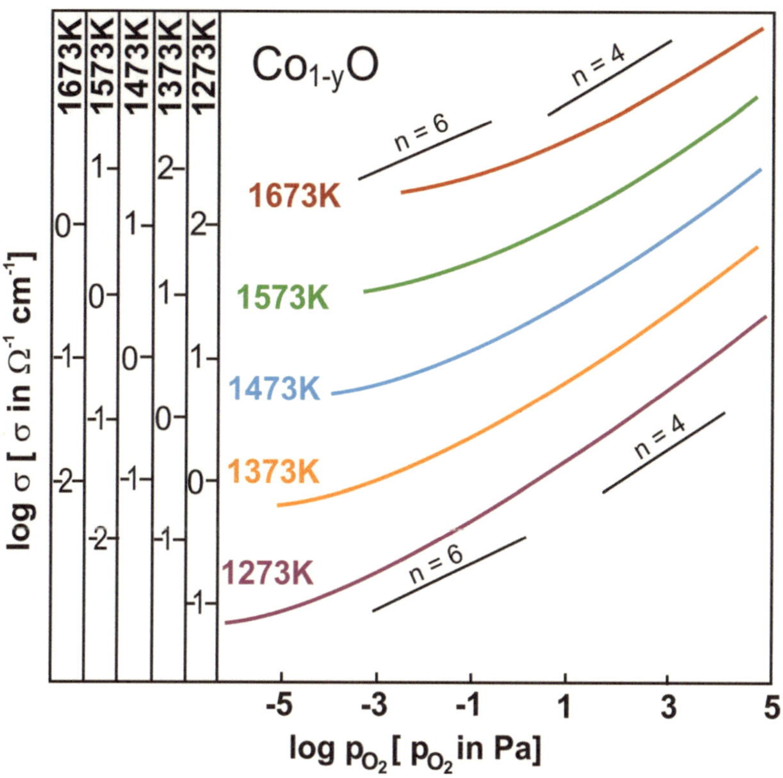

Figure 5. Electrical conductivity, σ, as a function oxygen partial pressure and temperature [20].

3.3. Creep of FeO

Iron monoxide, termed also as wüstite, has the NaCl-type structure. Compared to other binary oxides of iron-row metals, the wüstite phase exhibits the highest deviation from stoichiometry, varying from 4.3 at% at the $Fe/Fe_{1-y}O$ phase boundary up to 16.7 at% at the $Fe_{1-y}O/Fe_3O_4$ interphase (**Figure 6**) [4, 5, 29]. The non-stoichiometry occurs because of the ease of oxidation of Fe^{2+} to Fe^{3+}effectively replacing a small portion of Fe^{2+} with two-thirds their number of Fe^{3+}, which take up tetrahedral positions in the close packed oxide lattice. The considerable concentration of defects, resulting from nonstoichiometry, leads to their strong interactions and formation of complexes. On the basis of neutron diffraction studies, Roth [30] has proposed the formation of defect complexes composed of two iron vacancies and interstitial iron $\left[(V_{Fe}'')_2 Fe_i^{3+}\right]'$. This defect is similar to an element of the spinel structure of magnetite Fe_3O_4. Therefore, the Roth complexes can be formally considered as a magnetite-type defect or a submicrodomain of Fe_3O_4 in FeO [28, 31]. On the basis of X-ray studies, Koch and Cohen [32] have postulated that defects in $Fe_{1-y}O$ form associates even larger than Roth 2:1 complexes, composed of 13 iron vacancies and 4 interstitials (termed as cluster 13:4). The formation of the clusters was later confirmed by the neutron diffraction

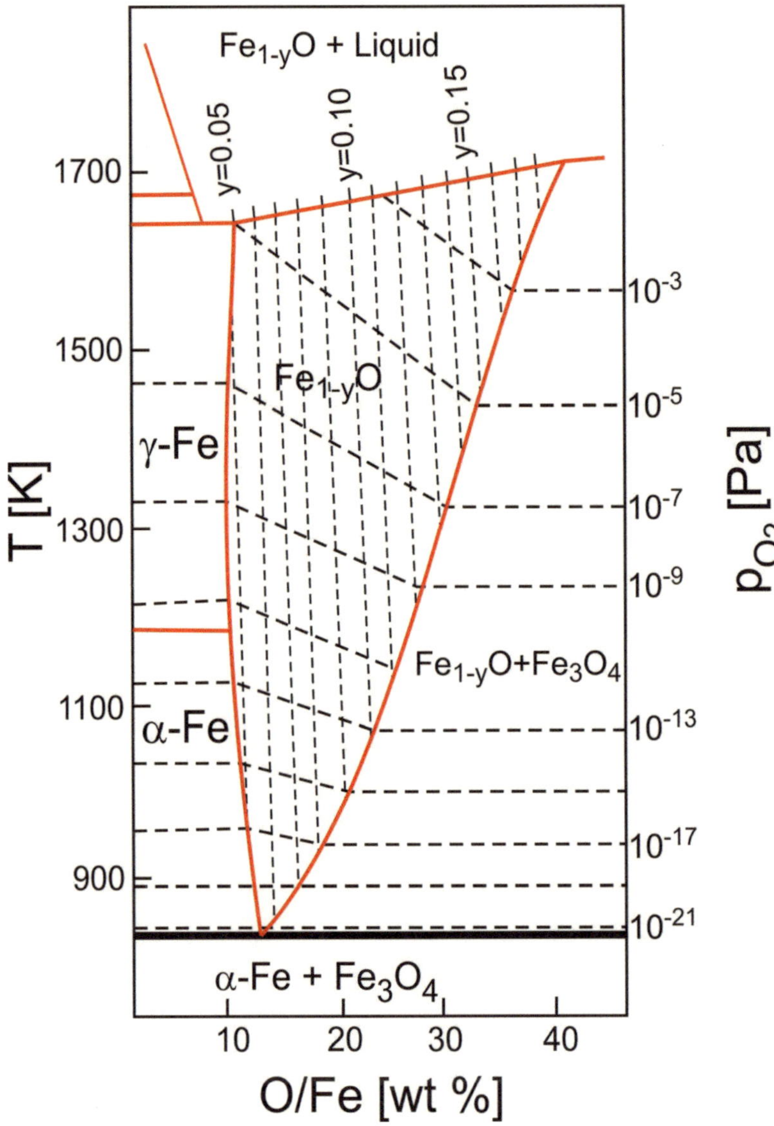

Figure 6. Stability range of $Feo_{1-y}O$ [4, 5, 29].

studies of Cheetham et al. [33, 34]. Catlow et al. [35, 36] using the computer simulation method have proposed various kinds of defect clusters such as 4:1, 6:2. 8:3, 10:4, 12:4 and 16:5. It has been postulated that the 4:1 cluster is a constructional unit in formation of higher order clusters. The cluster 16:5 resembles the magnetite structure and can be considered as a Fe_3O_4-type microdefect.

The creep studies of FeO single crystals were performed by Ilschner et al. [37] and by Jolles and Monty [38]. The required parameter $n_h = n = \frac{e}{k}\left[\frac{\partial\alpha}{\partial \ln p_{O_2}}\right]^{-1}$ (where α is Seebeck coefficient) needed

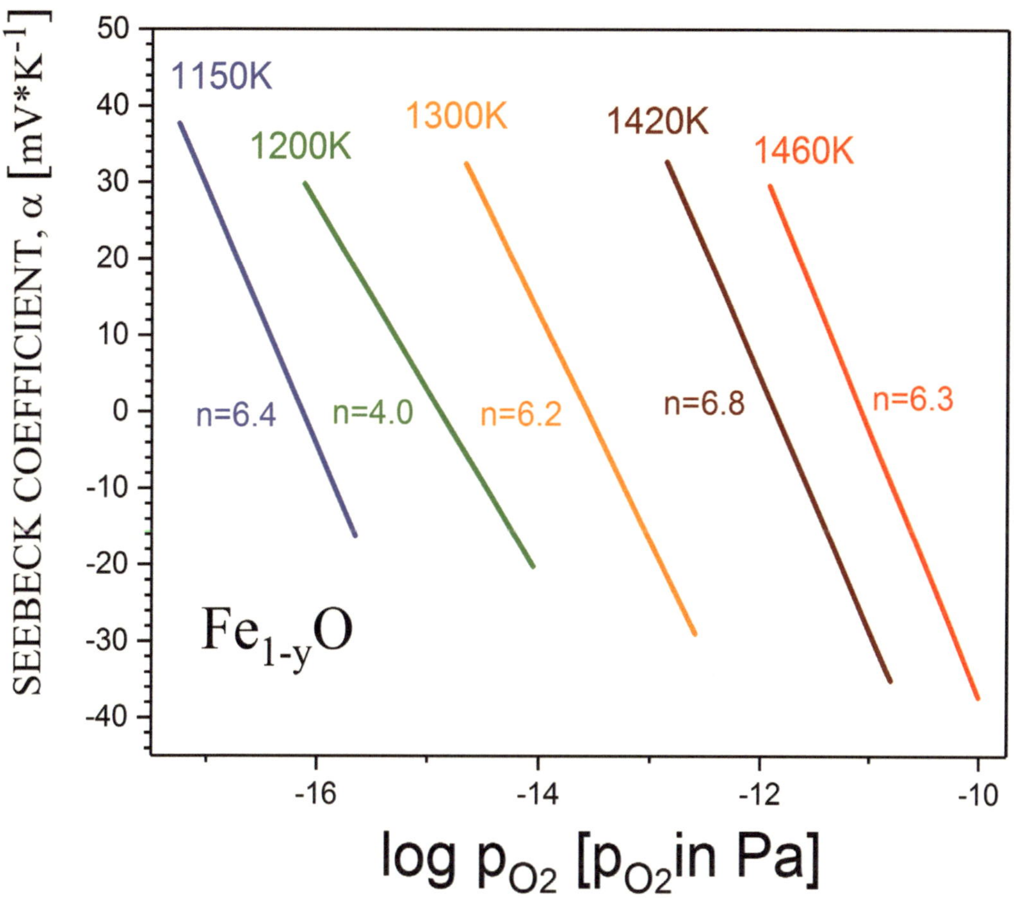

Figure 7. Seebeck coefficient, α, of $Fe_{1-y}O$ [31].

to interpretation of the creep rate versus p_{O_2} is presented in **Figure 7**. The results of the creep studies are summarized in **Table 1**.

4. Zirconia-based materials

Zirconium dioxide, ZrO_2 has three polymorphic modifications: monoclinic below 1440 K, tetragonal 1440–2640 K and cubic above 2640 K. The transition between them involves a large volume expansion cause a cracking upon cooling from high temperature, which is detrimental to the materials structural applications. The destructive phase transformation can be suppressed by total or partial stabilization of high temperature modifications (cubic or tetragonal form). This stabilization consists on the addition to ZrO_2 several mole % of MgO, CaO or Y_2O_3 among others. The most popular stabilizer is Y_2O_3 termed as yttria.

Yttria-stabilized zirconia (YSZ) ceramics is a strategic functional material which has found extensive applications in electrochemical devices, such as gas sensors, solid oxide fuel cells and electrolysis cells.

Also the YSZ well known as an excellent construction material due to such properties as: high hardness, low wear resistance, low coefficient of friction, high elastic modules, chemical inertness, low thermal conductivity and high melting point [39–42]. It is also recognized that the useful mechanical properties are obtained in multiphase material known as partially stabilized zirconia (PSZ). Garvie and Nicholson [43] have demonstrated that a fine-scale precipitate of monoclinic zirconia in a cubic stabilized matrix enhances the strength of PSZ. Very interested mechanical properties were discovered in fine-grained tetragonal ZrO_2 stabilized with 3 mol% Y_2O_3 termed as 3Y-TZP [44, 45]. The macroscopic and microscopic behavior of 3Y-TZP can be characterized as structural or micrograin superplasticity, according to the definition used by metallurgists. The 170% elongation reported by Wakai et al. [44] on 3Y-TZP has been considered as evidence of ceramic superplasticity in the ceramic materials. 3Y-TZP, termed as 'ceramic steel' [46, 47], is now considered to be the model ceramic system. The fine grade size leads to a very dense, non-porous ceramic with excellent mechanical strength, corrosion resistance, impact toughness, thermal shock resistance and very low thermal conductivity. Due to its characteristics Y-TZP is used in wear parts, cutting tools and thermal barrier coatings.

Also, electrical properties of 3Y-TZP are very interesting. At moderate temperatures below 970 K the grain interior of Y-TZP has higher conductivity [47, 48] than that of fully (YSZ) or partially stabilized (PSZ) zirconia [50]. However, the total conductivity of Y-TZP is lower due to the high contribution of grain boundary resistivity, known as the blocking effect [47, 49]. It was found that addition of Al_2O_3 leads considerable reduction of the blocking effect [50].

4.1. Defect structure of YSZ

The defect reactions in the yttria-stabilized zirconia can be written as:

$$Y_2O_3 \rightarrow \quad 2Y'_{Zr} + V_O^{\bullet\bullet} + 3\,O_O^x \tag{16}$$

$$O_O^x \Leftrightarrow V_O^{\bullet\bullet} + 2\,e' + \frac{1}{2}\,O_2 \tag{17}$$

$$nil \Leftrightarrow e' + h^{\bullet} \tag{18}$$

The majority defect oxygen in yttria-stabilized zirconia are oxygen vacancies $V_O^{\bullet\bullet}$ and yttrium ions Y^{3+} (Y'_{Zr} using defect notation) occupying zirconium positions the electroneutrality conditions is given by:

$$2\left[V_O^{\bullet\bullet}\right] = \left[Y'_{Zr}\right] \tag{19}$$

Ionic conductivity of ZrO_2 + 10 mol% Y_2O_3, termed as 10YSZ resulting from the Eq. (19) is illustrated in **Figure 8** [51].

Combining Eqs. (16)–(19) we can determine concentration of both electronic defects:

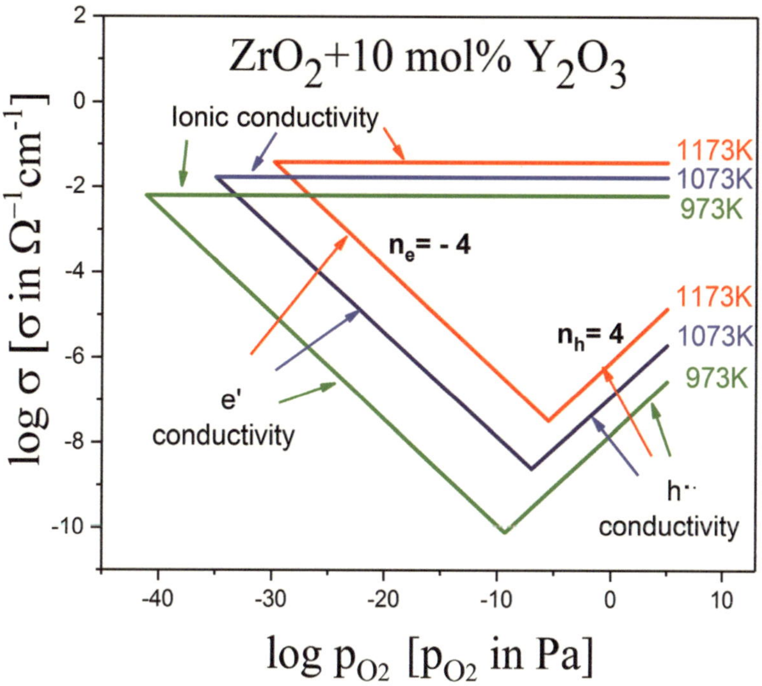

Figure 8. Both ionic and electronic components of the electrical conductivity of yttria-stabilized zirconia (10YSZ) in the range 973–1173 K as a function of p_{O_2}, according to Weppner [51].

$$[e'] = \left\{ \frac{2K_V}{[Y'_{Zr}]} \right\}^{1/2} p_{O_2}^{-1/4} \tag{20}$$

and

$$[h^{\bullet}] = \left\{ \frac{K_i^2 [Y'_{Zr}]}{2K_V} \right\}^{1/2} p_{O_2}^{1/4} \tag{21}$$

where K_V and K_i are equilibrium constants of reactions (17) and (18), respectively. Dependences of electrons and electron holes conductivities are presented in **Figure 8**.

The possible minority defects are: zirconium interstitials $Zr_i^{\bullet\bullet\bullet\bullet}$ and zirconium vacancies V''''_{Zr}. They occupy positions in zirconium sublattice and form according to the following reactions:

$$2O_O^x + Zr_{Zr}^x \Leftrightarrow Zr_i^{\bullet\bullet\bullet\bullet} + 4e' + O_2 \tag{22}$$

and

$$O_2 + 4e' \Leftrightarrow 2O_O^x + V''''_{Zr} \tag{23}$$

Applying the mass action law to reactions (22) and (23) gives:

$$[Zr_i^{\bullet\bullet\bullet\bullet}] = K_{22}[e']^{-4}p_{O_2}^{-1} \tag{24}$$

and

$$[V_{Zr}^{''''}] = K_{23}[e']^4 p_{O_2} \tag{25}$$

where K_{22} and K_{23} are the equilibrium constants of reactions (22) and (23), respectively.

4.2. Creep of single crystalline zirconia

Marcartney et al. [52] studied creep of single crystal zirconia stabilized with 12 mol% CaO within 1623–1723 K. They observed considerable climb and glide formation. Dominguez-Rodriguez et al. [53–56] examined cubic YSZ (9.4–18 mol% Y_2O_3) single crystals. The determined activation energies of the deformation process were 646.4; 800.8 and 944.7 kJ/mol for 9.4, 12 and 18 mol% of yttria, respectively. Lankford [57] studied the deformation of two single crystals of YSZ containing 2.8 and 12 mol% Y_2O_3.The temperature range of the experiments was 296–1423 K. The plastic flow with rapid decrease in strength with increasing temperature was observed. The deformation was attributed to dislocation activity alone. Corman [58] examined the creep of single crystal 9.5 mol% yttria partially stabilized zirconia, within temperature range 1923–2123 K and stress from 12.5 to 100 MPa.

The stress exponent, m, and activation energy Q (Eq. (1)) were m = 4 and Q = 436 kJ/mol. This value is close to the lattice diffusion of Zr (418 kJ/mol). So, he concluded that cation diffusion is controlled the process deformation.

4.3. Creep of polycrystalline zirconia

The dependence of steady state creep rate, $\dot{\varepsilon}$; on stress, σ; temperature, T and grain size, d of polycrystalline zirconia is given by a general relationship of the form:

$$\dot{\varepsilon} = \frac{ADGb}{kT}\left(\frac{b}{d}\right)^p\left(\frac{\sigma}{G}\right)^m \tag{26}$$

where A is a dimensionless constant, G is the shear modulus, b is the Burgers vector, k is Boltzmann's constant, p and m are constants termed the inverse grain size and stress exponent, respectively and D is the appropriate diffusion coefficient given by:

$$D = D_o\, exp\left(-\frac{Q}{RT}\right) \tag{27}$$

where D_o is a frequency factor, Q is the appropriate activation energy and R is the gas constant.

Creep of polycrystalline zirconia was recently intensively studied. The subjects of interest were cubic stabilized zirconia, CSZ [59–64], partial stabilized zirconia, PSZ [65–67] as well as tetragonal zirconia polycrystals, TZP [44, 68–73].

St Jacques and Angers [59] examined zirconia stabilized with 18 mol% CaO in the temperatre range 1470–1670 K and stress between 3.5 and 27.5 MPa. The parameter m in Eq. (26) was equal to 1, which suggested diffusionally controlled creep. The detailed study by Dimos and Kohlstedt [60] on a cubic 25YSZ indicated that the creep is controlled by the Nabarro-Herring diffusion mechanism with parameters $m \approx 1$ and $p \approx 2$. Wakai et al. [61, 63] reported that a stress exponent $m \approx 2$ in an 8YSZ. Sharif et al. [64] studying a static grain growth claimed that cubic zirconia is prone to extensive grain growth.

Evans [66] studied deformation of yttria- and scandia- partially stabilized zirconia, PSZ ($1436 < T < 1808$ K; $4.1 < \sigma < 7.1$ MPa). The activation energies were 373 and 360 kJ/mol for scandia and yttria, respectively, the parameter $p = 2$ was found, the parameter m was 1.5 for scandia stabilized zirconia and it assumed two values $m = 1$ and $m = 6$ for yttria. Seltzer, and Talty [67] studied Y-PSZ at high temperatures (up to 2270 K), they found $m = 1.5$ for fine-grained samples (d varied from 15 to 20 μm) and activation energy was 531 kJ/mol. Chevalier et al. [68] studied PSZ materials stabilized by MgO, they concluded that cavitations and microcracking by grain boundary sliding have been identified as the main creep mechanisms.

Wakai et al. [44] examined creep of 3Y-TZP within temperature range of 1423–1723 K. The stress exponents $m = 1.5$ at 1423 K and gradually increased with temperature reaching value $m = 1.9$ at 1723 K. The activation energy was 586 kJ/mol. Nauer and Carry [69] investigated the creep behavior TZP containing 2, 3 and 4 mol% Y_2O_3. At lower stresses ($\sigma < 10$ MPa) the parameters were: $m = 2.4$ and $p = 1$ and activation energy $Q = 590$ kJ/mol. At higher stresses ca. 100 MPa, the parameter $m = 1$. Authors proposed two different deformation regimes: an interface reaction controlled creep at low stresses and a grain boundary diffusion controlled creep at high stresses. Lakki et al. [70] studied 2Y-TZP material using two techniques the creep and the internal friction (it consists on applying a cyclic stress $\sigma = \sigma_o \cos(\omega t)$ to a sample and measuring its response in the form of strain, ε). Both techniques provide at low stresses (< 15 MPa) close values of the activation energies and lead to conclusion that grain boundary sliding is the main deformation mechanism. Owen and Chokshi [71] and Chokshi [72] investigated 3Y-TZP. The experimental data over a wide range of stresses revealed a transition in stress exponent. Deformation at low and high stress regions was associated with $m \approx 3$, $p \approx 1$ and $m \approx 2$, $p \approx 3$, respectively. The activation energy was $Q = 550$ kJ/mol for both regions. Authors postulated that the interface reaction is controlled at low stress region and the grain boundary sliding at high stress region. Ghosh et al. [73] examined effect of silica additions on creep behavior of 3Y-TZP. The determined results were verified by tracer diffusion measurements. The creep data at high stresses are consisted with Coble diffusion and creep at lower stresses is attributed to interface-controlled diffusion mechanism. Addition of silica has only minor effect on both grain boundary and lattice diffusion. Ghost and Chokshi [74] studied the creep of nanocrystalline 3Y-TZP who obtained results indicating the same deformation mechanism as for micro-sized materials. Liu et al. [75] examined creep behavior of zirconia stabilized with 2.5 mol%. They postulated the grain boundary sliding accompanied by intergranular dislocation due to existence of amorphous phases.

5. Summary

The emphasis of this review was to compile existing data in the creep behavior of two types of oxides: iron-row group $M_{1-y}O$ (M = Ni, Co, Fe) which properties are strongly related with the oxygen partial pressure (p_{O_2}) dependent departure from stoichiometry (y) and superionic ZrO_2-based materials. Two different approaches to creep methodology are used. In case of the transition metal oxides such as $M_{1-y}O$ the creep experiments required determination of the rate deformation ($\dot{\varepsilon}$) as a function σ, T, p_{O_2} and grain size (d) for polycrystalline materials. On the other hand, the oxide superionics such as stabilized zirconia concentration of ionic defects ($[V_O^{\bullet\bullet}]$) is fixed by the amount used stabilizer (usually Y_2O_3). Electronic defects (e' and $h^{\bullet}$) play role only in the extremely reduced or oxidizing conditions (usually very hard to achieve experimentally). Therefore, the creep rate is studied as a function σ, T and d in case of the polycrystals.

From this review several general trends in creep behavior emerge. Deformation of single crystals is typically controlled by dislocation glide on the most favorable slip system. On the other hand, deformation of polycrystalline oxides is controlled by diffusion of the slowest species along the fastest path, such as through the lattice, along grain boundaries or through a second phase formed at the grain boundaries. Dislocations play a more or less important role depending on the specific oxide.

Steady state creep of $M_{1-y}O$ appears to be controlled by oxygen diffusion through either oxygen vacancies or interstitials. On the other hand in case of stabilized zirconia it is controlled by cation diffusion (Zr^{4+} or Y^{3+}). In any cases diffusion, involving both lattice and grain boundary data are needed to verification of proposed mechanisms. Recently used secondary ion mass spectrometry (SIMS) is suitable for this purpose [73, 76–78].

Acknowledgements

This work was financially supported by the National Science Centre of the Republic of Poland, under Grant No 2016/23/B/ST8/00163. One of the author (MR) would like to thank to Professor J. Philibert- Head of the Laboratoire de Physique des Materiaux CNRS Bellevue and his coworkers: Dr. J. Castaing, Dr. C. Monty and Dr. E. Jolles for the opportunity to perform studies on creep of single crystals FeO in their labs.

Author details

Krystyna Schneider* and Mieczyslaw Rekas

*Address all correspondence to: kryschna@agh.edu.pl

AGH, University of Science and Technology, Krakow, Poland

References

[1] Ashby M. Materials. Oxford: Elsevier; 2014. p. 336

[2] Bretheau T, Castaing J, Rabier J, Veyssiere P. Dislocations motion and high temperature plasticity of binary and ternary oxides. Advances in Physics. 1979;**28**:835

[3] Philibert J. Creep and diffusion. Solid State Ionics. 1984;**12**:321-336

[4] Rekas M, Mrowec S. On defect clustering in the wustite phase. Solid State Ionics. 1987;**22**: 185-197

[5] Nowotny J, Rekas M. Defect structure and thermodynamic properties of the wustite phase ($Fe_{1-y}O$). Journal of the American Ceramic Society. 1989;**72**:1221-1228

[6] Dickmann R. Cobaltous oxide point defect structure and non-stoichiometry, electrical conductivity, cobalt tracer diffusion. The Journal of Physical Chemistry. (Munich. 1977;**107**:189-210

[7] Nowotny J, Rekas M. Seebeck effect of undoped and Cr-doped NiO. Solid State Ionics. 1984;**12**:253-261

[8] Kröger FA, Vink HJ. In: Seitz F, Turnbull D, editors. Solid State Physics. Vol. 3. New York: Academic Press; 1956. p. 307

[9] Bransky I, Tallan NM. High temperature defect structure and electrical properties of NiO. The Journal of Chemical Physics. 1968;**49**:1243-1249

[10] Farhi R, Petot-Ervas G. Thermodynamic study of point defects in single crystalline nickel oxide. Analysis of experimental results. Journal of Physics and Chemistry of Solids. 1978;**39**:1175-1179

[11] Domingues-Rodrigueez A, Castaing J, Philibert J. Plastic deformation of pure and doped Nio single crystals. Materials Science and Engineering. 1977;**27**:217-223

[12] Carrera-Caño J, Domingues-Rodrigueez A, Marquez R, Castaing J, Philibert J. Point defects and high-temperature creep of non-stoichiometric NaCl-type oxide single crystals. I, NiO. Philosophical Magazine A. 1982;**46**:397-407

[13] Jimenez-Melendo M, Cabrera-Cano J, Dominguez-Rodriguez A, Castaing J. Pure diffusional creep of NiO polycrystals. The Journal of Physical Chemistry Letters. 1983;**44**: L-339-L-343

[14] Krishinamachri V, Notis MR. High temperature deformation of polycrystalline NiO and CoO. Acta Metallurgica (pre 1990). 1977;**25**:1307

[15] Dubois C, Monty C, Philibert J. Oxygen self- diffusion in NiO single crystals. Philosophical Magazine A. 1982;**46**:419-433

[16] Dubois C, Monty C, Philibert J. Influence of oxygen pressure on oxygen self-diffusion in NiO. Solid State Ionics. 1984;**12**:75-78

[17] Choi JS, Moore WJ. Diffusion of nickel in single crystals of nickel oxide. The Journal of Physical Chemistry. 1962;**66**:1308-1311

[18] Lindner RL, Akerstro A. Diffusion of nickel-63 (NiO). Discussions of the Faraday Society. 1957;**23**:133-136

[19] O'Keeffe M, Moore WJ. Diffusion of oxygen in single crystal of nickel oxide. The Journal of Physical Chemistry. 1961;**65**:1435-1439

[20] Nowotny J, Rekas M, et al. Journal of the American Ceramic Society. 1989;**72**:1199-1207

[21] Stoneham AM, Tomlison SM, Catlow CRA, Harding JH. Clustering of defects. In: Adles D, Fritsche H, Ovshinsky SR, editors. Physics of Disordered Materials. New York: Plenum Press; 1985. p. 243-252

[22] Nowotny J, Rekas M. Defect structure of cobalt monoxide. II. The Debye-Hückel Model. Journal of the American Ceramic Society. 1989;**72**:1207-1214

[23] Nowotny J, Rekas M. Defect structure of cobalt monoxide. III. The cluster model. Journal of the American Ceramic Society. 1989;**72**:1215-1220

[24] Clauer AH, Selzer MS, Wilcox BA. Creep of CoO single-crystals. Journal of Materials Science. 1971;**4**:1379-1388

[25] Krishinamachri V, Jones JT. Compressive creep of CoO single crystals. Journal of the American Ceramic Society. 1974;**57**:506-507

[26] Nehring VW, Smyth JR, McGee TD. Compressive creep of CoO single crystals. Journal of the American Ceramic Society. 1974;**60**:328-332

[27] Routbort JL. The stoichiometry dependence of the deformation of Co1-δO. Acta Metallurgica. 1982;**30**:663-671

[28] Dominguez-Rodriguez A, Sanchez M, Marquez R, Castaing J, Monty C, Philibert J. Point defects and high-temperature creep of non-stoichiometric NaCl-type oxide single crystals. II CoO. Philosophical Magazine A. 1982;**46**:411-418

[29] Darken LS, Gurry RW. The system iron-oxygen II Equilibrium and thermo-dynamics of liquid oxide and other phases. Journal of the American Chemical Society. 1946;**68**:798-812

[30] Roth WL. Defects in the crystals and magnetic structure of ferrous oxide. Acta Crystallographica. 1960;**13**:140-149

[31] Nowotny J, Rekas M, Wierzbicka M. Defect structure and electrical properties of the wustite phase. Zeitschrift für Physikalische Chemie. 1982;**131**:191-198

[32] Koch FB, Cohen JB. Defect structure of $Fe_{1-y}O$. Acta Crystallographica Section B. 1969;**25**:275-287

[33] Cheetham AK, Fender BEF, Taylor RI. High temperature neutron diffraction study of $Fe_{1-y}O$. Proceedings of the Physical Society. London, Solid State Physics. 1971;**4**:2160-2165

[34] Battle PD, Cheetham AK. The magnetic structure in non-stoichiometric ferrous oxide. Journal of Physics C. 1979;**12**:337-345

[35] Catlow CRA. Defect clustering in non-stoichiometric oxides. In: Sørensen T, editor. Nonstoichiometric Oxides. New York: Academic Press; 1981. p. 61-98

[36] Stoneham AM, Tomlinson SM, Catlow CRA, Hardings JH. Defect clustering in rock-salt structured transition metal oxides. In: Simkovich G, Stubican VS, editors. Transport in Non-Stoichiometric Compounds. New York: Plenum Press; 1985. p. 243-252

[37] Ilschner B, Reppich B, Riecke E. High-temperature steady-state creep and atomic disorder in iron^{II} oxide. Discussions of the Faraday Society. 1964;**38**:243-250

[38] Jolles E, Monty C. High temperature creep of $Fe_{1-x}O$. Philosophical Magazine A. 1991;**64**:765-775

[39] Badwal SPS, Bannister MJ, Hannik RHJ, editors. Science and Technology of Zirconia V. Basel: Technocomic Publ. Comp.; 1993

[40] Somiya S, Yamamoto N, Yanagida H, editors. Science and Technology of Zirconia III. Westerville, OH (USA): American Ceramic Society Inc; 1988

[41] Claussen N, Ruhle M, Heuer AH, editors. Science and Technology of Zirconia, II. Columbus, OH (USA): American Ceramic Society, Inc; 1983

[42] Kisi E. Zirconia Engineering Ceramics. Switzerland: Trans.Tech. Publ. Ltd; 1998

[43] Garvie RC, Nicholson PS. Structure and thermo-mechanical properties of partially stabilized zirconia in the $CaO\text{-}ZrO_2$ system. Journal of the American Ceramic Society. 1972;**55**:152-157

[44] Wakai F, Sakaguchi S, Matsuno Y. Superplasticity of yttia-stabilized tetragonal ZrO_2 polyctrystals. Advanced Ceramic Materials. 1986;**1**:259-263

[45] Carry C. In: Kobayashi M, Wakai F, editors. Proceedings of the MRS International Meeting on Advanced Materials, Pittsburgh, PA (USA). 1989;**7**:199-215

[46] Garvie RC, Hannik RH, Pacoe RT. Ceramic steel. Nature. 1975;**258**:703-704

[47] Badwal SPS. Yttria tetragonal zirconia polycrystalline electrolytes for solid state electrochemical cells. Applied Physics A. 1990;**50**:449-462

[48] Badwal SPS, Drennan J. Grain boundary resistivity in Y-TZP materials as a function of thermal history. Journal of Materials Science. 1989;**24**:88-96

[49] Meyer D, Eisele U, Satet R, Rödel J. Codoping of zirconia with yttria and Scandia. Scripta Materialia. 2008;**58**:215-218

[50] Obal K, Brylewski T, Pedzich Z, Rekas M. Modification of yttria-doped tetragonal zirconia polycrystal ceramics. International Journal of Electrochemical Science. 2012;**7**: 6831-6845

[51] Weppner W. Electrochemical transient investigations of the diffusion and concentration of electrons in yttria sabilized zirconia-solid electrolytes. Zeitschrift für Naturforschung. 1976;**31a**:1336-1343

[52] Mecartney ML, Donlon WT, Heuer AH. Plastic deformation in CaO stabilized ZrO_2 (CSZ). Journal of Materials Science. 1980;**15**:1063-1065

[53] Dominguez-Rodriguez A, Lagerloff KPD, Heuer AH. Plastic deformation and solid-solution hardening of Y_2O_3-stabilized ZrO_2. Journal of the American Ceramic Society. 1986;**69**:281-284

[54] Martinez-Fernandez J, Jimenez-Melendo M, Dominguez-Rodriguez AA. High temperature creep of yttria-stabilized zirconia single crystals. Journal of the American Ceramic Society. 1990;**73**:2452

[55] Dominguez-Rodriguez A, Jemenez-Melendo M, Casting J. Plasticity of zirconia. In: Bradt RC, Brookes CA, Routbort JL, editors. Deformation of Ceramics. New York: Plenum Press; 1995. p. 31-41

[56] Gomez-Garcia D, Martinez-Fernandez J, Dominguez-Rodriguez A, Casting J. Mechanizm of high-temperature creep of full stabilized zirconia single crystals as a function of the yttria content. Journal of the American Ceramic Society. 1997;**80**:1668-1672

[57] Lankford J. Deformation and fracture of Yttria-stabilized zirconia single crystals. Journal of Materials Science. 1986;**21**:1981-1989

[58] Corman GS, High-temperature creep of yttria-stabilized zirconia single crystals. G.E. Research and Development, Technical Information Series. 1989. Ceramics Laboratory, Schenectady NY (USA) #89CRD084

[59] St Jacques RG, Angers R. Creep of CaO-Stabilized ZrO_2. Journal of the American Ceramic Society. 1972;**55**:571-574

[60] Dimos D, Kohlstedt DL. Diffusional creep and kinetic demixing in yttria-stabilized zirconia. Journal of the American Ceramic Society. 1987;**70**:531-536

[61] Chen I-W, Xue LA. Development of superplastic structural ceramics. Journal of the American Ceramic Society. 1990;**73**:2585-2609

[62] Wakai F, Nagano T. Effecxt of solute ion and grain size on superplasticity of ZrO_2 polycrystals. Journal of Materials Science. 1991;**26**:241-247

[63] Wakai F. Step model of solution-precopitation creep. Acta Metallurgica et Materialia. 1994;**42**:1163-1172

[64] Sharif AA, Imamura PM, Michell TE, Mecartney ML. Control of grain growth using intergranular silicate phase in cubic yttria-stabilized zirconia. Acta Materialia. 1998;**6**: 3863-3872

[65] Sudhir B, Chokshi AH. Compression creep characteristics of 8 mol% - yttria-stabilized zirconia. Journal of the American Ceramic Society. 2001;**84**:2625-2632

[66] Evans PE. Creep in yttria- and scandia-stabilized zirconia. Journal of the American Ceramic Society. 1970;**53**:365-369

[67] Seltzer MS, Talty PK. High temperature creep of Y_2O_3 – stabilized ZrO_2. Journal of the American Ceramic Society. 1975;**58**:124-130

[68] Chevalier J, Olagon C, Fantzyyi G, Gros H. Creep behavior of alumina, zirconia and zirconia-toughened alumina. Journal of the European Ceramic Society. 1997;**17**:859-864

[69] Nauer M, Carry C. Creep parameters of yttria doped zirconia materials and superplasticity deformation mechanisms. Scripta Metallurgica et Materialia. 1990;**24**:1459-1463

[70] Lakki A, Schaller R, Nauer M, Carry C. High temperature siperplastic creep and internal friction of yttria doped zirconia polycrystals. Acta Metallurgica et Materialia. 1993;**41**: 2845-2853

[71] Owen DM, Hokshi AH. The high temperature mechanical characteristics of super plastic 3 mol% yttria stabilized zirconia. Acta Materialia. 1998;**46**:667-679

[72] Chokshi AH. The role of diffusion creep in the superplastic deformation of 3 mol% yttria stabilized tetragonal zirconia. Scripta Materialia. 2000;**42**:241-248

[73] Ghosh S, Kilo M, Borchardt G, Chokshi AH. Diffusion and creep in silica-doped tetragonal zirconia. Journal of the American Ceramic Society. 2009;**92**:3004-3013

[74] Ghosh S, Chokshi AH. Creep in nanocrystalline zirconia. Scripta Materialia. 2014;**86**:13-16

[75] Liu E, Wang H, Xiao G, Yuan G, Shu X. Creep-related micromechanical behavior of zirconia-based ceramics investigated by nanoindentation. Ceramics International. 2015. DOI: 10.1016/j.ceramint 2015.06.136

[76] Bak T, Nowotny J, Prince K, Rekas M, Sorrell CC. Grain boundary diffusion of magnesium in zirconia. Journal of the American Ceramic Society. 2002;**85**:2244-2250

[77] Swaroop S, Kilo M, Argirusis C, Borchardt G, Chokshi AH. Lattice and grain boundary diffusion of cations in 3Y-TZP. Analyzed using SIMS. Acta Materialia. 2005;**53**:4975-4985

[78] Kowalski K, Obal K, Pedzich Z, Schneider K, Rekas M. Lattice and grain boundary diffusion of Al in tetragonal yttria-stabilized zirconia polycrystalline ceramics (3Y-TZP). Analyzed using SIMS. Journal of the American Ceramic Society. 2014;**97**:3123-3127

A Unified Creep-Fatigue Equation with Application to Engineering Design

Dan Liu and Dirk John Pons

Additional information is available at the end of the chapter

http://dx.doi.org/10.5772/intechopen.70877

Abstract

Background: Creep-fatigue damage occurs under cyclic loading at elevated temperature. The existing creep-fatigue models have limited ability to cover the full combination of creep and fatigue behaviours, except with extensive prior empirical testing. Consequently, they cannot effectively and efficiently be used for early engineering design.

Approach: We present a strain-based unified creep-fatigue formulation that overcomes these limitations. We validate this equation against empirical data for multiple materials, and shows it is able to cover full ranges from pure fatigue to pure creep. A simplified formulation is developed where the coefficients are extracted through simple creep-rupture tests. We show how the equation may be used in a design situation, by application to a representative gas turbine blisk. Included here is a demonstration of how the equation may be integrated into finite element analysis, which is an important practical consideration in the design work flow.

Outcomes: The results demonstrate that the unified equation evidences fidelity to empirical data for creep-fatigue behaviours for multiple metallic materials. The usefulness of this equation is the ability to identify candidate materials for creep-fatigue loading situations. The ability to achieve this at relatively early design stages is advantageous because of the economy and convenience provided.

Keywords: creep-fatigue, accuracy, economy, engineering design, turbine blisk

1. Introduction

Creep-fatigue failure results from the interaction of pure fatigue and creep, and is influenced by temperature, frequency and applied loading. The conventional strain-based creep-fatigue equations are quantitatively accurate in narrow areas of application, but suffer from poor ability to generalize to other materials and loading regimes. Consequently, those equations are ineffective and inefficient for engineering design. A recent development in the field is a unified formulation of creep-fatigue. The advantages of this unified creep-fatigue equation are that it accommodates the crucial parameters (temperature, frequency and loading), covers different materials, represents the full range of conditions from pure fatigue to pure creep and is economical in the testing regime to determine coefficients. This chapter describes the unified creep-fatigue equation and applies it to design. The theory can be applied to evaluate the fatigue capacity of candidate materials at the early stages of design.

2. Desirable characteristics of a creep-fatigue method for design

For the perspective of engineering design, the development of a creep-fatigue method should consider following four areas:

2.1. Unified characteristic

The unified characteristic is defined as the ability to predict fatigue life for multiple temperatures and cyclic times for multiple materials. In this case, the desirable formulation should accommodate all relevant variables (including temperature, cyclic time and applied loading), and should present small difference between predicted life and experimental result under multiple situations.

2.2. Integrated characteristic

The integrated characteristic refers to the ability to cover full range of conditions from pure-fatigue condition to pure-creep condition. In this case, the desirable formulation should have the ability to be transformed to the representations of pure fatigue and pure creep, and the transformed formulations should be consistent with the general understanding of fatigue and creep mechanisms.

2.3. Economy

A desirable engineering method for fatigue-life prediction should present good balance between accuracy and economy. The economy of this theory can be assessed by evaluating the life-prediction error with overall experimental cost. Generally, the higher sensitive a numerical formulation has, the more experimental data are needed and the poorer economy is presented.

2.4. Applicability to engineering design

An engineering-based creep-fatigue method should be effectively and efficiently applicable to practical design process, where an economical, convenient and accurate method is represented to engineers and designers. Such as, the creep-fatigue evaluation by using this formulation is applied to the initial stage of design to select material or optimize structure.

Overall, a desirable creep-fatigue formulation for engineering design should have unified and integrated characteristics, and present good economy and applicability to practical design.

3. Brief review of existing creep-fatigue models

Creep-fatigue behavior is generally influenced by temperature and frequency/cyclic time, wherein increasing temperature or decreasing frequency results in reduced fatigue capacity due to intensified creep damage. Normally, the existing creep-fatigue equations were derived from empirical data through curve fitting, and present the extension of the Coffin-Manson equation (Eq. (1)) [1, 2]:

$$\Delta\varepsilon_p/2 = \varepsilon_f'\left(2N_f\right)^{\beta} \tag{1}$$

where $\Delta\varepsilon_p$ is the plastic amplitude, ε_f' is the fatigue ductility coefficient, β is the fatigue ductility exponent and N_f is the cycles to failure. The development of creep-fatigue model was firstly attempted by Coffin, who proposed the frequency-modified Coffin-Manson equation (Eq. (2)) [3]:

$$\varepsilon_p = C\left(N_f f^{k-1}\right)^{-\beta_0} \tag{2}$$

where f is the frequency and k is a constant obtained from experiments and is given different values for different temperatures (temperature dependency is indirectly introduced). Then, through directly integrating with temperature dependence, the frequency-modified Coffin-Manson equation (Eq. (2)) was further developed to show the combined influence of temperature on creep-fatigue. For example, Solomon proposed a creep-fatigue equation (Eq. (3)) [4] for Sn40Pb solder:

$$\varepsilon_p = C_1(T)\left(N_f f^{k-1}\right)^{-\beta_0} \tag{3}$$

with

$$C_1(T) = 1.338 - 2\times10^{-4}T - 1\times10^{-5}T^2 - 2\times10^{-7}T^3$$

where T is the temperature, $\beta_o = 0.5$ and k is the constant and related to the frequency: $k = -0.42$ for 6×10^{-5} Hz $\leq f \leq 3\times10^{-4}$ Hz and $k = -0.84$ for 3×10^{-4} Hz $\leq f \leq 0.3$ Hz. In addition, based on the creep-fatigue tests on 63Sn37Pb solder, Shi et al. presented a creep-fatigue formulation (Eq. (4)) [5]:

$$\varepsilon_p = C_2(T)\left[N_f f^{k(T)-1}\right]^{-\beta_0(T)} \tag{4}$$

with

$$C_2(T) = 2.122 - 3.57\times10^{-3}T + 1.329\times10^{-5}T^2 - 2.502\times10^{-7}T^3$$

$$\beta_0(T) = 0.731 - 1.63\times10^{-4}T + 1.392\times10^{-6}T^2 - 1.151\times10^{-8}T^3$$

$$k_1(T) = 0.919 - 1.765\times10^{-4}T - 8.634\times10^{-7}T^2$$

$$k_2(T) = 0.437 - 3.753\times10^{-4}T - 8.04\times10^{-7}T^2$$

where $k_1(T)$ and $k_2(T)$ are the frequency-exponent functions for 10^{-3} Hz < f < 1 Hz and 10^{-4} Hz < f < 10^{-3} Hz, respectively.

Not all strain-based creep-fatigue models follow the pattern of the frequency-modified Coffin-Manson equation. For example, Jing et al. proposed a temperature-modified Coffin-Manson equation (Eq. (5)) [6], wherein the temperature dependence is incorporated into the coefficient and exponent of the Coffin-Manson equation, respectively.

$$\Delta\varepsilon_p/2 = C_3\left(2N_f\right)^{\beta} \tag{5}$$

with

$$C_3(T) = 68.79 - 0.34T + 250.56/\sqrt{T}$$

$$\beta(T) = 1.29 - 0.0053T + 2.5/\sqrt{T}$$

In addition, the creep-fatigue model (Eq. (6)) [7] developed by Engelmaier presents the influence of both temperature and frequency on creep-fatigue, where a logarithmic relationship between temperature and frequency is included.

$$\varepsilon_p = C_4 N_f^{-\beta_0\left(\overline{T},f\right)} \tag{6}$$

with

$$\beta_0\left(\overline{T},f\right) = 0.442 + 6\times10^{-4}\overline{T} - 1.74\times10^{-2}\ln\left(1 + 43200f\right)$$

where $\overline{T}$ is the mean temperature and f is the cyclic frequency ($1 \le f \le 1000$ cycles/day).

Besides temperature and frequency, applied loading also contributes to creep damage. This factor was considered by Wong and Mai, and then they proposed a unified creep-fatigue equation (Eq. (7)) [8] which accommodates temperature, frequency and applied loading.

$$\varepsilon_p = C_0 s(\sigma) c(T,f) N_f^{-\beta_0 b(T,f)} \tag{7}$$

with

$$s(\sigma) = \begin{cases} 1 & \textit{when creep is dormant} \\ \exp\left[-\left(\sigma_{yield}\varepsilon_p^{n'}\right)/A\right] & \textit{when creep is active} \end{cases}$$

$$c(T,f) = 1 - c_1\left(T - T_{ref}\right) - c_2\log\left(f/f_{ref}\right)$$

$$b(T,f) = 1 - b_1\left(T - T_{ref}\right) - b_2\log\left(f/f_{ref}\right)$$

where n' is the cyclic hardening index, A, c_1, c_2, b_1 and b_2 are the positive constants, T_{ref} is the reference temperature below which creep becomes dormant and f_{ref} is the reference frequency above which creep becomes dormant.

Overall, the existing creep-fatigue formulations were developed through limited empirical data for specific materials, and not all models accommodate both temperature and frequency dependencies. Consequently these derivations and models may not be extended to predict fatigue life for multiple temperatures, frequencies or different materials, thus cannot present a unified characteristic. Theoretically, the existing models may show the unified characteristic by recalculating the coefficients for each new material encountered. However, this would merely provide a numerical model without a fundamental physical theory. Furthermore such an approach would be poor economy due to the empirical effort involved.

In addition, these existing models only describe creep-fatigue behavior, and cannot cover the full range of conditions from pure fatigue to pure creep. Taking Solomon's equation as an example, the pure-fatigue condition is presented by letting $C_1(T)f^{\beta(1-k)} = C_0$. When the extreme frequency is imposed, $f \rightarrow \infty$, the functions $C_1(T)$ becomes infinitely small, which causes $T \rightarrow \infty$. This does not agree with the general understanding of pure fatigue, where the temperature should be lower than 35% of the melting temperature [9]. The pure-creep condition is presented by putting $\varepsilon_p = 0$. This is satisfied by letting $C_1 = 0$, which returns $T = 172°C$. This implies creep-rupture only occur when the temperature closes to melting temperature (186°C for 60Sn40Pb), which does not agree with general understanding that creep is active at much lower temperature [9].

Furthermore, the existing creep-fatigue equations were derived from the method of curve fitting, where more coefficients were introduced to achieve high quality of fitting to empirical data. When these models are applied to describe creep-fatigue behavior for another material, a large amount of empirical effort is involved to get high fitting accuracy. In this case the outcomes are strongly sensitive to the quality and quantity of empirical data.

Finally, per Section 2, these disadvantages are significant in the case of engineering design, where decisions (such as material selection) must be made on incomplete information. Consequently, the existing methods are often not applicable to practical design process except in narrowly defined areas.

These limitations have been recently improved by the *strain-based unified creep-fatigue* equation [10, 11]. The next sections describe the advantages of this new model, and present a case study illustrating its applicability to engineering design.

4. Description of the strain-based unified creep-fatigue equation

The strain-based unified creep-fatigue equation (Eq. (8)) [10, 11] is based on the underlying physical mechanisms of fatigue and creep. It provides a linear relationship between temperature and applied loading, based on the observation of creep-diffusion phenomenon [12], and it includes a power-law relation between number of cycles and applied loading which is consistent with crack-growth behavior [13]. Structurally, this formulation presents an extension of the Coffin-Manson equation, and the creep effect is numerically incorporated based on the concept of *fatigue capacity*. This equation includes the variables of temperature, cyclic time and applied loading, and proposes that the full fatigue capacity is gradually consumed by the elevated temperature and prolonged cyclic time. The strain form is:

$$\varepsilon_p = C_0 c(\sigma, T, t_c) N^{-\beta_0} \tag{8}$$

with

$$c(\sigma, T, t_c) = 1 - c_1(\sigma)\left(T - T_{ref}\right) - c_2 \log\left(t_c/t_{ref}\right)$$

$$T - T_{ref} = \begin{cases} T - T_{ref} & for \quad T \geq T_{ref} \\ 0 & for \quad T \leq T_{ref} \end{cases}$$

$$t_c/t_{ref} = \begin{cases} t_c/t_{ref} & for \quad t_c \geq t_{ref} \\ 1 & for \quad t_c \leq t_{ref} \end{cases}$$

where ε_p is the plastic strain, N is the creep-fatigue life, T is the temperature, t_c is the cyclic time, T_{ref} is the reference temperature and is defined as 35% of melting temperature where the creep is active [9], t_{ref} is the reference cyclic time and is suggested as a small value, C_0 and β_0 are the constants which reflect the full fatigue capacity and $c_1(\sigma)$ and c_2 are the creep-related function and constant, respectively.

Function $c_1(\sigma)$ and constant c_2 are derived from creep-rupture tests which give the relationship of the Manson-Haferd parameter [14] against applied loading (σ), and the convergence point ($\log t_a$, T_a) of all $\log t$-T lines at different stresses. Then, the pure-creep condition ($\varepsilon_p = 0$) gives the formula of constant c_2 (Eq. (9)) through letting $T = T_{ref}$, and suggests function $c_1(\sigma)$ (Eq. (10)) through letting $t_c = t_{ref}$.

$$c_2 = \frac{1}{\log\left(t_a/t_{ref}\right)} \tag{9}$$

$$c_1(\sigma) = -\frac{c_2}{P_{MH}(\sigma)} \tag{10}$$

For the same amplitude of applied loading, the creep damage caused by constant stress is significantly larger than the damage resulting from reversed loading [15], thus a moderating factor f_m is introduced to compress the constant stress, then this presents an equivalent creep damage for the cyclic situation. This moderating factor is determined by the shape of loading wave, and is defined as the ratio of the average level to the peak value of applied cyclic loading. In this case, f_m is normally given as 0.6366 for the sinusoidal wave and 0.5 for a triangular wave. Then, with the strain-stress relation, function $c_1(\sigma)$ is converted into the form in terms of plastic strain (Eq. (11)):

$$c_1(\sigma) = -\frac{c_2}{P_{MH}(\sigma)} \rightarrow -\frac{c_2}{P_{MH}(\varepsilon_P)} = -\frac{c_2}{P_{MH}\left[f_m \cdot K(T, t_c) \cdot \varepsilon_P{}^{n(T,t_c)}\right]} \tag{11}$$

where $K(T, t_c)$ and $n(T, t_c)$ are the strength coefficient and strain hardening exponent for cyclic-loading situation, respectively, and they are functions of temperature and cyclic time.

Finally, creep-fatigue data are applied to obtain the magnitudes of C_0 and β_0 through minimizing the error (Eq. (12)) between the predicted creep-fatigue life ($N_{pre,\,ij}$) and experimental results ($N_{\exp,\,ij}$):

$$error = \sum_{i,j} \left(\log N_{pre,\,ij} - \log N_{\exp,\,ij} \right)^2 \tag{12}$$

5. Evaluation of creep-fatigue models for design

Next we show that the strain-based unified creep-fatigue equation is applicable to multiple situations, covers the full range of fatigue-to-creep, and provides an economical method for engineering design. Comparing with the existing models, this new model shows significant advantages on these three areas.

5.1. The unified characteristic

Generally, the existing creep-fatigue models shown in Section 3 present good ability of fatigue-life prediction in the field where they were derived. However, the accuracy reduces when these models are extended to other materials at multiple temperatures and cyclic times. This limitation is improved by the strain-based creep-fatigue equation, which presents a general formulation based on physical mechanisms of fatigue and creep. Theoretically, the existing creep-fatigue models may also present unified characteristic through recalculating the coefficients for different materials, but the economy becomes poor which is not desired for engineering design.

The unified characteristic for the strain-based unified creep-fatigue equation is validated on multiple materials (including low melting temperature material: 63Sn37Pb solder, and high melting temperature materials: stainless steel 304, Inconel 718 and GP91 casting steel) at multiple temperatures and cyclic times. The coefficients of the unified formulation for these materials are obtained by the method shown in Section 4, and the results are presented below.

5.1.1. 63Sn37Pb solder

The coefficients of the unified formulation for 63Sn37Pb solder is shown in **Table 1** based on the creep-rupture data [16] and creep-fatigue data [5]. This material was validated in our previous research [11], but the coefficients are recalculated here since the derivation method was further improved in the present work.

C_0	β_0	c_2	T_{ref} (K)	t_{ref} (s)	f_m	Convergent point	Average error
7.894	0.825	0.1215	160	1	0.6366	(160 K, 8.232)	0.003301
$c_1(\sigma)$	$9.9586 \times 10^{-4} + 1.01122 \times 10^{-4} \cdot f_m \cdot \sigma + 8.09657 \times 10^{-7} \cdot {f_m}^2 \cdot \sigma^2$						

Table 1. Coefficients of the unified formulation for 63Sn37Pb solder.

5.1.2. *Stainless steel 304*

The coefficients of the unified formulation for stainless steel 304 is shown in **Table 2** based on the creep-rupture data [17] and creep-fatigue data [18].

5.1.3. *Inconel 718*

The coefficients of the unified formulation for Inconel 718 is shown in **Table 3** based on the creep-rupture data [19] and creep-fatigue data [20].

5.1.4. *GP91 casting steel*

The coefficients of the unified formulation for GP91 casting steel is shown in **Table 4** based on the creep-rupture data [21] and creep-fatigue data [22].

The validation on the above four materials shows that the strain-based unified creep-fatigue equation has the ability to cover multiple materials and presents good fatigue-life prediction at multiple temperatures and cyclic times. Structurally, Wong and Mai's equation also provides a general form, thus may have chance to present unified characteristic. In particular, the coefficients are not fixed across multiple materials, hence have to be determined in each case. The investigation indicates that Wong and Mai's equation presents smaller average errors on the materials of 63Sn37Pb solder (0.000284), stainless steel (0.002810) and Inconel 718 (0.001585) than the unified formulation. Normally, the equation with more coefficients has better fitting for empirical data. This is the main reason for smaller average error shown by using Wong and Mai's equation.

C_0	β_0	c_2	T_{ref} (K)	t_{ref} (s)	f_m	Convergent point	Average error
0.8524	0.578	0.0666	600	1	0.5	(600 K, 15.01)	0.004257
$c_1(\sigma)$	$8.5843 \times 10^{-4} + 9.74017 \times 10^{-6} \cdot f_m \cdot \sigma - 1.86542 \times 10^{-8} \cdot f_m^{\,2} \cdot \sigma^2$						

Table 2. Coefficients of the unified formulation for stainless steel 304.

C_0	β_0	c_2	T_{ref} (K)	t_{ref} (s)	f_m	Convergent point	Average error
0.5658	0.608	0.0782	560	1	0.6366	(560 K, 12.78)	0.008643
$c_1(\sigma)$	$2.7679 \times 10^{-3} - 4.12347 \times 10^{-6} \cdot f_m \cdot \sigma + 3.91221 \times 10^{-9} \cdot f_m^{\,2} \cdot \sigma^2$						

Table 3. Coefficients of the unified formulation for Inconel 718.

C_0	β_0	c_2	T_{ref} (K)	t_{ref} (s)	f_m	Convergent point	Average error
0.6879	0.667	0.0547	610	1	0.6366	(610 K, 18.28)	0.00876
$c_1(\sigma)$	$9.51808 \times 10^{-4} + 1.20344 \times 10^{-5} \cdot f_m \cdot \sigma - 1.75045 \times 10^{-8} \cdot f_m^{\,2} \cdot \sigma^2$						

Table 4. Coefficients of the unified formulation for GP91 casting steel.

Theoretically, the equations of Coffin's, Solomon's, Shi's, Engelmaier's and Jing's may also have opportunity to present the unified characteristic through transforming them into general formulations. For example, Solomon's equation is rewritten as:

$$\varepsilon_p = C_1(T)\left(N_f f^{k-1}\right)^{-\beta_0} \tag{13}$$

with

$$C_1(T) = \mathrm{a} + \mathrm{b}T + \mathrm{c}T^2 + \mathrm{d}T^3$$

where a, b, c, d, k and β_0 are the constants obtained from experiments. Then, this modified formulation gives small errors (such as 0.000529 for 63Sn37Pb and 0.00103 for stainless steel 316), and even present better fatigue-life prediction than the unified formulation at some situations.

However, we cannot conclude that Wong and Mai's equation and those modified formulations are better than the strain-based unified creep-fatigue equation. To be specific, it is significant that Wong and Mai's equation and those modified formulations are introduced as many as coefficients to get good quality of fitting (such as seven independent coefficients for Wong and Mai's equation), thus the representation of creep-fatigue behavior is a numerical-based method and the accuracy highly relies on the number of empirical data. This means that more empirical data gives more accurate prediction, and thus poor economy is presented. However, only two independent coefficients in the unified formulation are derived by curve fitting. This implies less experimental effort is needed, and then a more economical method for fatigue-life prediction is given. This will be discussed in Section 5.3.

Overall, comparing with other existing creep-fatigue models, the strain-based unified creep-fatigue equation provides a better method for fatigue-life prediction at multiple situations, wherein the small average errors are given and the unified characteristic is proved.

5.2. The integrated characteristic

The integrated characteristic refers to the ability to cover full range of conditions from pure-fatigue condition to pure-creep condition. According to the concept of fatigue capacity, the full fatigue capacity (the pure-fatigue condition) is continuously consumed by the increasing creep effect to the condition of creep-fatigue, and finally to the pure-creep condition as the fatigue capacity is completely consumed. The comparison between different creep-fatigue models on the integrated characteristic is collected in **Table 5** [23].

Normally, the creep effect presents the dependence on temperature, cyclic time and applied loading, thus the creep-fatigue model should accommodate these relevant variables to show a good description of creep-fatigue. However, creep effect in the existing creep-fatigue models shown in Section 2 cannot be totally presented, specifically, only the temperature and cyclic time are included into the equations proposed by Solomon, Shi et al., and Engelmaier, only the cyclic time is presented in Coffin's equation and only the temperature is shown in Jing's equation.

Creep-fatigue models	Pure fatigue	Creep-fatigue	Pure creep
Coffin's equation (Eq. (2))	X	√	X
Solomon's equation (Eq. (3))	X	√	X
Shi's equation (Eq. (4))	X	√	X
Jing's equation (Eq. (5))	X	√	X
Engelmaier's equation (Eq. (6))	√	√	X
Wong & Mai's equation (Eq. (7))	√√	√√	X
Unified creep-fatigue equation (Eq. (8))	√√	√√	√√

√√: This equation can well describe the phenomena of this condition.
√: This equation can partly describe the phenomena of this condition.
X: This equation cannot describe the phenomena of this condition.

Table 5. The capacity of integrated characteristic.

While, Wong and Mai's equation and the unified formulation accommodate these three variables, thus they are believed to have a better fatigue-life prediction at the creep-fatigue condition (accuracy has been shown in Section 5.1). Further investigation of stress-related functions in these two equations shows that function $c_1(\sigma)$ in the unified formulation is directly derived from creep-rupture behavior, but is not for Wong and Mai's equation. Therefore, the strain-based unified creep-fatigue equation theoretically has better presentation of creep effect.

In addition, a good creep-fatigue equation also should be capable of covering two ends of creep-fatigue condition: pure-fatigue condition and pure-creep condition. However, the equations proposed by Coffin, Solomon, Shi et al., Jing, Engelmaier, and Wong and Mai do not totally satisfy this condition. On the one hand, the condition of pure fatigue is presented by letting $C_1(T)f^{\beta(1-k)} = C_0$($C_0$ is the ductility coefficient at pure-fatigue condition) for Solomon's equation, $C_2(T)f^{\beta[1-k(T)]} = C_0$ for Shi's equation and $\beta_0(\overline{T},f) = \beta_0$ (β_0 is the ductility exponent at pure fatigue condition) for Engelmaier's equation. When the extreme frequency is imposed, $f \to \infty$, the functions $C_i(T)$ and $\beta_0(\overline{T},f)$become infinitely small, which causes $T \to \infty$. In addition, the temperature component is not shown in Coffin's equation, thus the activation of creep effect in terms of temperature is ignored in this equation. These do not agree with the general understanding of pure fatigue, where the temperature should lower than 35% of melting temperature [9]. For Jing's equation, the derivation of function $C_3(T)$ is based on the material of 80Au/20Sn solder, thus a low temperature (where the creep effect is dormant) may return a reasonable value to describe full fatigue capacity. While, when this equation is applied on steel, function $C_3(T)$ yields to an impossible negative value since the temperature at the pure-fatigue condition is much higher than the situation for solder. However, the pure-fatigue condition can be well-presented by both Wong & Mai's equation and the strain-based unified creep-fatigue equation through letting $T = T_{ref}$ *and* $t = t_{ref}$, where they can be restored to the Coffin-Manson equation.

On the other hand, the condition of pure creep is presented by letting $\varepsilon_p = 0$. This condition cannot be satisfied by the equation of Coffin, because the coefficient C and exponent β_0 are constant, which leads to impossible zero time or zero frequency. In addition, for the equation of

Engelmaier, the pure-creep condition is satisfied by letting $\beta_0(\overline{T},f)\to+\infty$, which implies $T\to+\infty$. The pure-creep condition for the equations proposed by Solomon, Shi et al. and Jing et al. is satisfied by letting $C_1 = C_2 = C_3 = 0$, which returns T = 172, 198 and 249°C, respectively. The Solomon's and Jing's equations imply that creep-rupture only occur when the temperature closes to melting temperature (186°C for 60Sn40Pb solder and 280°C for 80Au/20Sn solder), while Shi's equation suggests that creep-rupture only occur above melting temperature (183°C). The creep activation temperatures obtained from these four creep-fatigue equations do not agree with general understanding of creep, where creep is active at 35% of melting temperature. Although Wong and Mai's equation has a good description for pure fatigue, it cannot present pure-creep condition entirely. This is because letting $s(\sigma)c(T_R,f_R)=0$ cannot deduce any well-known time-temperature parameters, such as the Larson-Miller parameter [24], the Sherby-Dorn parameter [25] and the Manson-Haferd parameter [14]. However, the Manson-Haferd parameter can be well-presented through letting $c(\sigma, T, t_c)=0$ for the strain-based unified creep-fatigue equation.

Overall, comparing with other existing creep-fatigue models, the strain-based unified creep-fatigue equation can cover the full range of conditions from pure fatigue to pure creep. Specifically, this equation has capability to model pure fatigue at $c(\varepsilon_p,T,t_c) = 1$ where this equation is restored to the Coffin-Manson equation, model pure creep-rupture at $c(\varepsilon_p,T,t_c) = 0$ where this equation is transformed to the Manson-Haferd parameter, and model creep-fatigue damage at $0 \leq c(\varepsilon_p,T,t_c) \leq 1$ where the creep effect is influenced by temperature, cyclic time and applied loading.

5.3. The economy

Engineering design of fatigue-life evaluation always requires good performance on both accuracy and economy [26]. The accuracy has been shown in Section 5.1 and the economy will be discussed in this section, where Wong and Mai's equation is select to compare with the strain-based unified creep-fatigue equation in terms of economy. The results show that the unified formulation provides a more economical method for fatigue-life prediction because the minimum experimental effort is involved.

Normally, the accuracy to predict or fit a distribution (or a trend) by a numerical formulation partly depends on the number of coefficients, which implies that the formulation with more coefficients presents better fitting [27]. However, this factor is not isolated, and is strongly related to the number of data points. Specifically, the more coefficients are introduced, then the more data points are needed to pick up, and then the better fitting can be gotten. This does not mean that a good formulation should contain as many as coefficients because this may lead to redundancy on the empirical effort. Therefore, is very important for deriving a formulation to keep a good balance between accuracy and consumption. In this case, the strain-based unified creep-fatigue equation satisfies this requirement.

As mentioned in Section 5.1, Wong and Mai's equation has seven independent coefficients which are totally obtained through numerical method, but only two independent coefficients are included in the unified formulation. This suggests that more empirical data are needed to give better fitting for Wong & Mai's equation, but this result in poor economy since performing a large number of creep-fatigue experiments is expensive and time-consuming. This significantly is not a good choice for engineering design due to the undesirable high cost. However, when less creep-fatigue tests are involved to control the total cost, the accuracy for fatigue-life prediction remarkably reduces. This limitation is improved by using the strain-based unified creep-fatigue equation, where a better balance between accuracy and economy is given. The comparison between Wong and Mai's equation and the unified formulation on the accuracy of prediction in terms of empirical-data number is shown in **Table 6**, where the materials of 63Sn37Pb and stainless steel 304 are investigated.

Taking 63Sn37Pb solder as an example, **Table 6** shows that Wong and Mai's equation gives better accuracy on fatigue-life prediction than the unified formulation when all eight groups of creep-fatigue data are imposed, wherein the average errors are 0.000284 for Wong and Mai's equation and 0.003301 for the unified formulation. Then, three groups of creep-fatigue data (T = 233 K, t_c = 1 s; T = 298 K, t_c = 10s; T = 298 K, t_c = 1000s) were selected to extract the coefficients, and the numerical method still yields to quite small errors for fitting. However, when Wong and Mai's equation with the coefficients obtained at this stage is extended to predict fatigue life at total eight creep-fatigue situations, a poor accuracy is given, where the average error dramatically worsens to 0.2445 from 0.000284. This undesirable result is significantly improved by the strain-based unified creep-fatigue equation, wherein the difference between the average errors of fatigue-life prediction under two situations is very small (specifically, 0.003301 for the coefficients obtained from eight groups of creep-fatigue data and 0.00355 for the coefficients obtained from three groups of data). This implies that less creep-fatigue experiments may be involved to obtain the coefficients of the unified formulation, and the accuracy of fatigue-life prediction by using Wong and Mai's equation is more sensitive on the number of empirical data. Consequently, the unified equation provides a more economical method for engineering design because of the reduced number of creep-fatigue experiments. This result is further demonstrated by the material of stainless steel 304, and the average errors under two situations are shown in **Table 6**.

Overall, it is clear that the better fatigue-life prediction provided by Wong and Mai's equation (also other modified formulations from the original equations of Coffin's, Solomon's, Shi's, Engelmaier's and Jing's) is obtained through sacrificing economy. However, the strain-based

Materials	**Number of data to derive coefficients**	**Average errors on predict fatigue life**	
		Wong & Mai's equation	**Unified creep-fatigue equation**
63Sn37Pb	8 groups of data	0.000284	0.003301
	3 groups of data	0.2445	0.00355
Stainless steel 304	7 groups of data	0.002810	0.004257
	3 groups of data	0.4684	0.007199

Table 6. Accuracy of prediction in terms of empirical-data number.

unified creep-fatigue equation provides a more practical method of fatigue-life evaluation for engineering design, wherein a good balance between accuracy and economy is achieved.

The research conducted by Manson indicated that the Coffin-Mansion equation can be modified to a universal slops formulation (Eq. (14)) through introducing the ductility (D) based on the observation of 47 materials [28]:

$$\Delta\varepsilon_p = 0.547D^{0.43}{N_f}^{-0.5} \tag{14}$$

Significantly, the coefficients of this equation can be directly derived from material properties, thus no fatigue test is involved. This remarkably reduces the cost spent on experiments, and thus could be introduced to simplify the unified creep-fatigue equation to eliminate creep-fatigue test. According to the empirical data on the materials of Stainless steel 316, Stainless steel 304, Inconel 718 and GP91 casting steel, the simplified formulation is presented as (Eq. (15)):

$$\Delta\varepsilon_p = 0.8965D^{0.3998}c(\sigma, T, t_c)N_f^{\ \ 0.629} \tag{15}$$

This simplified form provides an easy way to extract the coefficients without any creep-fatigue tests, and only creep-rupture tests are needed. Therefore, a more economical method of fatigue-life prediction is proposed. This formulation could be further modified and improved if more materials are investigated. However, the solder materials are not included in this simplified form because solder materials present quite different material properties from carbon-based materials, where they has more significant deformation at break and higher full fatigue capacity [29, 30]. Therefore, the solder materials should be separately discussed to develop a simplified formulation.

6. Design application: case study for gas turbine blisk

Evaluation shown in Section 5 indicates that the strain-based unified creep-fatigue equation presents good characteristics on unifying, integration and economy. This provides an effective and efficient creep-fatigue method for engineering design at initial stage, such as material selection, structure optimization and boundary-condition determination. Generally, during the process of design, the coefficients of this unified formulation can be extracted through minimum experimental effort. Then, the coefficients can be applied to numerically evaluate creep-fatigue under service condition. In addition, finite element analysis (FEA) is frequently applied to conduct numerical analysis of creep-fatigue in different engineering components, such as heat exchanger [31, 32], turbine blade [33–35] and electronic package [36, 37]. In this case, the coefficients obtained from the unified equation also can be imported into FEA software to reduce the complexity of analysis.

A gas turbine blisk works under high temperature, and experiences repeated starting up and shutting off. This is a typical creep-fatigue situation, where damage is caused by the combination of creep effect and fatigue effect. In this case study, the theory of highly accelerated life test is accepted to make a selection of material. In particular, the strain-based creep-fatigue equation is used to obtain the creep-fatigue-related parameters for the evaluation of creep-fatigue damage.

6.1. Methodology for fatigue-based design

This case study is based on the theory of highly accelerated life test, which is used at the initial stage of design to improve the reliability of product, such as the selection of material, the optimization of structure, and the decision of manufacturing processes. Normally, the accelerated life test is done through intensifying the influence of stress-related factors on life. Therefore, the prediction of fatigue life and creep-rupture time in this case study cannot present the real value for the failure at the operation condition, but can provide effective guidance for engineering design, for example, differentiating between different candidate solutions.

6.1.1. Loading characteristics

Generally, the total damage is accumulated by the fatigue damage caused by repeated process of starting up and shutting off, the creep damage due to elevated temperature and the damage results from vibration. Therefore, the creep-fatigue evaluation based on each operational unit is divided into three parts: regular-loading fatigue partition (repeated process of starting up and shutting off), creep partition (elevated temperature) and irregular-loading partition (vibration) (**Figure 1**).

The partition which includes vibration is not considered because the amplitude of this irregular-loading is so small that can be ignored. Therefore, this evaluation focuses on the regular-loading partition and creep partition. The fatigue damage is caused by the repeated process of starting up and shutting off, and the maximum loading appears at the 100% rotational velocity. The waveform for the loading is triangular. The cyclic time is arbitrarily

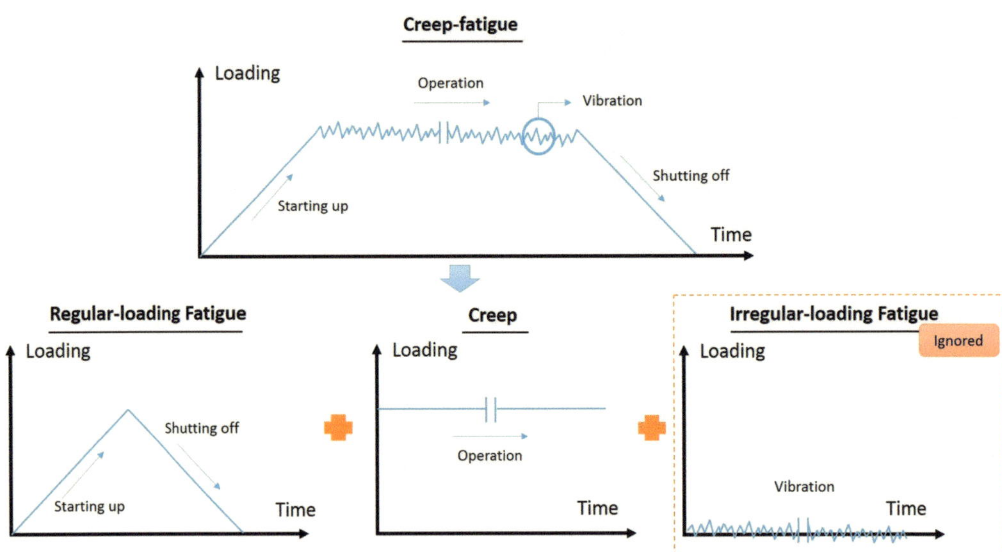

Figure 1. The partition of creep-fatigue.

defined as 1 s, which is also defined as the reference cyclic time for the strain-based creep-fatigue equation. Significantly, this small cyclic time is much smaller than the real situation, where the gas turbine blisk is impossible to start up and then shut off within 1 s. However, the choice of frequency does not influence the result of stress/strain distribution for FEA. In addition, the constant loading for creep damage at elevated temperature is defined as the applied loading at the maximum rotational velocity.

6.1.2. *FEA approach*

FEA methods such as ANSYS can accommodate creep, fatigue and creep-fatigue but only with data. Normally, FEA-based creep-fatigue evaluation [33, 34, 38] is based on the exploration of creep and fatigue behavior by experiments, where the relationships of creep strain versus time and applied loading versus fatigue life are imported into FEA software as engineering data to simulate creep-fatigue behavior. Then, the simulation is conducted under cyclic loading and elevated temperature. Generally, this is a complex process. This is firstly because the amount of empirical experiment required, especially the need to redo the creep test when the applied loading is changed. In addition, the introduction of a creep effect may result in non-convergence for FEA, and then the analysis settings may need to be repeatedly modified to get solution convergence. This makes it difficult to apply FEA to early design stages where the creep-fatigue material properties are not yet established empirically.

However, the complexity can be improved through using the strain-based unified creep-fatigue equation (Eq. (8)). Specifically, this new formulation is used to get the creep-fatigue-related coefficients under the service condition, then these coefficients are introduced into FEA as the engineering data. The key concept is that equation provides a method to transfer the creep effect into the creep-fatigue-related coefficients. Therefore, FEA can be conducted without the need to obtain explicit creep data. This significantly reduces the complexity of simulation. This process of substitution is described below.

The process is that the unified equation gives the creep-fatigue-related coefficients for the candidate materials under consideration. Then, the coefficients were input into ANSYS as engineering data to evaluate fatigue life. The next stage was to take the maximum stress and strain obtained from FEA and substitute into Morrow's equation. This provides another estimate of the fatigue life. Finally, the results obtained by the two methods were compared, and design implications identified. The FEA stress and strain results were also used to evaluate the creep damage using the Manson-Haferd parameter. This gives an estimate of the creep-rupture time for the creep part of the loading. The evaluation was conducted on Inconel 718 and GP91 casting steel, and the ideal material should have good performance on both fatigue and creep.

6.2. Selection of material

6.2.1. *Geometry*

The geometry (**Figure 2**) of gas turbine blisk was structured by Solidworks [39], and one of blades was selected to conduct FEA by using ANSYS WORKBENCH. The key dimension is shown in **Table 7**.

Figure 2. Geometry.

No.	Items	Dimensions
1	R_1 (the distance from center to tip of the blade)	469 mm
2	R_2 (the distance from center to root of the blade)	295 mm
3	Number of blades	24
4	The outlet anger for mean cross section	30°
5	The area of mean cross section	3036.45 mm^2

Table 7. Key dimension of geometry.

6.2.2. *Service condition*

The turbine blisk is working under cyclic loading and elevated temperature, and the service condition is tableted in **Table 8**:

6.2.3. *Coefficients for creep-fatigue condition and material properties*

According to the coefficients of the strain-based unified creep-fatigue equation for the materials of Inconel 718 shown in **Table 3** and GP 91 casting steel shown in **Table 4**, the fatigue-related coefficients at 811 K which can be input into ANSYS as engineering data are presented in **Table 9**:

In addition, the material properties of Inconel 718 [40] and GP91 casting steel [22] at 811 K are shown in **Table 10**:

6.2.4. *Loading*

The centrifugal force, tangential force and axial force to blades are the main loadings which cause creep-fatigue damage at elevated temperature during the repeated process of starting up and shutting off. The maximum loading is calculated at the situation of 100% rotational velocity.

Rotational velocity (n)	Temperature (T)	Mass flow rate ($\dot{M}$)	Nozzle angle (α)
3000 rpm	811 K (538°C)	31.6 kg/s	14°

Table 8. Service condition.

Materials	Strain-life relation $\Delta\varepsilon_p/2 = C(2N_f)^{\beta}$		Stress-life relation $\Delta\sigma/2 = \sigma(2N_f)^{b}$		Strain–stress relation $\Delta\sigma/2 = K(\Delta\varepsilon_p/2)^{n}$	
	C	β	σ	b	K	n
Inconel 718	0.4828	−0.636	851.6	−0.09178	946	0.1443
GP91 casting steel	0.5093	−0.662	382	−0.0422	398.5	0.0638

Table 9. The creep-fatigue-related coefficients for Inconel 718 and GP91 casting steel.

Material	Density (kg/m^3)	Yield stress (MPa)	Tensile stress (MPa)	Young's modules (MPa)
Inconel 718	8220	1069	1276	179,000
GP91 casting steel	7700	342	402	161,500

Table 10. Material properties of Inconel 718 and GP91 casting steel.

6.2.4.1. *Centrifugal force*

The centrifugal force can be obtained through Eq. (16) [35]:

$$F_c = 0.5\rho A\omega^2\left({R_1}^2 - {R_2}^2\right) \tag{16}$$

where F_c is the centrifugal force, ρ is the density, A is the area of cross section and ω is the rotational velocity in rad/s. With the rotational velocity (3000 rpm), the magnitude of ω is given by Eq. (17):

$$\omega = 2\pi n/60 = 2\pi \times 3000/60 = 314.2\ rad/s \tag{17}$$

Then, the centrifugal force is presented by Eq. (18):

$$F_c = 0.5 \times 8220 \times 0.00303645 \times 314.2^2 \times \left(0.469^2 - 0.295^2\right) = 163780.776\ N \tag{18}$$

6.2.4.2. *Tangential force and axial force*

The tangential force and axial force are calculated through velocity triangle (**Figure 3**) [33], which is a well-accepted method for the force calculation of blade.

The mean cross section is selected to calculate the tangential force and axial force, and the results in velocity triangle are shown in **Table 11**.

Then, the tangential force (F_t) is give as (Eqs. (19) and (20)):

$$F_t = \dot{M}\Delta v_x = 31.6 \times (240.1667 + 236.17) = 15052.23972\ N \tag{19}$$

$$F_t/blade = 15052.23972/24 = 627.177\ N \tag{20}$$

and the axial force (F_a) is presented as (Eqs. (21) and (22)):

$$F_a = \dot{M}\Delta v_y = 31.6 \times (67.0645 - 59.8878) = 226.78372\ N \tag{21}$$

$$F_a/blade = 226.78372/24 = 9.449\ N \tag{22}$$

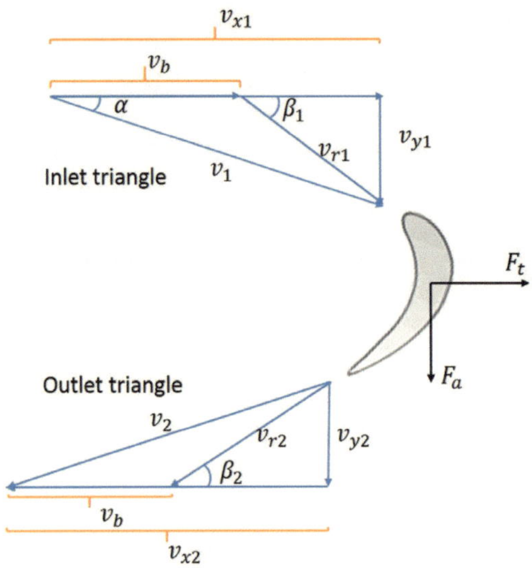

Figure 3. Velocity triangle for blade.

In the inlet triangle						**In the outlet triangle**			
v_b **(m/s)**	v_1 **(m/s)**	β_1	v_{r_1} **(m/s)**	v_{x_1} **(m/s)**	v_{y_1} **(m/s)**	v_{r_2} **(m/s)**	β_2	v_{x_2} **(m/s)**	v_{y_2} **(m/s)**
120.01	247.55	26.519^0	134.129	240.197	59.888	134.129	30^0	67.065	236.17

Table 11. The results in velocity triangle.

6.2.5. The evaluation of fatigue damage

6.2.5.1. Finite element analysis results

The finite element analysis is conducted through nonlinear analysis method [34, 35, 38]. The creep-fatigue-related coefficients from **Table 9** and the material properties from **Table 10** are input into ANSYS WORKBENCH as engineering data. The boundary condition (force) presented in Section 6.2.4 is imposed. Finally, the maximum stress, maximum total strain and fatigue life are given. The results are tabulated in **Table 12**, and fatigue-life prediction is shown in **Figure 4** for Inconel 718 and **Figure 5** for GP91 casting steel.

6.2.5.2. Predicted fatigue life through Morrow's equation

Based on the stress and strain obtained from FEA, the fatigue life can be calculated through Morrow's equation (non-zero-mean-stress condition) (Eq. (23)) [38, 41]:

http://dx.doi.org/10.5772/intechopen.70877

$$\varepsilon_t = 3.4\left(\frac{\sigma_s - \sigma_m}{E}\right)N_f^{-0.12} + C_0^{0.6}N_f^{-0.6} \tag{23}$$

where ε_t is the total strain, σ_s is the tensile stress, E is the Young's modulus, C_0 is the fatigue ductility coefficient based on $\Delta\varepsilon_p - N$ relation and σ_m is the mean stress which is defined as the half of the maximum stress obtained from FEA.

Substituting Eq. (23) with the values obtained from FEA gives the results: N_f=2440 for Inconel 718 and N_f=348 for GP91 casting steel.

Materials	Maximum stress	Maximum total strain	Minimum fatigue life
Inconel 718	1311.8 MPa	0.011597	2350 cycles
GP91 casting steel	755.27 MPa	0.029661	327 cycles

Table 12. FEA results for Inconel 718 and GP91 casting steel.

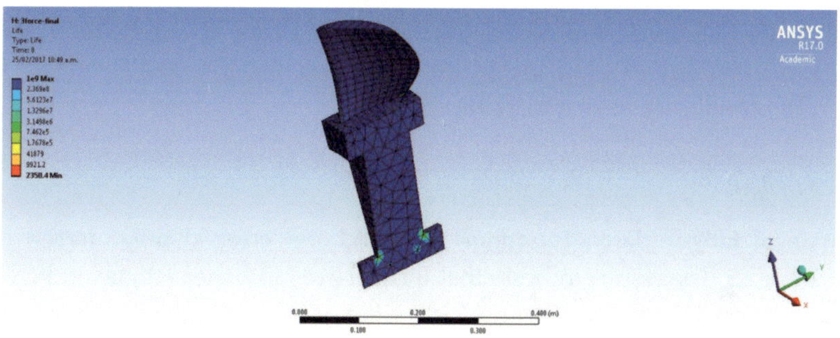

Figure 4. The fatigue life distribution for Inconel 718.

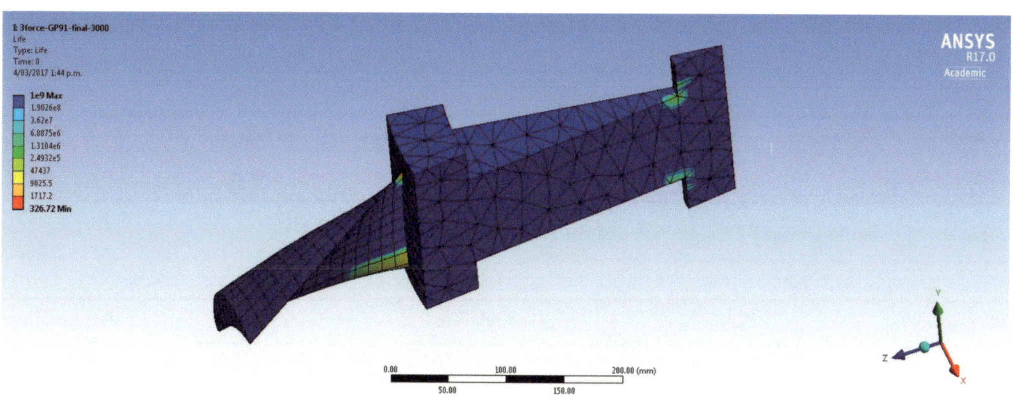

Figure 5. The fatigue life distribution for GP91 casting steel.

6.2.5.3. *Comparing the results*

The fatigue damage is evaluated through both FEA and the Morrow's equation, and the results are provided in **Table 13**. The error between these two results is given by Eq. (24).

$$error = \frac{N_{f-Morrow} - N_{f-FEA}}{N_{f-FEA}} \times 100\% \tag{24}$$

As shown in **Table 13**, the fatigue life obtained through Morrow's equation is higher than the result obtained from FEA. This may be because the influence of mean stress is removed from the second term (plastic strain), and thus reduces the mean-stress effect on fatigue capacity and obtains a longer fatigue-life prediction. The FEA results are more conservative, and most designers would want to proceed with them rather than the Morrow's results.

6.2.6. *The evaluation of rupture time due to creep damage*

According to the FEA results, the maximum plastic strain (0.005406 for Inconel 718 and 0.02498 for GP91 casting steel) can be obtained, and it could then be transformed to stress (445.4 MPa for Inconel 718 and 314.9 MPa for GP91 casting steel) by the stress-strain relation (**Table 9**). Finally, the stress is used to get the rupture time through the Manson-Haferd parameter (P_{MH}) obtained during the process of extracting the coefficients of the unified formulation. The results are shown in **Table 14**.

6.2.7. *The design decision*

The evaluation of fatigue damage (fatigue partition) and creep damage (creep partition) is collected in **Table 15**. The results indicate that Inconel 718 has better fatigue and creep capacity than GP91 casting steel. Therefore, Inconel 718 would be recommended as the material to take forward into the next stages of design. In safety critical situations, it would still be prudent to conduct the empirical creep-fatigue tests to validate the selection. However, it will be appreciated that conducting such tests for one material for the purposes of validation is more efficient than conducting the same tests for all candidate materials—some of which will not be taken forward.

Materials	FEA-based results	Morrow-based results	Error
Inconel 718	2350 cycles to failure	2440 cycles to failure	3.83%
GP91 casting steel	327 cycles to failure	348 cycles to failure	6.42%

Table 13. The results obtained through FEA and the Morrow's equation.

Materials	P_{MH}	$1/P_{MH}$– σ relation	Rupture time
Inconel 718	−0.0218	$-5.0 \times 10^{-8}\sigma^2 + 5.27 \times 10^{-5}\sigma - 3.5375 \times 10^{-2}$	5588 hours
GP91 casting steel	−0.0549	$3.2 \times 10^{-7}\sigma^2 - 2.2 \times 10^{-4}\sigma - 1.74 \times 10^{-2}$	4791 hours

Table 14. Creep damage for Inconel 718 and GP91 casting steel at 811 K.

Materials	Evaluation of fatigue damage	Evaluation of creep damage
Inconel 718	2350 cycles to failure	5588 hours to failure
GP91 casting steel	327 cycles to failure	4791 hours to failure

Table 15. The evaluation of fatigue damage and creep damage.

7. Limitation

The strain-based unified creep-fatigue equation was developed for the situation of zero-hold-time cyclic loading. In this case, fatigue makes more contribution than creep on failure [42], because total time is too small to produce remarkable creep damage. However, for the cyclic loading with hold time, the creep effect gradually intensifies as the hold time increases. Then, creep damage makes more contribution and the failure finally occurs due to a creep effect. Therefore, at the situation with short hold time, the unified formulation may still present a reasonable prediction of fatigue life, but the accuracy of this prediction may become worse when the hold time is prolonged.

8. Conclusion

The unified creep-fatigue equation (Eq. (8)) presents an effective and efficient method for engineering design. The advantages of this equation are:

1. Better fatigue-life prediction for multiple temperatures and cyclic times for multiple materials.
2. Ability to cover the full range of conditions from pure fatigue to pure creep.
3. Good economy since the coefficients may be determined through minimum experimental effort. Especially, a simplified form is presented where the coefficients can be obtained from simple creep-rupture tests without need for any creep-fatigue tests.
4. Applicability to engineering design and FEA.

Conflicts of interest

The authors declare no conflict of interest. The research was conducted without personal financial benefit from any funding body, and no such body influenced the execution of the work.

Author contributions

The work was conducted by DL and supervised by DP. The validation on different materials, the integrated characteristic and the economy are analyzed by DL. The case study on gas turbine blisk was conducted by DL with guidance from DP.

Author details

Dan Liu* and Dirk John Pons

*Address all correspondence to: dan.liu@pg.canterbury.ac.nz

Department of Mechanical Engineering, University of Canterbury, Christchurch, New Zealand

References

[1] Coffin LF Jr. A Study of the Effects of Cyclic Thermal Stresses on a Ductile Metal. New York, USA: Knolls Atomic Power Lab; 1953

[2] Manson SS. Behavior of Materials under Conditions of Thermal Stress. Ohio, US: Lewis Flight Propulsion Lab.; 1954

[3] Coffin L. Fatigue at High Temperature. Fatigue at Elevated Temperatures. Pennsylvania, USA: ASTM International; 1973

[4] Solomon H. Fatigue of 60/40 solder. IEEE Transactions on Manufacturing Technology. 1986;**9**:423-432

[5] Shi X, Pang H, Zhou W, Wang Z. Low cycle fatigue analysis of temperature and frequency effects in eutectic solder alloy. International Journal of Fatigue. 2000;**22**:217-228

[6] Jing H, Zhang Y, Xu L, Zhang G, Han Y, Wei J. Low cycle fatigue behavior of a eutectic 80Au/20Sn solder alloy. International Journal of Fatigue. 2015;**75**:100-107

[7] Engelmaier W. Fatigue life of leadless chip carrier solder joints during power cycling. IEEE Transactions on Components, Hybrids, and Manufacturing Technology. 1983:232-237

[8] Wong E, Mai Y-W. A unified equation for creep-fatigue. International Journal of Fatigue. 2014;**68**:186-194

[9] Ashby MF, Shercliff H, Cebon D. Materials: Engineering, Science, Processing and Design. Oxford, UK: Butterworth-Heinemann; 2013

[10] Liu D, Pons D, Wong E-H. The unified creep-fatigue equation for stainless steel 316. Metals. 2016;**6**:219

[11] Liu D, Pons D, Wong E-H. Creep-integrated fatigue equation for metals. International Journal of Fatigue. 2016;**98**:167-175

[12] Poirier J-P. Creep of Crystals: High-Temperature Deformation Processes in Metals, Ceramics and Minerals. Cambridge, UK: Cambridge University Press; 1985

[13] Weertman J. Fatigue crack propagation theories. In: Fatigue and Microstructure. Metals Park, Ohio: ASM; 1979. p. 279-206

[14] Manson S, Haferd A. A Linear Time-Temperature Relation for Extrapolation of Creep and Stress-Rupture Data. Lewis Flight Propulsion Lab; 1953

[15] Halford G. Cyclic creep-rupture behavior of three high-temperature alloys. Metallurgical Transactions. 1972;**3**:2247-2256

[16] Shi X, Wang Z, Yang Q, Pang H. Creep behavior and deformation mechanism map of Sn-Pb eutectic solder alloy. Journal of Engineering Materials and Technology. 2003;**125**:81-88

[17] Fritz LJ, Koster W. Tensile and Creep Rupture Properties of Uncoated and Coated Engineering Alloys at Elevated Temperatures. 1977

[18] Kanazawa K, Yoshida S. Effect of temperature and strain rate on the high temperature low-cycle fatigue behavior of austenitic stainless steels. Creep and Fatigue in Elevated Temperature Applications International conference sponsored by the Institution of Mechanical Engineers, American Society of Mechanical Engineers, American Society for Testing Materials, Philadelphia, 23-27 September 1973 and Sheffield, 1-5 April 1974; Vol. 11975

[19] Sugahara T, Martinolli K, Reis DA, de Moura Neto C, Couto AA, Neto FP, et al. Creep Behavior of the Inconel 718 Superalloy. Defect and Diffusion Forum. Trans Tech Publications; 2012. p. 509-514

[20] Fournier D, Pineau A. Low cycle fatigue behavior of Inconel 718 at 298 K and 823 K. Metallurgical Transactions A. 1977;**8**:1095-1105

[21] Tabuchi M, Hongo H, Li Y, Watanabe T, Takahashi Y. Evaluation of microstructures and creep damages in the HAZ of P91 steel weldment. Journal of Pressure Vessel Technology. 2009;**131**:021406

[22] Mroziński S, Golański G. Low cycle fatigue of GX12CrMoVNbN9-1 cast steel at elevated temperature. Journal of Achievements in Materials and Manufacturing Engineering. 2011;**49**:7-16

[23] Liu D, Pons D. Development of a unified creep-fatigue equation including heat treatment. Fatigue & Fracture of Engineering Materials & Structures. 2017. DOI: http://dx.doi.org/10.1111/ffe.12670

[24] Larson FR, Miller J. A time-temperature relationship for rupture and creep stresses. Trans ASME. 1952;**74**:765-775

[25] Dorn J, Orr RL, Sherby O. Creep correlations of metals at elevated temperatures. AIME TRANS. 1954;**200**:71-80

[26] Shigley JE, Mischke CR. Mechanical Engineering Design. New York, USA: McGraw-Hill; 2003

[27] Chapra SC, Canale RP. Numerical Methods for Engineers. New York, USA: McGraw-Hill; 2006

[28] Manson S. A modified universal slopes equation for estimation of fatigue characteristics of metals. Journal of Engineering Materials and Technology. 1988;**110**:55

[29] Kanchanomai C, Miyashita Y, Mutoh Y. Low-cycle fatigue behavior of Sn-Ag, Sn-Ag-Cu, and Sn-Ag-Cu-Bi lead-free solders. Journal of Electronic Materials. 2002;**31**:456-465

[30] Shohji I, Yoshida T, Takahashi T, Hioki S. Tensile properties of Sn-Ag based lead-free solders and strain rate sensitivity. Materials Science and Engineering: A. 2004;**366**:50-55

[31] Patil R, Anand S. Thermo-structural fatigue analysis of shell and tube type heat exchanger. International Journal of Pressure Vessels and Piping. 2017;**155**:35-42

[32] Zhao X, Zhou Y, Yuan K. Creep-fatigue damage evaluation of Ni-based superalloy inconel 617 based on finite element analysis. 23rd International Conference on Nuclear Engineering: Nuclear Power - Reliable Global Energy, ICONE 2015, May 17–21, 2015. Chiba, Japan: American Society of Mechanical Engineers (ASME); 2015. p. et al.; GLSEQ, LLC/SCI Technologies. Inc; Hitachi-GE Nuclear Energy, Ltd.; Mitsubishi Heavy Industries, Ltd. (MHI); Toshiba Corporation; Westinghouse Electric Company

[33] Kumar RR, Pandey K. Static structural and modal analysis of gas turbine blade. International Conference on Advanced Material Technologies; 2016

[34] Madhu P. Stress analysis and life estimation of gas turbine Blisk for different materials of a jet engine. International Journal of Science and Research. 2016;**5**:1103-1107

[35] Tulsidas D, Shantharaja M, Kumar K. Design modification for fillet stresses in steam turbine blade. International Journal of Advances in Engineering & Technology. 2012;**3**: 343-346

[36] Che F. Material characterization and low cycle fatigue model of low Ag content lead-free solder. IEEE. Transactions on Device and Materials Reliability. 2017

[37] Hsieh MC. Modeling correlation for solder joint fatigue life estimation in wafer-level chip scale packages. Microsystems, Packaging, Assembly and Circuits Technology Conference (IMPACT), 2015 10th International: IEEE; 2015. pp. 65–8

[38] Khan SBaARA. Fatigue and creep interaction in steam turbine bladed disk. International Journal of Innovative Research in Science, Engineering and Technology. 2014;**3**

[39] Vava G. Simple Rotor Blisk. 2016; Available from: https://grabcad.com/library/simple-rotor-blisk-1

[40] Metals S. Inconel alloy 718. Publication Number SMC-045 Special Metals Corporation. 2007

[41] Dowling NE. Mechanical Behavior of Materials: Engineering Methods for Deformation, Fracture, and Fatigue. London, UK: Pearson; 2012

[42] Hattori H, Kitagawa M, Ohtomo A. Effect of grain size on high temperature low-cycle fatigue properties of inconel 617. Tetsu to Hagane. 1982;**68**:2521-2530

Chapter 6

Review of Long-Term Durable Creep Performance of Geosynthetics by Constitutive Equations of Reduction Factors

Han-Yong Jeon

Additional information is available at the end of the chapter

http://dx.doi.org/10.5772/intechopen.72330

Abstract

In an elastic solid the strain stays constant with time and is constant and the stress decays slowly with time. The increase in strain is not linear, and the curve becomes steeper with time and also as the stress-rate is increased. The slope of the curve tends to decrease with time, but it is steeper for higher strain rates. The variation of both strain and stress with time is linear for constant stress- and strain-rate tests upon elastic materials. The final comment about the compressive creep test and data interpretation is as follows: (1) Description of the creep mechanism of the geosynthetics (exactly not compression but perpendicular compression) is very important because the creep mechanism of tension and compression is quite deferent. (2) To reduce the specimen-to-specimen, many ramp-and-hold (in the case of tension creep: 1 h) tests are recommended. (3) Loading rate is also important because it make initial strain value. To check the nonaffected loading rate, prior to the main creep test, some kind of short-term test is needed. (4) The method to assess the reduction factor by creep also will be reviewed because the value will be changed according to the applied load.

Keywords: time dependence, creep loading, stress- and strain-rate tests, creep mechanism, geosynthetics, ramp-and-hold, reinforcements, reduction factor by creep

1. General time-dependent behavior

The exact nature of the time dependence of the mechanical properties of a polymer sample depends upon the type of stress or straining cycle employed. The variation of the stress and strain with time t is illustrated schematically in **Figure 1** for a simple polymer tensile specimen subjected to four different deformation histories [1, 2]. In each case, the behavior of an elastic material is also given as a dashed line for comparison. During creep loading, a constant stress is applied to the specimen at t = 0 and the strain increases rapidly at first, slowing down over longer time periods. In an elastic solid, the strain stays constant with time. The behavior of an elastic material during stress relaxation is shown in **Figure 1(b)**. In this case, the strain is held constant, and the stress decays slowly with time, whereas in an elastic material at a constant stress-rate is shown in **Figure 1(c)**. The increase in strain is not linear, and the curve becomes steeper with time and also as the stress-rate is increased. If different constant strain rates are used, the variation of stress with time is not linear. The slope of the curve tends to decrease with time, but it is steeper for higher strain rates. The variation of both strain and stress with time is linear for constant stress- and strain-rate tests upon elastic materials.

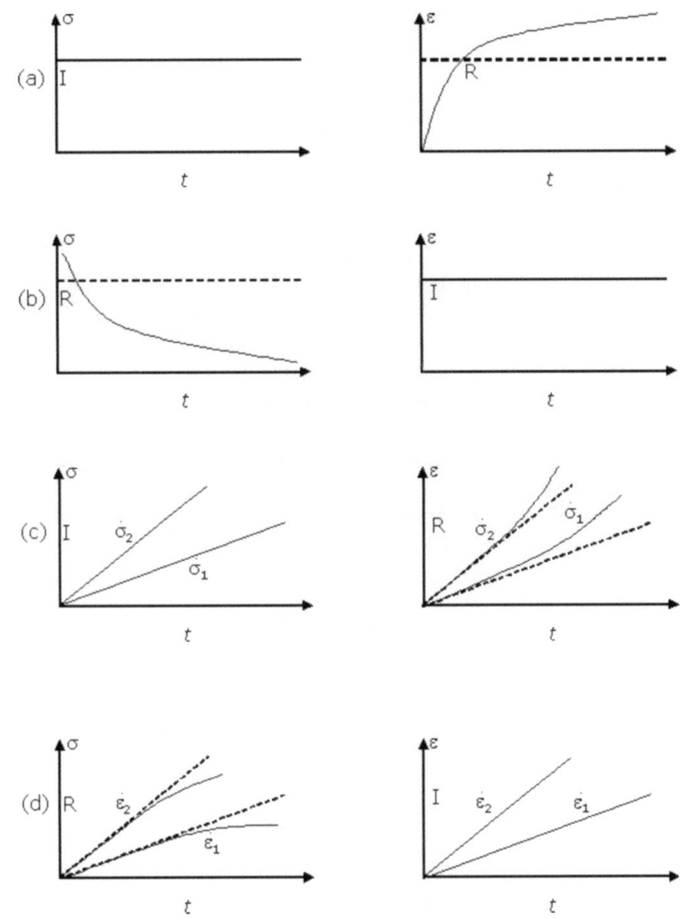

Figure 1. Schematic representation of the variation of stress and strain with time indicating the input (I) and responses (R) for different types of loading: (a) creep, (b) relaxations, (c) constant stressing rate, and (d) constant straining rate.

http://dx.doi.org/10.5772/intechopen.72330

2. Creep deformation

Creep behavior is typically time-dependent elongation property for geosynthetics, which are used to reinforce and improve the long-term performance and sustainable durability in the soil-related structure. Also, creep behavior is changed under temperature and service life after installation period

Figure 2 shows a curve of creep strain versus time of the creep test results, and slope of this curve is creep rate [3]. In general, creep phenomena can be divided into three stages.

i. First stage – primary creep, start at a rapid rate and slows with time.

ii. Second stage – secondary creep has a relatively uniform rate.

iii. Third stage – tertiary creep has an accelerating creep rate and terminates by failure of material at time for rupture (**Figure 2**).

Creep analysis in designing with geosynthetic reinforcement is considered, and the tertiary creep is very important because creep fracture can be occurred within this area.

As geosynthetics are generally manufactured from polymer materials, they exhibit a visco-elastic behavior, time, load, and temperature-dependent, under a sustained constant load. The creep behavior of geosynthetics is generally presented as strain versus time or (log time) curves and strain-rate versus time as shown in **Figure 2**. At time t, the total strain ε(t) can be expressed from the following equation:

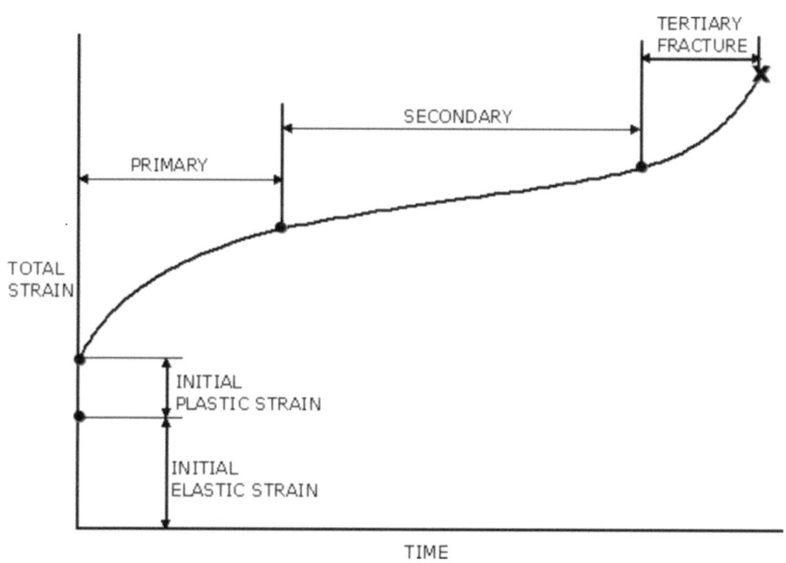

Figure 2. Time-dependent strain curve with primary, secondary, and tertiary stages.

$$\varepsilon(t) = \varepsilon_0 + \varepsilon_I + \varepsilon_{II} + \varepsilon_{III} \quad (1)$$

where ε0 is the instantaneous strain and εI, εII, and εIII are the primary, secondary, and tertiary creep strains, respectively.

The ε0 is composed from both recoverable (elastic) and irrecoverable (plastic) instantaneous stain. However, the final sudden creep strain increase before rupture does not necessarily occur: it is typical for PE (polyethylene) and PP (polypropylene) not for PET (polyester) at standard load levels and temperatures. Based on the dominant creep stage, there are three types of creep generation: dominant primary creep, dominant secondary creep, and dominant tertiary creep as shown in **Figure 3** [4].

As already very well known, the creep behavior of geosynthetics depends from the following factors:

a. polymer nature and polymer structure;

b. geosynthetics structure (nonwoven, woven, and integral structure);

c. loading conditions;

d. temperature conditions;

e. soil environment conditions, and so on.

Therefore, to investigate more specifically the long-term tensile creep of geosynthetics, it is necessary to know the effect of each of the above-mentioned factors on load-strain behavior.

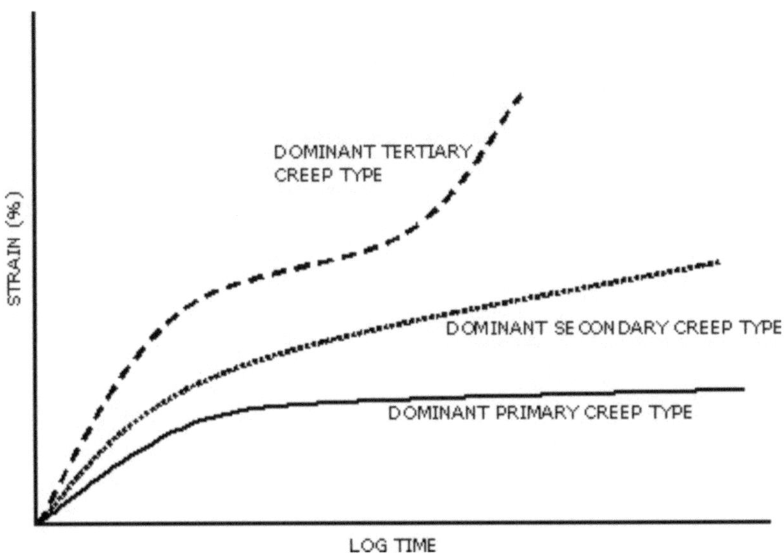

Figure 3. Type of creep behaviors generation.

Nevertheless, polymer structure parameters such as molecular weight, molecular orientation, and crystalline volume, degree of branching, and draw molecular ratio are also very important to evaluate creep sensitivity of polymers.

3. Time-temperature superposition principle

Laboratory tests can only be conveniently conducted for about a week or so. Consequently, one must use a method of extrapolating short-time tests over several decades of time so that a lower limit of the modulus can be determined for use in design. On the other hand, it is also sometimes difficult to obtain very short time-scale data, and a method is needed to extrapolate data obtained under practicable experimental conditions to these short-time scales. An empirical extrapolation method is available for amorphous polymer systems and, in general, for systems where structure does not change during the period of testing. Time and temperature have essentially equivalent effects on the modulus values of amorphous polymers, in that the shapes of the curves are similar and the modulus values in each region are about equal. Because of the equivalent effect of time and temperature, data at different temperature can be superposed by shifting individual curves one at a time and consecutively along the log t axis on the reference temperature T_0. This time-temperature superposition procedure has the effect of producing a single continuous curve of modulus values extending over many decades of log t for reference temperature. This constructed plot is known as the master curve.

Before the curves can be shifted to make the master curve, the modulus values should be corrected for density and temperature to obtain *reduced modulus* values, $E(t)_{reduced}$

$$E(t)_{reduced} = \left(\frac{T_0}{T}\right)\left(\frac{\rho_0}{\rho}\right) E_r(t) \tag{2}$$

where T_0 = reference temperature, K; ρ_0 = density at T_0; ρ = density at T.

The density correction (ρ_0/ρ) is not usually very large. The temperature correction (T_0/T) is suggested by a rubber-elasticity theory. This procedure asserts that the effect of temperature on viscoelastic properties is equivalent to multiplying the time scale by a constant factor at each temperature. In mathematical terms, it is expressed as

$$E_T[a(T)] = E_{T_0}(t) \tag{3}$$

where the parentheses indicate functional dependence and the brackets indicate multiplication. The quantity *a(T)* is called the *shift factor*, and it must be obtained along the log time scale necessary to match the curve. The parameter *a(T)* is chosen as unity at the reference temperature and is a function of temperature alone, decreasing with increasing temperature.

An important empirical correlation has been developed by which the shift factor can be computed for temperature between **T_g and T_g + 100°C**. This correlation was developed by Williams, Landel, and Ferry, and the so-called WLF equation is expressed as

$$\log a(T_0) = \frac{-C_1(T-T_0)}{C_2+T-T_0} \quad (4)$$

for $T_0 = T_g$, $C_1 = 17.44$ and $C_2 = 51.6$, while for $T_0 = T_g + 45°C$, $C_1 = 8.86$ and $C_2 = 101.6$.

3.1. TTS principle in polyethylene

The modeling of the creep results has been carried out by many researchers in the past three decades. A creep model consisting of two thermally activated Eyring-rate processes has been proposed by Ward and coworkers, particularly for ultra-high molecular weight polyethylene fibers. It has been seen that even in isotropic PE, this model could describe the creep response successfully. The corresponding Arrhenius plot of the dynamic data (log a_T vs. 1/T) is shown in **Figure 4** [5]. It is evident that the irrecoverable creep can be approximated by a single temperature-activated process with the activation energy of 118 kJ/mol over the entire region of stress and temperature experimentally covered. Therefore, there should be a linear relationship between log a_T and 1/T that may be described by the Arrhenius equation:

$$a_{T/Tref} = \exp\left[\frac{E}{R}\left(\frac{1}{T}-\frac{1}{T_{ref}}\right)\right] \quad (5)$$

where E = Arrhenius activation energy (118 kJ/mol), R = gas constant (8.314 $Jmol^{-1}$ K^{-1}), T = absolute temperature (K), and T_{ref} = reference temperature (K).

3.2. Relation between activation energy and creep load for geosynthetics

- The modern development of the theory of reaction rates may have come from the proposal made by S. Arrhenius to account for the influence of temperature on the rate of inversion of sucrose. He suggested that an equilibrium existed between inert and active molecules of the reactant and that the latter only were able to take part in the inversion process. By applying the reaction isochrones to the equilibrium between inert and active species, it can be readily shown that the variation of the specific rate of the reaction with temperature should be expressed by an equation of the form

$$\ln K = \ln A - \frac{E}{RT} \quad (6)$$

where E is the difference in heat content between the activated and inert molecules. An A is a quantity that is independent of or varies relatively little with temperature. Subsequently, the Arrhenius Eq. (6) was written in the equivalent form

$$K = Ae^{-E/RT} \quad (7)$$

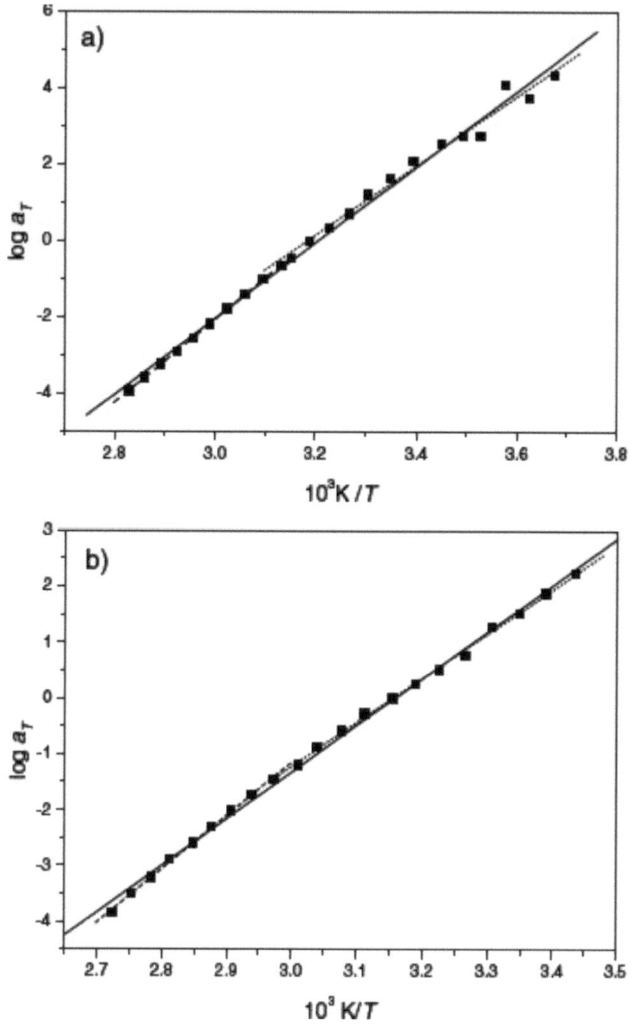

Figure 4. Arrhenius plot of the temperature shift factor from dynamic mechanical storage modulus. Squares: experimental results; solid line: fitting considering all points; dashed line: fitting in the high-temperature region: dotted line: fitting in the low-temperature region: (a) results for PEc and (b) results for PEs.

and it is now generally accepted that a relationship of this kind represents the temperature dependence of the specific rates of most chemical reactions, and even of certain physical processes, provided that the temperature range is not large, the quantities *A* and *E* can be taken as constant.

- Most of the methods to calculate the activation energy are about chemical reaction and phase transitions. The method about the geosynthetics viscoelastic properties is as follows.

The relation between temperature *T* and creep strain-rate ($\dot{\varepsilon}$) can be expressed in the mathematical expression, the Arrhenius Equation, in the form:

$$\dot{\varepsilon} = Ae^{\frac{-E}{RT}} \tag{8}$$

where $\dot{\varepsilon}$ = creep strain-rate, t^{-1}, A = a pre-exponential rate constant, t^{-1}, E = experimental activation energy (cal/mol), R = universal gas constant (= 1.987 cal/mol-K), and T = absolute temperature (°K).

Eq. (8) shows that creep strain-rate increase with temperature and with the decrease in the activation energy, providing that all other factors affecting creep are kept constant. Eq. (8) is usually used in predicting creep rate for longer time intervals. Although time t is not an explicit parameter in the equation, the ratio between $\dot{\varepsilon}_1$ at temperature T_1 and $\dot{\varepsilon}_2$ at temperature T_2 presents a multiplier coefficient of the kinetics μ that can be used in shifting strain rates along the time axis. This relationship is expressed in the form:

$$\ln\mu = \ln(\dot{\varepsilon}_1/\dot{\varepsilon}_2) = \frac{E}{R}\left(\frac{1}{T_2} - \frac{1}{T_1}\right) \tag{9}$$

One of the inherit problems in the used Arrhenius equation arises from the difficulty in determining the activation energy E from creep tests as the strain-rate is constantly changing during the test. However, a common and simple technique for the estimation of E consists of applying rapid change in temperature during creep under constant load. Creep strain rates $\dot{\varepsilon}_1$ and $\dot{\varepsilon}_2$ are measured before and after the change in temperature from T_1 and T_2, respectively, and E can be determined from Eq. (9). The evaluation of the activation energy E, according to the reference, is as follows:

The activation energy at various creep loads was assessed through testing HDPE geogrid specimens at various elevated temperatures and loading levels. The HDPE geogrid specimens were placed inside the test ovens and were tested at various temperatures ranging from 32 to 71°C. A hydraulic system applied creep loads ranging from about 18 to 40% of $T_{max.}$ The first set of tests on the geogrid consisted of increasing the temperature incrementally every 20 h while maintaining the load constant. The slopes of strains before and after the change in temperature represented creep strain rates at these temperatures and loading levels. The application of Eq. (9) resulted in an estimation of the activation energy for each loading level.

4. Stepped isothermal method (SIM) for geosynthetics

According to the former working draft of the ASTM standard test procedure, SIM is defined as "a method of exposure that uses temperature steps and dwell times to accelerate creep response of a material being tested under load." It departs from traditional accelerated time-temperature superposition (TTS) work in that it uses a single specimen per test and employs temperature steps throughout the test duration. While the data is shifted using principles well established by TTS work.

The single specimen/multiple temperature steps approach allows for the following three advantages over conventional elevated temperature testing.

http://dx.doi.org/10.5772/intechopen.72330

a. Reduces uncertainty in selecting appropriate shift factors to construct master curves.

b. Facilitates the reduction of uncertainty associated with inherent variability of multispecimen tests.

c. Can develop 100-year master curves in a relatively short period of testing time (16–32 h).

A description of the SIM method and its application to geonet creep testing has been discussed by Thoronton et al. [6]. The influence of temperature on creep is illustrated hypothetically in **Figure 5**; creep strain increases with increasing temperature. Stated differently, a given value of material strain can be reached within shorter times by using higher test temperatures. This principle is used in Stepped Isothermal Method (SIM) to obtain creep information within time periods significantly lower than those required for conventional creep testing. A typical SIM experiment is explained graphically in **Figures 5** and **6**. A series of successively higher temperature increments are implemented on a test specimen at a constant stress. Strain is measured during each temperature increment.

For example, temperature step in **Figures** 7 and 8 can be 7°C, and a total of six such steps are utilized. Each temperature step can be 1000 s long, that is, each temperature is maintained constant for 1000 s. The resulting strain is illustrated in **Figure** 7. Each strain increment plots linear on log time scale provided some restrictions and conditions specific to each material are observed. These conditions relate to material failure or rupture strength and melt or softening temperature.

In general, the lower the temperature and stress, the better behaved the relationship between time and strain. The result of rescaling the raw SIM data is shown **Figure 8** as a plot of creep modulus versus log (t − t′). The term t′ is actually a set of rescaling terms that create matching slopes between the isothermal dwells in log time space. The result of shifting the rescaled data is the master creep modulus curve of **Figure 6**. Note that the modulus curve extends beyond the 114y (106 h).

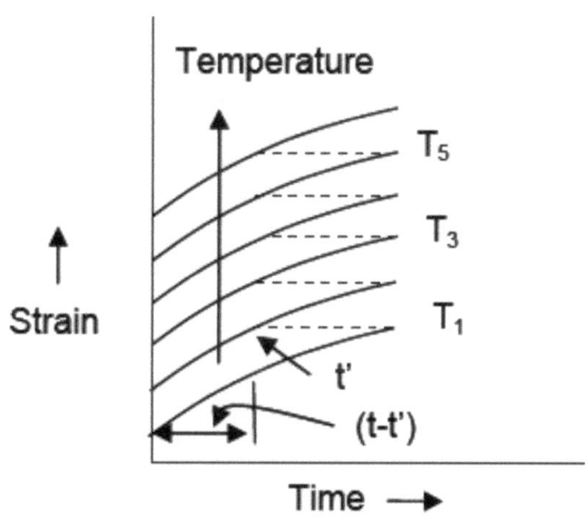

Figure 5. Effect of stress and temperature curves on creep behaviour.

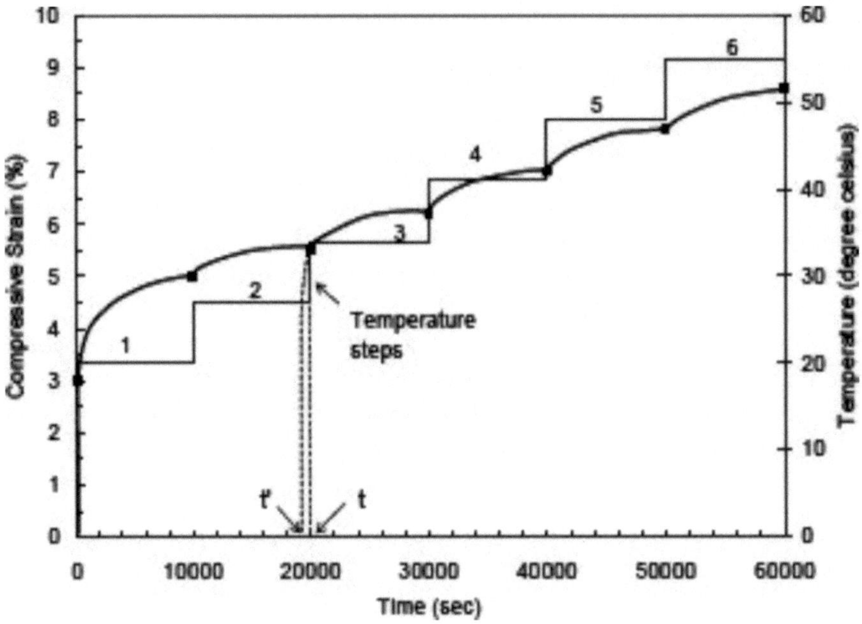

Figure 6. Schematic of SIM method.

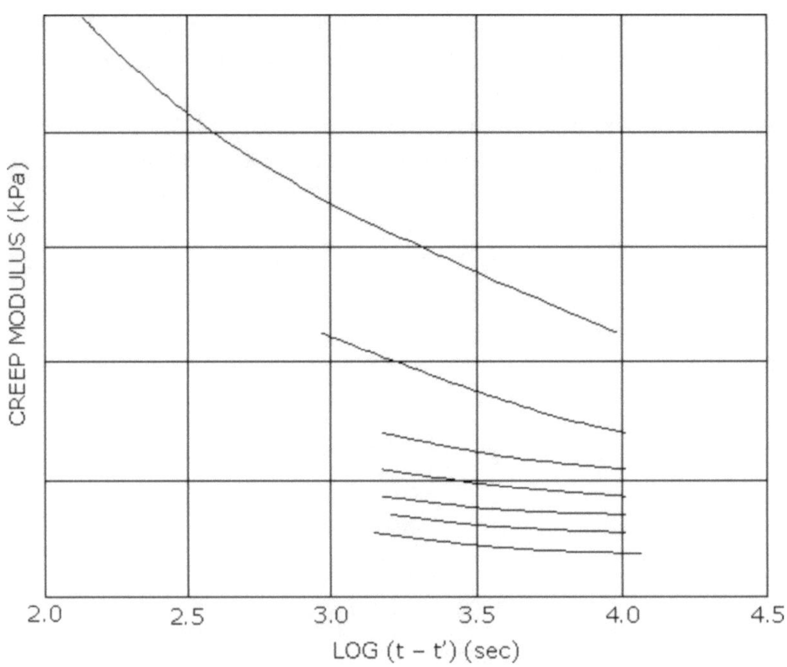

Figure 7. Creep modulus versus log (t – t′) after rescaling.

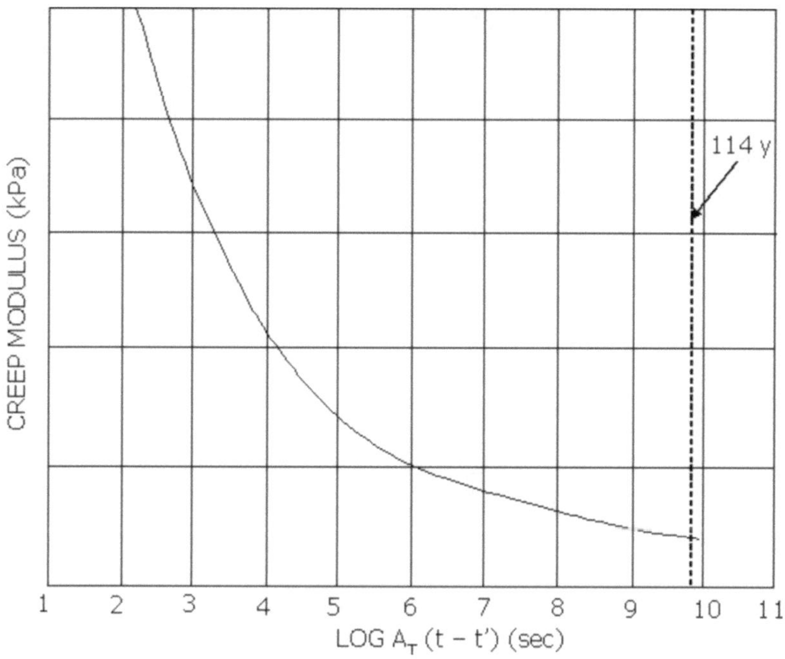

Figure 8. Creep modulus versus log AT (t − t') (sec).

The creep behavior from SIM on HDPE geogrids may be different than that from the TTS method because the low glass transition temperature (T_g) and melting temperature (T_m) of HDPE resin. When performing SIM on PET geogrids, the test temperature range is below the T_g (~70°C), and the material is in a glassy state. On the contrary, HDPE geogrid is tested in its rubbery state and relatively close to the T_m (124°C) at the same test temperature range. Plastic deformation would be a governing factor in the creep behavior of HDPE, particularly at the high testing temperatures.

The creep mechanism of SIM and TTS of HDPE is evaluated using the activation energy, which is obtained by plotting the shift factor against reciprocal test temperatures according to Arrhenius equations.

$$a_T = A \exp\left(\frac{-E}{RT}\right) \tag{10}$$

where E is the activation energy for the creep deformation, R is gas constant, and T is the test temperature

Table 1 shows the activation energy values of six creep tests based on SIM and TTS procedures in the referenced study [7]. For the same loading, the energy values resulting from SIM and TTS are very similar. The activation energy decreased linearly with stress, as shown in **Figure 9**.

Y. G. Hsuan [7]		Lothspeich and Thornton [10]		Farrag [14]	
Stress/method	**E (kJ/mol)**	**Stress/method**	**E (kJ/mol)**	**Stress/method**	**E (kJ/mol)**
10% UTS/SIM	302	37% UTS/SIM	230	19% UTS/TTS	91
20% UTS/SIM	257				
30% UTS/SIM	227 ± 10			23% UTS/TTS	102
40% UTS/SIM	198				
10% UTS/TTS	271			31% UTS/TTS	96
20% UTS/TTS	219				

Table 1. Activation energies with shift factor and strain-rate.

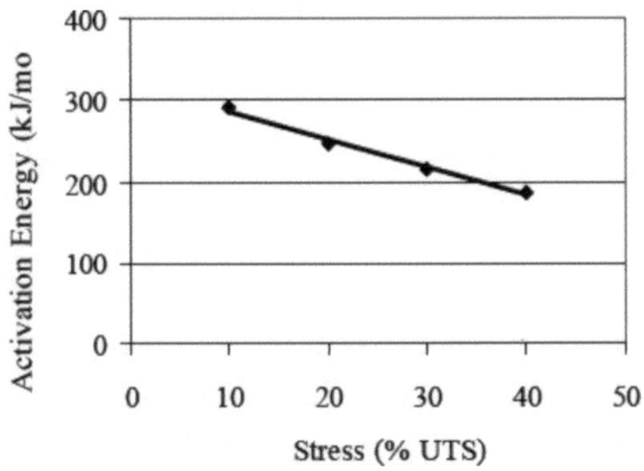

Figure 9. Activation energy versus applied stress.

This similar result for PET yarns and PP woven geotextile were also observed, respectively, [6, 8] and this behaviour was accounted as stress decreases the thermal energy requiring for polymer flow [9].

However, the activation energies resulted from this study as well as those calculated from data published is much higher than other published values [10]. It was found an activation energy value of 118 kJ/mol for PE fibers and an energy value of 85 ± 30 kJ/mol was obtained by studying the stress relaxation of an oriented HDPE [11, 12]. Creep performance tests using TTS with test durations of 1,000 hours were done at each temperature [13, 14]. The activation energies calculated from their shift factor at different applied load are presented in **Table 1**, and the values are closer to the published range for polyethylene materials. The high activation energy values from this study may result from the short test duration, which is only 10,000 s (2.7 h) for each temperature step. The discrepancy in activation energy values suggests that the creep mechanism is different between the short-term SIM and long-term TTS or conventional creep tests. Caution must be carried out when SIM and short-term TTS procedures are applied to new geosynthetics that do not have long-term creep data to be compared. However, once the verification is confirmed, SIM or short-term TTS can be extremely useful to evaluate similar geosynthetics with the same polymer type.

5. Comparison between SIM and conventional creep tests

5.1. Tensile creep test results

Figures 10–13 show the examples of creep curves derived from the results of the conventional long-term creep tests and the SIM tests [15]. The SIM tests are shown as thick lines, and the conventional tests are shown as thin lines. In this reference, the criteria used for comparison are the strain after 1 h (the initial strain), the strain at the end of the conventional test (final strain), and the difference between these strains (the time-dependent strain). The total strain at any time is the sum of the initial strain and the time-dependent strain. At the risk of over-simplification, the strain developed at 1 h could be taken to be that occurring during construction and that after 1 h as the strain generated in service.

Figure 10 shows the derived creep curves for polypropylene fabric. At a load of 56 kN/m (27% of tensile strength), the SIM data give a higher strain than the measured values. For example, at 122,374 h, the results of the conventional test give a total strain of 7.92% and a time-dependent strain of 3.21%, while for the equivalent time, the SIM data predict strains of 9.61 and 4.95%, respectively. Thus, the time-dependent strain is over predicted by 1.74% strain or 54% as a proportion of the actual strain. At a load of 114 kN/m (55% of tensile strength), the initial strain (i.e., at 1 h) recorded in the conventional test was 8.62%, while the SIM data predicted a value of 6.73%: this wide variability is not an unusual phenomenon in creep testing. The variability can be reduced by performing additional "ramp-and hold" tests ,and indeed the tests were used to adjust the initial strain of the conventional tests.

Nonetheless, the two creep curves are reasonably parallel until close to the rupture point of the specimen. At 7967 h, the point at which the rate of strain in the conventional test began to

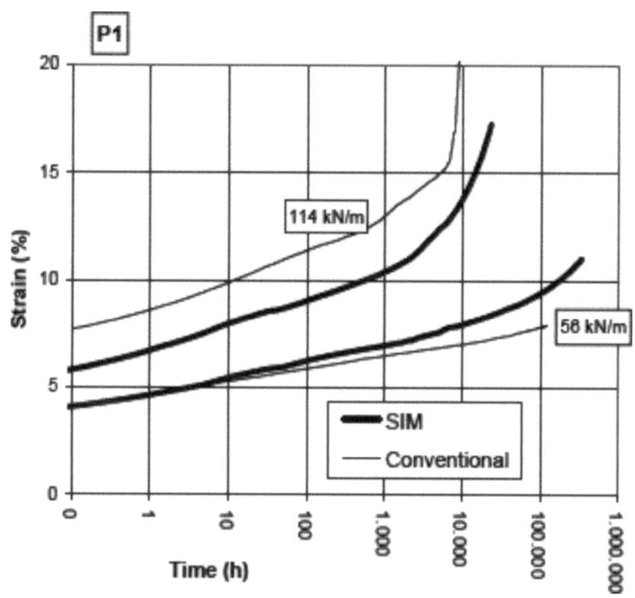

Figure 10. Comparison between results of conventional and SIM tests on material (polypropylene fabric Terram W20/4).

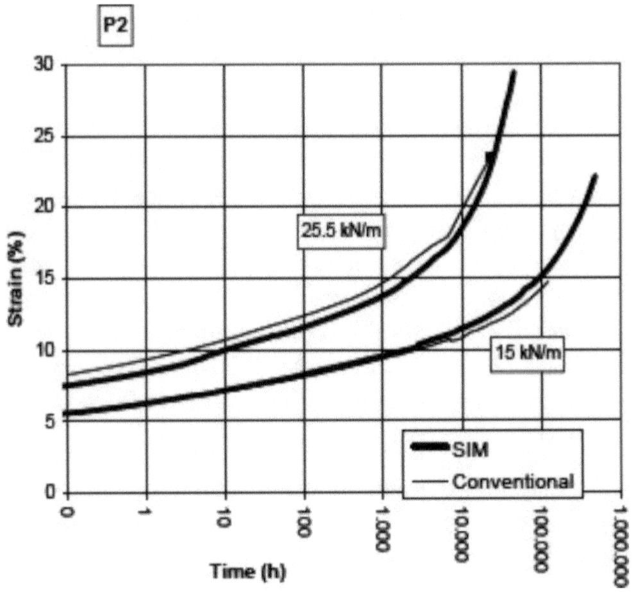

Figure 11. Comparison between results of conventional and SIM tests on material (polypropylene fabric Lotrak 45/45).

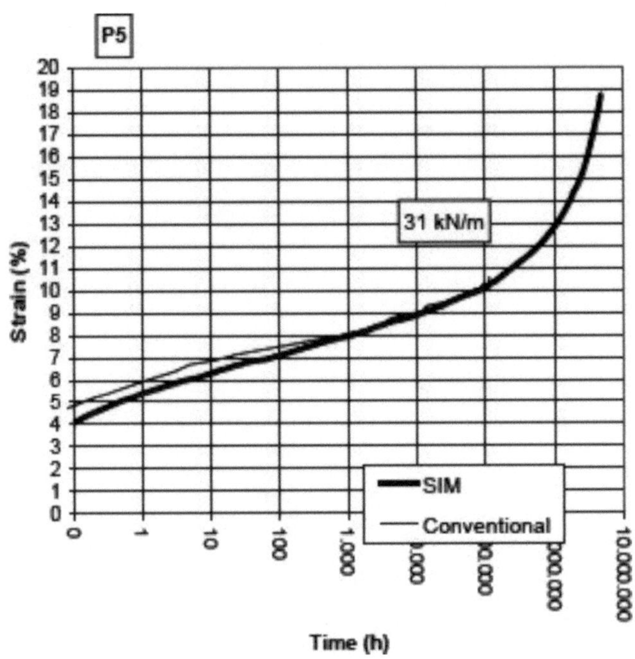

Figure 12. Results of conventional and SIM tests for polyethylene grid. Tensile strength 77.2 kN/m; tests at 31 kN/m (40%).

increase substantially, the total measured strain was 17.08%, while the predicted value was 13.19%. The time-dependent strains at this point were 8.46 and 6.46% respectively; thus, this strain was overpredicted by 2.0% or by 24% proportionally. The specimen in the conventional

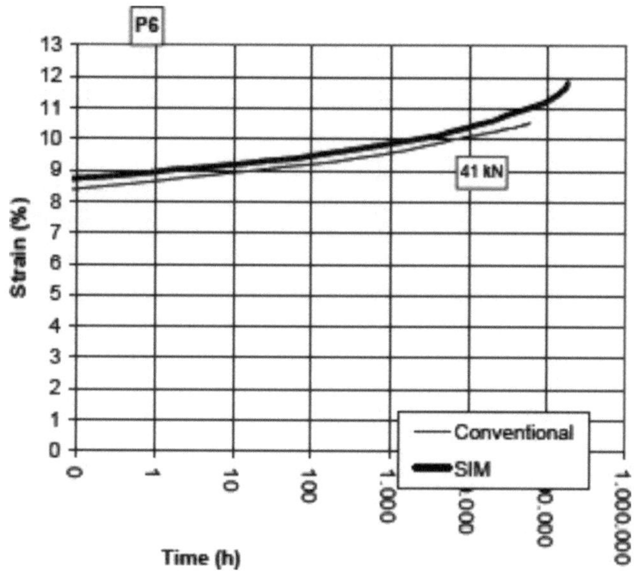

Figure 13. Results of conventional and SIM tests for polyester strip. Tensile strength 58.5 kN/m; tests at 41 kN/m (70%).

test ruptured after 9825 h at about 19% strain. The SIM test was terminated after the equivalent of 23,340 h at which point the strain was 17.3%.

Figure 11 shows the creep curves for polypropylene fabric. As shown in **Figure 11**, the creep curves derived from both types of test are in good agreement. At a load of 15 kN/m (31% of tensile strength), the total strain at the end of the conventional test (121,175 h) was 14.71%, and for the equivalent time, the SIM data predicted a strain of 15.62%. The time-dependent strains were 8.48 and 9.36%, respectively; thus, the difference was 0.88% strain – or 10% proportionally. At a load of 25.5 kN/m (52% of tensile strength), in the conventional test, the total strain after 6617 h was 17.90%, and for the equivalent time, the SIM data predict a value of 17.13%. For the same duration, the time-dependent strains are in excellent agreement at 8.71 and 8.69%, respectively. The specimen in the conventional test ruptured after 23,109 h at 23% strain, and the specimen in the SIM test ruptured after the equivalent of 49,000 h at 31% strain.

Figure 12 shows the creep curves for material polyethylene grid Tensar at 31 kN/m (40% of ultimate tensile strength). There is excellent agreement between the creep curves derived from the results of the conventional test and that predicted by the SIM data. The strain measured at the end of the conventional test (at 113,179 h) was 10.25% and that predicted by the SIM data was 10.23%. For this time, the time-dependent strains were 4.34 and 4.85%, respectively – a difference of 0.51 or 12% proportionally.

The principal objective of the tests undertaken on material Paraweb strip was to compare the creep-rupture characteristics, and so the tests were not necessarily performed at the same load levels. However, the results of a pair of tests that were undertaken at the same load (of 40.95 kN/strip, or 70% of tensile strength) are shown in **Figure 13**. The data from the conventional test indicate a strain of 10.64% at the end of a test (at 69,558 h) and a time-dependent strain of 2.00%. The predicted strains derived from the SIM test were 11.11 and 2.17%, respectively.

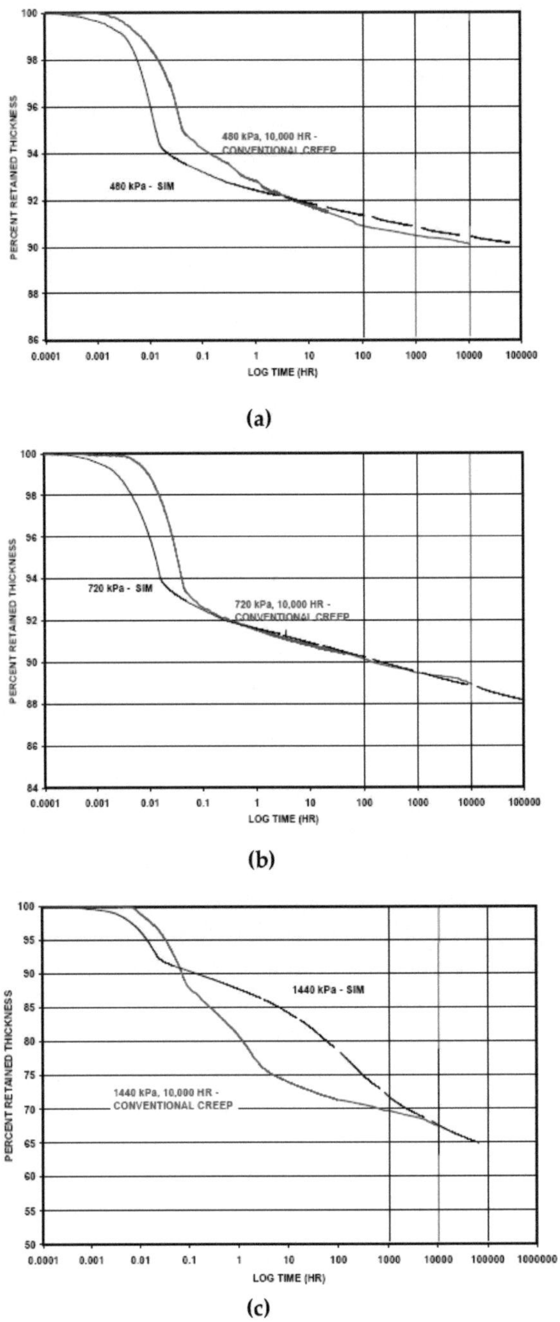

Figure 14. Comparison of 10,000-h creep response. (a) at 480 kPa, (b) at 480 kPa, and (c) at 1440 kPa.

The high initial strain and low time-dependent strain are typical of polyester. The difference in the time-dependent strains is small at 0.17% strain or at 9% proportionally.

As noted earlier, there can be quite a wide variability in the measurements of the initial strain in creep tests. In the conventional tests, the creep curve was adjusted using the following procedure:

- Perform two additional creep tests at the same load, each lasting just 1 h.
- Measure the strains after 1 h.
- Calculate the average of the strains after 1 h for the long-term test and the two additional tests.
- Subtract the strain after 1 h for the long-term creep test from this average.
- Add this difference to all strains measured for the long-term creep test.

In effect, the above shifts the entire long-term creep curve such that it passes through the average strain measured after 1 h, and the time-dependent strain is unchanged. This procedure has been named "ramp-and-hold" [6]. The data from the SIM tests were not adjusted. The agreement in the time-dependent strains determined by the conventional and SIM tests is no better than that between the final strains, but it would have been worse had the initial strains of the conventional tests not been corrected. This shows that better estimates of the initial and final strains can be obtained by undertaking these additional ramp-and-hold tests.

5.2. Tensile creep test results

The examples of the 10,000-h creep response and the comparison to the creep response at the same load generated via SIM testing are shown in **Figure 14(a)–(c)** [16]. The 10,000-h creep reduction factors were determined. Also, the calculated reduction factors for predicted 10,000 h and 35-year service lives from both conventional and accelerated creep testing are summarized in **Tables 2** and **3**. According to the results of this reference, while 10,000 h and 35-year creep reduction factors at 480 and 720 kPa were substantially similar as a result of SIM and conventional testing, they were observed to differ by as much as 66% at 1440 kPa. However, the cause of this difference was found to be related to the retained thickness at the relatively short duration of 100 h. As seen in **Figure 14(c)**, significant differences in the thickness response to loading and the onset of creep at 1440 kPa were manifest in less than 10 h of test time. This is clearly unrelated to the elevated temperatures used during SIM testing.

Load (kPa)	10,000 h RF_{CR} at 20°C	10,000 h RF_{CR}-SIM	Difference
240	NA	1.038	NA
480	1.056	1.041	−1.4%
720	1.057	1.071	+1.3%
960	NA	1.135	NA
1200	NA	1.288	NA
1240	1.323	1.963	+48%

Table 2. SIM versus conventional testing at 10,000 h.

Load (kPa)	35 Year RF_{CR} at 20°C	35 Year RF_{CR}-SIM	Difference
240	NA	1.049	NA
480	1.065	1.075	+0.9%
720	1.113	1.118	+0.4%
960	NA	1.281	NA
1200	NA	1.601	NA
1240	1.652	2.739	+66%

Table 3. SIM versus conventional testing at 35 years.

6. Review of creep for geosynthetics

The behavior of geosynthetics may be classified into three groups: (1) temperature-independent, stress-dependent deformation, asymptotic to a final creep value (PET geotextiles), (2) temperature- and stress-dependent deformation, only under a threshold mainly of stress deformation are asymptotic to a final creep value (HDPE-extruded geogrid), and (3) temperature- and stress-dependent deformation (woven PP-slit film geotextiles). The creep behavior of PET non-woven and PP woven geotextiles is not modified by the confining pressure; however, the total strain of PP non-woven geotextiles is strongly reduced with a confining pressure of 50 kPa and no more with higher confining pressure. The strain reducing increases with the level of load. From these results it seems that a confining pressure on a geotextile does not modify the structural creep but can strongly modify the polymer creep. This modification is effective with low values of the confining pressure. It makes one feel sorry that the reason why was not explained.

In this study, the performance limit strain was about 15%, but they do not use these values because it is seen that there was no evidence of the onset of rupture occurring at strains of less than 16% and the strain time and Sherby-Dorn curves show that the grids behave in a predictable manner, and there are no sudden changes in performance. Thus it can be decided that the performance limit strain of 10%, safe, if conservative, design criteria for the ultimate load condition of that product. However, it is seen that the sudden change near the 15% and the importance is the reason why how to generate these Sherby-Dorn curves. From this time, it can be that the use of the 10% limit performance strains to the most of geosynthetic reinforcements. This is very sad to the geosynthetics research field. Another thing is the application and analysis of time-temperature superimposition principle. it is seen that the shifted the raw strain data to the time scale without any explanation of theory or shift mechanism.

The purpose of this study was to make sure about the various aspects of the test procedure, such as the type of clamping, the loading time, the minimum length of the specimens, the required testing time, the extrapolation procedure, and the level of the accuracy of the prediction. And as suggested the time-temperature superposition, using the creep modulus curves, has been used to extrapolate the behavior of integral geogrids to 106 h. The time shift between the curves at 20 and 30°C is equal to 100 or 102.0, while the time shift between 20 and 40°C is equal to 2000 h or 103.301. This means, practically, that 1 h of creep testing at 30°C produces the same elongation than 100 h at 20°C, for the same testing condition; or 104 h at 30°C corresponds to 106 h at 20°C and similarly for the 40°C testing, with a 2000 factor.

Using the new SIM, over 15 load specific master creep modulus curve were generated, some of which extended to over 100-year design lifetimes, each from a single specimen in a test that was completed in less than 18 h. The results of the conventional and SIM procedures were found to be an equivalent for the polyester products examined. Recently, the compressive creep behavior of a geonet was also evaluated using SIM and the compressive creep test about the EPS Geofoam was evaluated using SIM.

7. Conclusions

The final comment about the compressive creep test and data interpretation as follows:

1. Description of the compressive mechanism of the geonet (exactly not compression but perpendicular compression) is very important because the creep mechanism of tension and compression are quite deferent.
2. To reduce the specimen-to-specimen, many ramp-and-hold (in the case of tension creep: 1 h) tests are recommended.
3. Loading rate is also important because it makes initial strain value (εi). To check the non-affected loading rate, prior to the main creep test, some kind of short-term test is needed.
4. The loading level may be a decision according to the site-specific conditions.
5. If the time-temperature superposition will be used, the exact answer about that what shift factor value is used in this study. How these values are used? Why the method to calculate the shift factor was used? Is strongly needed to the qualified paper?
6. The method to assess the RF_{creep} also will be reviewed because the value will be changed according to the applied load. Maybe to find the applying load value that satisfies the performance limit strain at the design time.

Author details

Han-Yong Jeon

Address all correspondence to: hyjeon@inha.ac.kr

Department of Applied Organic Materials Engineering, Inha University, Incheon, South Korea

References

[1] Koerner RM. Designing with Geosynthetics, 5th ed. NJ: Prentice Hall Inc.; 2005. Chapter 5

[2] Fannin RJ. Factors influencing the long-term flow capacity of geonets. 3rd International Conference on Geosynthetics. 1995. pp. 267-280

[3] ASTM D2990-1. Standard Test Methods for Tensile, Compressive, and Flexural Creep and Creep-Rupture of Plastics. West Conshohocken, PA, United States: ASTM International; 2001

[4] Cazzuffi D, Corbet S. In: Proceedings of 5th International Landfill Symposium. 2001. pp. 476-491

[5] Mano JF, et al. Polymer. 2001;**42**:6187

[6] Thornton JS, Paulson JN, Sandri D. Conventional and stepped isothermal methods for characterizing long-term creep strength of polyester geogrids. In: 6th International Conference on Geosynthetics, Atlanta, GA, USA. 1998. pp. 691-698

[7] Yeo SS, Hsuan YG. The short and long term compressive behavior of HDPE geonets and geocomposites under inclined conditions. Journal of Geotextiles and Geomembranes. 2007;**14**(3):154-164

[8] Baker LT, Thornton JS. Comparison of results using the stepped isothermal and conventional creep tests on a woven polypropylene geotextile. In: Geosynthetics Conference 2001, Portland, OR, USA. 2001. pp. 729-740

[9] Sherby OD, Dorn JE. Anelastic creep of polymethyl methacrylate. Journal of the Mechanics and Physics of Solids. 1958;**6**:145-162

[10] Lothspeich SE, Thornton JS. Comparison of different long-term reduction factors for geosynthetic reinforcement materials. In: 2nd European Geosynthetics Conference, Italy. 2000. pp. 341-346

[11] Govaert LE, Bastiaansen CWM, Leblans PJR. Stress-strain analysis of oriented polyethylene. Polymer. 1993;**34**(3):534-540

[12] Cembrola RJ, Stein RS. Crystal orientation relaxation studies of polyethylene. Journal of Polymer Science: Polymer Physics Edition. 1980;**18**:1065-1085

[13] Farrag K, Shirazi H. Development of an accelerated creep testing procedure for geosynthetics – Part I: Testing. Geotechnical Testing Journal. 1997;**20**(4):414-422

[14] Farrag K, Shirazi H. Development of an accelerated creep testing procedure for geosynthetics – Part II: Analysis. Geotechnical Testing Journal. 1998;**21**(1):38-44

[15] Greenwood JH, Voskmp W. Predicting the long-term strength of a geogrid using the stepped isothermal method. In: 2nd European Geosynthetics Conference, Italy. 2000. pp. 329-331

[16] Allen SR. The use of an accelerate test procedure to determine the creep reduction factors of a geosynthetic drain. Geotechnical Special Publication, Geo-Frontiers Conference. 2005. pp. 3297-3309

Chapter 7

Creep Lifing Models and Techniques

Zakaria Abdallah, Karen Perkins and Cris Arnold

Additional information is available at the end of the chapter

http://dx.doi.org/10.5772/intechopen.71826

Abstract

The deformation of structural alloys presents problems for power plants and aerospace applications due to the demand for elevated temperatures for higher efficiencies and reductions in greenhouse gas emissions. The materials used in such applications experience harsh environments which may lead to deformation and failure of critical components. To avoid such catastrophic failures and also increase efficiency, future designs must utilise novel/improved alloy systems with enhanced temperature capability. In recognising this issue, a detailed understanding of creep is essential for the success of these designs by ensuring components that do not experience excessive deformation which may ultimately lead to failure. To achieve this, a variety of parametric methods have been developed to quantify creep and creep fracture in high temperature applications. This study reviews a number of well-known traditionally employed creep lifing methods with some more recent approaches also included. The first section of this paper focuses on predicting the long-term creep-rupture properties which is an area of interest for the power generation sector. The second section looks at pre-defined strains and the re-production of full creep curves based on available data which is pertinent to the aerospace industry where components are replaced before failure.

Keywords: creep, creep lifing methods, long-term creep behaviour, stress rupture, creep prediction

1. Introduction

The drive towards more efficient gas turbines and the associated reductions in greenhouse emissions require the existing gas turbines to operate under higher severe temperatures. However, this aim is restricted by the limitation of the materials used in such harsh environments which may, eventually, lead to the deformation and failure of these components. In order to avoid such catastrophic failures and increase the efficiency, future designs must utilise novel or improved alloy systems with an enhanced temperature capability. One key material property that governs the life of many components within the gas turbine is creep. A detailed understanding of the creep behaviour of materials is seen as an essential requirement. However, understanding and predicting the creep properties is a very important challenge for researchers, which is the basis of this chapter. Therefore, the current research will thoroughly concentrate and investigate the long-term creep predictions of materials as well as their behavioural attributes under the applied stresses and temperatures.

Creep is defined as the plastic deformation of materials under the effect of high stresses and temperatures for long durations of time which, eventually, leads to fracture. Generally speaking, problems of creep failure and excessive distortion are experienced at temperatures equal, or just above, to the half of the melting temperature, TM, of a material. It might be possible to avoid creep problems by either selecting materials of high melting temperatures or maintaining the operation temperatures far away from those at which creep could take place, typically less than the third of the melting point of a material. However, these simple solutions do not provide a comprehensive and convincing answer to the problem. On one hand, materials of high melting temperatures can be developed and employed but will still show creep deformation under the high stresses and temperatures encountered in such applications. On the other hand, if temperatures are lowered to less than the third of the melting temperature, this will, in return, lower the efficiency, which is undesirable in these applications. Therefore, the design stage is the crucial part of the industrial process where decisions should be taken so as to avoid the long-term creep failures [1].

During the design stage, a comprehensive study and analysis of a material's behaviour should be made before this material is considered for a particular application. For certain applications, this might be adequate but for fundamental studies of creep behaviour, full creep curves must be available. For this purpose, creep tests can be carried out at different stresses and temperatures in order to provide the designer with the necessary information to study and analyse the long-term behaviour of materials under the applied stresses and temperatures. Since it is impractical to perform creep tests for the entire lifetime of some real applications, particularly when lifetimes can range, for instance, from 20,000 to 120,000 h as in the power generation applications, determining a conservative and an acceptable method for extrapolating the short-term measurements is a significant goal. Alternatively, for aerospace applications, where the time to a certain percentage strain is more desirable, this method should also provide accurate predictions of the creep behaviour based on this criterion. Starting from this point, many extrapolation techniques were devised for the purpose of predicting the long-term creep behaviour of materials without the need to carry out practical tests which could last for many years before being able to size and manufacture the required components. Minimising the scale of these larger tests will, in return, reduce the cost and save the time needed for such long-term tests. Hence, these predictions require short-term data to be available from the various types of creep tests at the same conditions as the actual application. Extrapolation methods must be taken into consideration that creep is a critical function of stress and temperature, that is, a relatively small change in either of these quantities can drastically affect the material's lifetime. These methods are being used to predict both creep-rupture and creep-deformation behaviours, in which the former has received a greater attention than the latter as a result of the more drastic consequences of brittle failures, that is, sudden rupture, compared with ductile failures, that is, excess deformation [2].

2. Review of creep parametric methods

Many approaches were proposed in an effort to predict the long-term creep properties based on short-term creep measurements so as to reduce the time scales and costs required to obtain such long-term data. Each of these approaches represents a technique through which the short-term creep-rupture data can be extrapolated using a time-temperature parameter. This concept is based on the assumption that all creep-rupture data, for a given material, can be superimposed to produce a single 'master curve' wherein the stress is plotted against a parameter that contains and combines time and temperature. Based on this master curve, that can only be constructed using available short-term measurements; extrapolation to longer times can then be obtained. These parametric methods play a key role during the design stage in which the high temperature components are designed to codes that are intended to assure a specific life. Generally, these design codes define a maximum allowable stress that can exist in a component during the anticipated design life. This allowable design stress, which is a combined function of time, temperature and material, is usually based on the rupture stress required to give the expected design life. It is tempting to infer that the plant will give a satisfactory service up to, but not much beyond, the design life. For this reason, two distinct parts of the service life can be defined, namely: (a) the original design life which can typically be 100,000 h and (b) the safe economic life. Although the latter is normally outside the influence of the design codes, it can be considered as a significant fraction of the overall service life. Moreover, due to the time-dependent nature of materials' properties at high temperatures and the fact that ultimate failure is, thus, implicit, consideration must be always given to a 'beyond design' end-of-life criterion. Since the time required for a crack to grow can be very short, life extension is only safe within the time scale for crack initiation unless defect growth is being monitored [3].

In general, current methods normally involve two approaches, namely: (1) those which involve the acquisition and monitoring of operational parameters, the use of standard materials data and the life fraction rule and (2) those based on post-service examination and testing which require direct access to the component being examined for sampling and measurement [3]. These parametric methods have a great advantage, at least in theory, of requiring only a relatively small amount of data to establish the required master curve. Some of these approaches proved their validity for creep predictions by providing satisfactory results whereas others failed to give precise long-term predictions.

2.1. Review of the power law

The power law represents a combination of the temperature and stress dependences of creep rate which are described by, respectively, Arrhenius's and Norton's laws. In these two laws, the secondary strain rate, $\acute{\varepsilon}_s$, is used to describe the creep rate of materials, as follows [1]:

- **Arrhenius Law:** As the strain rate, $\acute{\varepsilon}_s$, increases with increasing the temperature, T, a straight line relationship can be obtained when plotting ($\ln \acute{\varepsilon}_s$) against (1/T), as shown in **Figure 1(a)**. Thus;

$$\acute{\varepsilon}_s \, \alpha \, \exp\left(-Qc/RT\right) \tag{1}$$

where Qc is the activation energy for creep and R is the gas constant.

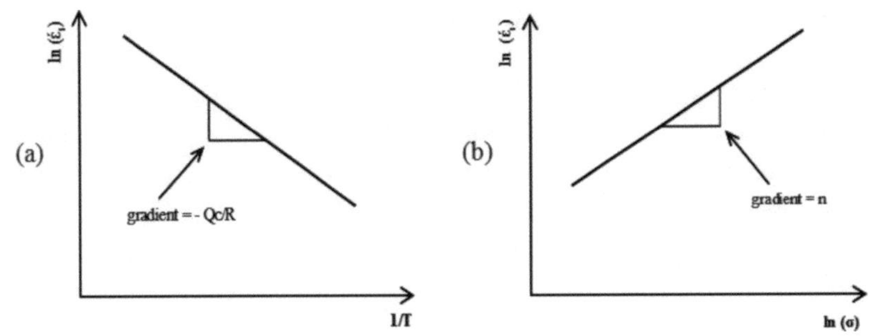

Figure 1. The secondary creep rate dependence of (a) temperature and (b) stress, respectively.

- **Norton's Law:** As the strain rate, $\acute{\varepsilon}_s$, increases with increasing the stress, σ, another straight line relationship can be obtained when plotting ($\ln \acute{\varepsilon}_s$) against ($\ln \sigma$), as shown in **Figure 1(b)**. Thus;

$$\acute{\varepsilon}_s \, \alpha \, \sigma^n \tag{2}$$

where n is the stress exponent. Combining these two laws together, that is, Eqs. (1) and (2), gives the power law equation as [1]:

$$\acute{\varepsilon}_s = A\,\sigma^n \exp\left(-Qc/RT\right) \tag{3}$$

where A is a constant.

It was also assumed that the value of Qc and n is constant but, in fact, after further research, it was found that their values vary according to the creep mechanism in different stress and temperature regimes [4]. The value of Qc is related to temperature, according to Eq. (1) and **Figure 2(a)**, such that Q_1 and Q_2 represent the value of Qc at high temperatures (due to vacancy flow through the lattice) and low temperatures (due to vacancy flow along grain boundaries), respectively [1]. On the other hand, the value of n is related to stress, according to Eq. (2) and **Figure 2(b)**, such that n_1 and n_2 represent the value of n at high stresses (due to dislocation creep) and low stresses (due to diffusional creep), respectively [1].

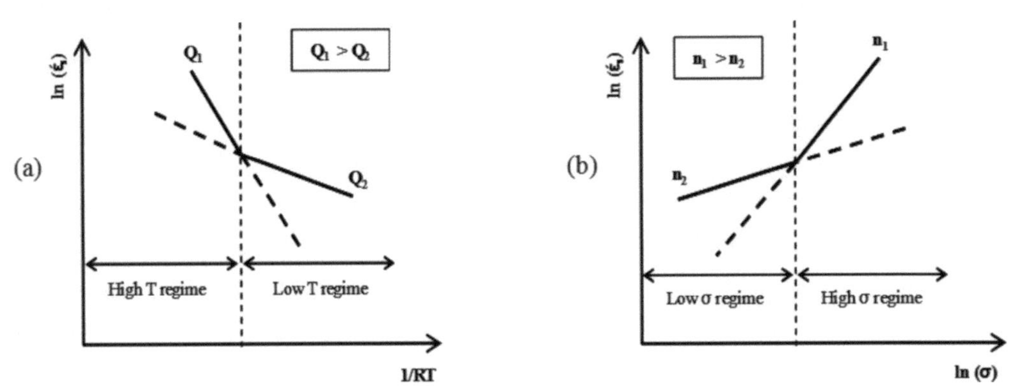

Figure 2. Transition of (a) Qc and (b) n, relative to temperature and stress, respectively.

According to Wilshire and Scharning [5, 6], when creep tests were carried out on the 9–12% chromium steels, it was found that the value of Qc and n was changing with increasing the temperature and decreasing the stress. Therefore, it can be deduced that there is a variation in the value of Qc and n used in the power law equation depending, respectively, on temperature and stress regimes during the creep process. For this reason, and since these values vary in an unpredictable manner, the power law equation does not allow accurate estimation of the long-term rupture strengths by extrapolating the short-term measurements [5, 6]. Furthermore, using these relationships for extrapolation will overestimate the actual long-term performance, **Figure 3**, which might lead to considerable errors in the prediction of creep behaviour and thus, catastrophic consequences. If a certain method is unable to accurately predict the creep behaviour, the consequences will be less severe if the method underestimates the actual measurements rather than overestimating them, as underestimation will keep the component life within the safe operational conditions.

2.2. Review of the Larson-Miller (LM) methodology

This parametric approach is one of the methods used to predict the stress rupture data of metals. It has been originally derived from Arrhenius relation at a constant stress and thus, a constant stress exponent n, but at a variable value of T and Qc, which gave the final form of this relation as [7]:

$$P_{LM} = f(\sigma) = T\,(C_{LM} + \log t_f) \tag{4}$$

where C_{LM} and P_{LM} are the Larson-Miller constant and parameter, respectively. The parameter, P_{LM}, can be used to superimpose the family of rupture curves into a single master curve [2]. The constant, C_{LM}, includes the Monkman-Grant constant M [8], which is a function of Qc that was proved elsewhere [2, 5, 6] to be a function of stress. Plotting log t_f against 1/T at constant

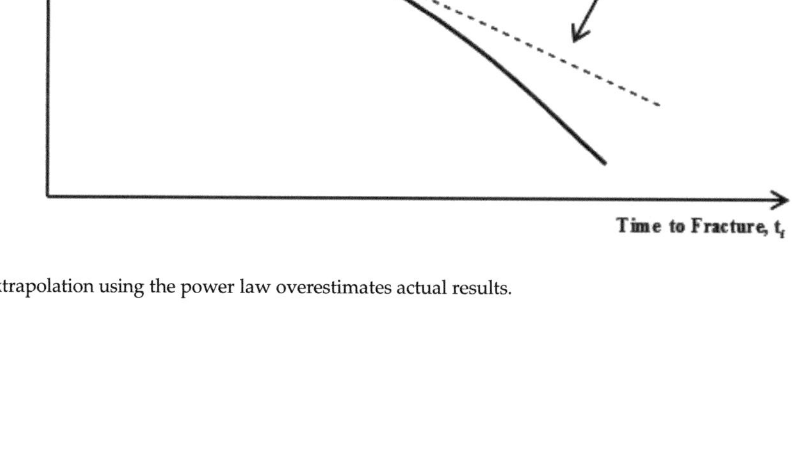

Figure 3. Extrapolation using the power law overestimates actual results.

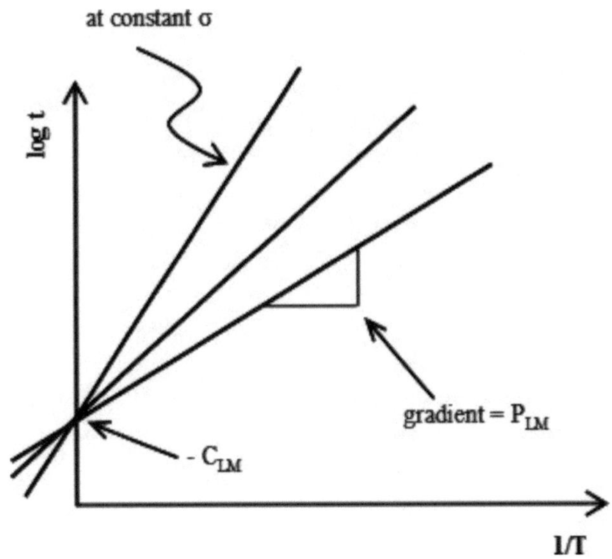

Figure 4. Determination of the Larson-Miller constant.

stresses, **Figure 4**, for some experimental data gave straight lines of slope P_{LM} and an intercept of - C_{LM} [9].

This method was further studied by Krivenyuk and Mamuzic [10], who described the constant C_{LM}, as:

$$C_{LM} = (T/\Delta T)\ m' \log(\sigma_1/\sigma_2) \tag{5}$$

where σ_1 and σ_2 are the corresponding stresses at a constant time value from two rectilinear stress rupture (SR) curves tested at T_1 and T_2 (where $T_2 = T_1 + \Delta T$), and m′ is the reciprocal of the slope, at the selected time value, of the SR curve at temperature T_1. When the value of C_{LM} was estimated based on the data of two rectilinear SR curves at temperature T_1 and T_2, it was found that the value of C_{LM} depends on the position of the two curves relative to each other. In other words, if the curves were parallel then, this means that, C_{LM} is constant. But, if the slope changed from one curve to another then, as the time to rupture increases, the value of the logarithm in Eq. (5) also increases leading to a significant dependence of C_{LM} on time. Hence, for equidistant curves, the time dependence of the constant C_{LM} is weak, whereas it might become sharp for curves that are distinguished by their slopes [10].

Larson and Miller took one step further in their original proposal, suggesting that the value of the constant C_{LM} could be taken as 20 for many metallic materials [7, 11]. However, it was found that the value of this constant varies from one alloy to another and is also influenced by factors such as cold-working, thermo-mechanical processing, phase transitions and/or other structural modifications [11]. Moreover, most applications of the Larson-Miller parameter are made by first calculating the value of C_{LM} that provides the best fit of the raw data, which means that C_{LM} is treated as a 'fitting constant' based on a 'trial and error' method instead of being a physically meaningful constant. For instance, a certain study [11] showed that the value of this constant for specific aluminium alloys ranged from about 13 to 27.

In studies of refractory and heat-resistant steels and alloys by Krivenyuk and Mamuzic [10], calculations often gave rather lower values of the constant C_{LM} than the commonly used value of 20. In these investigations, the difference in the values of this constant was mainly a result of the time dependence of this constant. In addition, the refractory metals were primarily studied at short loading times whereas the heat-resistant materials were investigated at longer loading times which led to higher values of C_{LM} for the latter, according to formula (5). In agreement with these findings, Cipolla and Gabrel [12] found a huge stress overestimation when the Larson-Miller equation was used on the high chromium steel (Grade 91) at all temperatures, especially at 600°C. Therefore, the requirement of a physical realism of extrapolation was not completely fulfilled by this method which is less conservative and seems to be less able to describe the strong curvature between the low and the high stress regimes.

The equation of Larson and Miller was reviewed by Wilshire and Scharning [5, 6] on the 9–12% chromium steels. Although it was generally accepted that C_{LM} should be taken as 20, the data fit with the curving LM plots was, frequently, better with other values, where, in the case of chromium steels, the best fit was obtained when C_{LM} was 36 instead of 20. This difference in the value of C_{LM} was attributed to the fact that it is a function of Qc which is, itself, a 'variable'. Thus, Larson and Miller's results were only in agreement with the theoretical equation for low temperature deformation, and could not accurately describe the high temperature properties.

A very logical explanation was given by Larke and Inglis [13] who assumed that if two different materials were tested at the same temperature, T, and fractured at the same time, t_f, then if the value of C_{LM} was the same for both materials, Eq. (4) would give the same value for P_{LM}, even though, as would in general be so, the stress to cause fracture is different for each material. Therefore, if the value of C_{LM} is considered 'presumably' as 20, as Larson and Miller suggested, then this suggestion will imply that, for the same conditions of testing, the fracture time would be the same for all materials, which is apparently unacceptable. In addition, this suggestion also means that if, for a given material, a set of stress rupture curves at different temperatures are established, then, over the same temperature range, these curves would be valid for any other material provided that only the stress scale is altered [13].

The graphical method, **Figure 4** recommended by Larson and Miller for determining the numerical value of C_{LM} was proved to be quite unsatisfactory [13]. This was based on the fact that, at least, one pair of lines intersects at a significantly different value of log t_f than the other pairs, and this, coupled with the fact that personal choice enters into the drawing of the curves associated with the basic log σ/log t_f data, increases the doubts on the acceptability of this method for determining the value of C_{LM} [13]. Another critical assessment of this method documented in Murry [14] concluded that the different curves which represent the variations of the Larson-Miller parameter with the initial stress, at different temperatures, very rarely coincided. It was also observed that the value of C_{LM} could vary from 2 to 55, very often in relation to the initial stress. In agreement with this assessment, another study also documented in Murry [14] found that the constant C_{LM} varied with the material, the test temperature and the initial stress. Along with these studies, another extensive work carried out by Penny and Marriott [2] on the Larson-Miller method stated that this method stands alone as the least accurate of all methods, both in correlation and extrapolation, where errors resulting from its use are significant even when good quality data are available.

Therefore, this parametric formula could only be used to a very limited extent to extrapolate time, temperature, stress and elongation since the value of C_{LM} was found to be variable. Moreover, the unknown curvature of the parametric plots of the Larson-Miller equation makes data extrapolation unreliable. Hence, even when tests lasting up to 30,000 h have been completed, this parametric method does not allow unambiguous determination of the 100,000 h rupture strengths.

2.3. Review of the Manson-Haferd (MH) methodology

Manson and Haferd [15] developed a linear time-temperature relationship for extrapolating creep and stress rupture data. The Manson-Haferd (MH) methodology was developed in order to eliminate the errors introduced by the Larson-Miller technique which assumed a fixed value of the constant used in its equation that led to inaccuracies in predicting the creep life [2, 15]. This technique assumes the same starting point of steady-state creep dominated by a power law behaviour but considers, later on, that the logarithm of the time varies linearly with the test temperature at a constant initial stress, according to [14, 15]:

$$\log t = a - P_{MH}\, T \tag{6}$$

where t is the time (either the time to fracture, t_f, or to a certain strain level, t_ε), $a = \log t_a + P_{MH}\, T_a$ (where t_a, P_{MH} and T_a are the Manson-Haferd time, parameter, and temperature constants, respectively), T is the absolute creep test temperature, and the point (T_a, t_a) is the point of intersection of the straight lines corresponding to the various iso-stress lines. Therefore, the Manson-Haferd parameter, P_{MH}, determines two constants compared to the Larson-Miller parameter that involves only one constant. Rearranging Eq. (6) gives [14, 15]:

$$P_{MH} = f(\sigma) = (\log t - \log t_a)/(T - T_a) \tag{7}$$

According to Manson and Haferd's suggestion, the parameter P_{MH} can, thus, be derived graphically from the intersection point of the extrapolated iso-stress lines when plotting log t_f against T. Moreover, plotting P_{MH} versus stress, σ, will force all creep data to collapse onto a single 'master curve'. The equation of this curve can then be determined by a curve fitting technique, which yields an equation relating time to a given percent creep, temperature and stress [15].

In agreement with Manson and Haferd, it was postulated elsewhere [14] that the parameter P_{MH} was derived from the approximately linear relationship found experimentally between log t_f and T as well as from the trend of the data that converge at a common point (T_a, t_a). This parameter, therefore, measures the slopes of the straight lines obtained for given values of stress. Values of T_a and log t_a which best fit the data vary for different materials [14]. Manson and Haferd showed that the values of T_a for most materials ranged from 0°F (−17.78°C) to 200°F (93.3°C) whereas the values of t_a varied appreciably [15]. Although single values of

T_a and log t_a might be found and universally agreed and used with satisfactory results, this possibility has not as yet been demonstrated. They also added that accurate results could be expected with this parameter, as with the LM parameter, only if the proper values of the constants were used for each material. However, the variation in the value of T_a and log t_a introduced many errors in extrapolating the short-term data, as it was found with the LM approach. Murray and Truman [16] also reviewed the MH technique and obtained new values of T_a and log t_a which accurately fitted the data of the austenitic steels used in the experiment. They also found that the values of the constants obtained were different from the standard values proposed by Manson and Haferd. Along with Murray and Truman, different values of these two constants were obtained elsewhere [17, 18] when experiments were carried out on different steels.

An advantage of the MH parameter is that it can be used for various materials and different times which could be either the time to a certain percent creep strain or the time to rupture. However, the numerical values of the MH constants read from the plots of log t_f against T are not precise enough unless very comprehensive experimental data are available. Furthermore, by using this technique, predicting the stress and the time values outside the temperature range on which the magnitudes of the constants are based can lead to significant errors [13]. An assessment carried out by Pink [19] stated that none of the methods had a consistent physical basis and that the apparent success of a certain procedure has only resulted from its applications in just circumstantial conditions. Furthermore, it was added that on one hand, the method of Larson and Miller, for instance, shows better consistency with the deformation processes occurring at low temperatures and thus, offers better results in the extrapolation of this type of data. Whereas on the other hand, the method of Manson and Haferd does not present any physical meaning, but coincidentally describes the complex pattern of deformation controlled by several mechanisms and is, thus, more reliable for long-term predictions of data generated at higher temperatures.

All of these methods were only proposed to analyse creep testing data since there is no mention in the literature of using the hot-tensile testing data, for example, in the analysis using these techniques [18]. Therefore, and based on these facts, the validity of this method is limited based on the conditions according to which the test is being carried out and thus, further research should be done in order to improve its capability of predicting the long-term creep properties before adopting its results.

2.4. Review of the Orr-Sherby-Dorn (OSD) methodology

The Orr-Sherby-Dorn (OSD) technique [20] involves a time-temperature parameter based on the parallelism of the iso-stress lines of a slope that represents the Orr-Sherby-Dorn constant, C_{OSD}. In this methodology, the assumptions of the Larson-Miller technique have been interchanged. In other words, the constant of the Larson-Miller equation, C_{LM}, became a function of stress whereas the parameter, P_{LM}, became a constant [2, 14]. Based on these new assumptions, the LM relation (Eq. (4)) can be re-arranged to give the OSD equation as [20]:

$$P_{OSD} = f(\sigma) = \log t_f - C_{OSD}/T \qquad (8)$$

where P_{OSD} and C_{OSD} are the Orr-Sherby-Dorn parameter and constant, respectively, T is the absolute creep test temperature and t_f is the time to fracture. The basis of the OSD life prediction methodology is that the activation energy, Qc, remains constant over the entire creep curve, with relatively sparse supporting data [20]. However, since the constant C_{OSD} includes the activation energy, Qc, then any variations in Qc will, thus, ensure that the superimposed parametric plots will be non-linear [5, 6]. Indeed, there is evidence that in some cases, the creep activation energy seems to increase systematically through the primary region [21].

In order to prove the variation in the value of C_{OSD}, tests were carried out by Murray and Truman [16] and graphs of log t_f against 1/T at constant stress values were plotted. The gradients of these plots, that is, the values of C_{OSD}, were also calculated. Eventually, it was found that in spite of the difference between the values of C_{OSD} obtained experimentally and the values proposed by Orr, Sherby and Dorn, the data were fitted with reasonable accuracy [16]. Since the slope of the resulting log t_f against 1/T line will be the numerical value of C_{OSD}, it was proposed by Orr, Sherby and Dorn that the adjacent log σ/log t_f curves will be equidistant from each other along the time scale [13]. Therefore, in principle, only one line of log t_f against 1/T at a constant stress needs to be drawn in order to determine the value of the constant C_{OSD}, although in practice, the average slope of lines corresponding to different stress levels would be determined. However, it was found quite impracticable to obtain such lines and, in consequence, another method for determining the value of C_{OSD} has been employed elsewhere [13]. A paper published by Mullendore [22] revealed certain limitations in methods that employ only a single time-temperature parameter, as with the OSD method, and this became particularly obvious in cases where structural instabilities were involved. It was also added that due to the multiplicity of rate processes affecting the creep strength of complex alloys at high temperatures, it is absolutely impossible for a single parameter to describe precisely all creep properties involved. A review was also carried out on some high temperature alloys in which it was observed that the criterion of a constant slope of the lines specified by the ODS methodology was even less accurate than the assumption of the LM technique [22]. Another critical assessment documented in Murry [14] and carried out by Garofalo [8, 23] revealed that at each test temperature, a separate curve could have been found in relation to the initial stress, which represents the variations of this method as well as the other two methods of Larson-Miller and Manson-Haferd. This leads to the conclusion that the parameters studied were not only functions of stress, but also of other parameters involved in the process. Therefore, this method is found to be indirect and not taking sufficient account for longer tests [24]. According to Brozzo [25], a plot of the logarithm of the minimum creep rate against the reciprocal of the absolute temperature, at constant stresses, should give a series of straight lines. The same results should be obtained if the logarithm of the time to fracture is plotted against the same variable, since it is linearly related to the minimum creep rate. Therefore, it was possible to interpret the ODS and the LM parameters in terms of these plots. However, appreciable deviations from the claimed linearity were generally exhibited, except possibly for a limited range of temperatures. The reasons behind the failure of the rate-process equation in solving this problem can be readily recognised from the possibility of the metal, or the alloy, to deform according to various creep mechanisms accompanied by different activation energies and the likelihood of occurrence of some metallurgical changes during creep. Along with these findings, direct evidence has been obtained by many investigators that metals and solid solution alloys can undergo a plastic deformation in different ways depending on the temperature and straining-rate conditions [25].

Therefore, based on these investigations, this methodology needs to include more materials and different processes in order to construct a complete and a comprehensive agreement about the value of its constants and the linearity of the plots that its equation implies.

2.5. Review of the Manson-Succop (MS) methodology

The Manson and Succop (MS) methodology [26] is identified by the analysis of the iso-stress lines in the plot of log t_f versus T. The Manson-Succop parameter, P_{MS}, was based on the parallelism of these lines of a slope that represents the Manson-Succop constant, C_{MS}, and is given by [26]:

$$P_{MS} = f(\sigma) = \log t_f + C_{MS}T \tag{9}$$

This method, in addition to other methods, was reviewed by Zharkova and Botvina [27] who confirmed that during long-term creep tests, fracture mechanisms changed according to the applied stress and the loading time. In this regard, they stated that fracture under high applied stresses was purely intergranular, under medium applied stresses it was also intergranular but resulted from wedge cracks formation and was also intergranular under low stresses but resulted from the formation and development of pores along grain boundaries. The change of fracture mechanisms was responsible for the appearance of the kink points in the long-term strength curves [27]. The known time-temperature parametric methods such as the Larson-Miller, Dorn, Manson-Succop, Manson-Haferd and many others, were based on relations with fixed values of constants in a wide range of temperatures and fracture durations which, in return, ignored the changes of fracture mechanisms and led to many errors and overestimations of the long-term creep life. For this reason, these methods are not necessarily reliable for creep life predictions [27].

2.6. Review of the Manson-Brown (MB) methodology

In general, as generated data do not necessarily show a linear trend in their behaviour, it is then necessary to use more complex functions to fit these data. The simplest function with an adjustable degree of curvature is the power function. Consequently, it is actually not surprising to find a generation of non-linear parameters containing the functional forms of the previous linear parameters raised to some power. The parameter which best illustrates this progression in complexity is the Manson-Brown parameter, P_{MB}, of the form [28, 29]:

$$P_{MB} = f(\sigma) = (\log t - \log t_a)/(T - T_a)^q \tag{10}$$

In this expression, there are three constants (t_a, T_a and the exponent q) which can be determined by a 'trial and error' graphical method. This equation represents the general form of the previously mentioned linear parameters such that, it represents [29]: (a) Manson-Haferd equation when q = 1, (b) Larson-Miller equation when q = −1 and T_a = 0, (c) Orr-Sherby-Dorn equation when log t_a and $1/T_a$ are both taken to be arbitrarily very large numbers with the condition that T_a log t_a = Qc and (d) Manson-Succop equation when q = 1 and log t_a and T_a are both taken to be arbitrarily very large numbers such that log t_a/T_a = − C_{MS}. This generalised technique is very beneficial and much better than the individual proposed methods such that the data would dictate the specific form of the equation instead of trying to force any equation to fit the data. Later on, Manson along with Roberts and Mendelson proposed a generalised parameter of the form [30]:

$$P_{Manson} = f(\sigma) = \sigma^V (\log t - \log t_a)/(T - T_a)^q \quad (11)$$

where v is an additional stress exponent constant. This equation presents a more generalised form of the previous methods where more linear parameters can be derived just with a slight change in the values of the constants involved. These generalised equations, that is, Eqs. (10) and (11), provide better techniques to predict the creep behaviour since they encompass most of the known parametric approaches under different test conditions.

2.7. Review of the Monkman-Grant (MG) methodology

The Monkman-Grant (MG) parametric method [8] uses the minimum strain rate, $\acute{\varepsilon}_{min}$, as a key variable to assess the time to fracture, t_f [31]. Monkman and Grant [8] noticed that the rupture time in the long-term creep tests could be related to the minimum strain rate by a power function of the form [8, 31]:

$$C_{MG} = \acute{\varepsilon}_{min}\, t_f \quad (12)$$

where C_{MG} is the Monkman-Grant constant and m is the time to fracture exponent. This equation suggests that the mechanisms that control creep deformation and creep rupture are, to a great extent, the same [8]. The constant, C_{MG}, in this relation usually depends on temperature [31]. The practical advantage of the Monkman-Grant rule is that the minimum strain rate, $\acute{\varepsilon}_{min}$, can be measured early in a creep test which, in return, facilitates the prediction of the long-term time to fracture, t_f. In other words, if the value of C_{MG} is determined, which is possible from short-term tests, the lifetime of a long-term test can be predicted once the minimum strain rate has been reached and recorded [31]. On the other hand, another study which was carried out by Borisenko et al. [18] argued that the product of the minimum creep rate and the time to fracture is a constant value, C_{MG}, which is independent of stress and temperature. They also added that the value of this constant ranges between 0.03 and 0.3 for all materials and that the value of m should be 1.0, which eliminates the exponent from this equation. But later, and after some experiments that were carried out on tungsten, they found that the relation must be of the exponential form.

Another interpretation presented by Davies and Wilshire [32], which was based on experiments carried out on pure nickel, suggested that the constant, C_{MG}, was only independent of stress and temperature under high temperature creep conditions, that is, above $0.45T_m$, where T_m is the absolute melting temperature of a material, whereas higher values of this constant were recorded at temperatures below $0.45T_m$. Moreover, they found that the value of the exponent m was not varying appreciably from unity and thus, can be ignored.

Baldan and Kaftelen [33] observed that proportionality was generally found between t_f and $\acute{\varepsilon}_{min}$ when the material was strained. This observation was based on the long-term creep tensile tests where it was found that the time to fracture was inversely proportional to the power function of the minimum creep rate for relatively simple alloys such as pure metals and single phase alloys. Their equation is given by [33]:

$$C_{MG} = \acute{\varepsilon}_{min}{}^{m}\, t_f \tag{13}$$

where the value of the exponent m ranged between ~0.8 and ~0.95. Besides, it was found that the value of the constant, C_{MG}, ranged from ~ 2 to ~ 15, depending on the material and the microstructural variables as this constant represents the contribution of the secondary creep strain to the total failure strain [33]. This equation was based on when the material was strained, cavities and cracks grew, linked-up and led, eventually, to an intergranular creep fracture. Assuming that creep fracture is actually controlled by the creep growth of cavities at grain boundaries, this result would then be consistent with the Monkman-Grant equation as, from the very beginning, the fracture process is always linked to the creep process [33].

Dobes and Milicka [34] argued that the value of C_{MG} and m changed according to the applied stress in contrast to the studies of Davies and Wilshire [32] and Chih-Kuang Lin [31] who previously found that the value of C_{MG} was dependent on stress and/or temperature. Therefore, Dobes and Milicka modified the Monkman-Grant relation into the form [34]:

$$C_{MG}\, \varepsilon_f = \acute{\varepsilon}_{min}{}^{m}\, t_f \tag{14}$$

where ε_f is the fracture strain recorded at t_f. This relation accounts for a possible stress dependence of the product ($\acute{\varepsilon}_{min}{}^{m}\, t_f$) due to changes in the fracture strain, ε_f, according to the applied stress. However, this modification of the equation does not improve the prediction capability since, instead of only one long-term creep parameter, that is, t_f, their relationship requires also the knowledge of the second long-term parameter, that is, ε_f. This is actually impractical since having known the values of these two parameters eliminates, in return, the need for any predictions which is mainly the aim of such approaches [35].

Some other studies [36] added that if continuous nucleation occurs, a modelling of the fracture process might lead to the Monkman-Grant relationship provided that diffusive and plastic coupling of cavity growth and cavity interactions are considered. Besides, this relationship might offer the possibility of long-term extrapolation if the same creep-deformation mechanism operates during the whole creep life [37]. A research done by Menon [38] on silicon nitride examined the applicability of the Monkman-Grant relationship in predicting the stress rupture life. The data showed that the Monkman-Grant lines relating the rupture life to the minimum creep rate were stratified with respect to temperature. For this reason, a modification to the known expression of the Monkman-Grant equation was proposed to accommodate this temperature dependence. Following this modification, another generalised form of the equation was proposed by Evans [39] who stated that the standard Monkman-Grant relation has the advantage of the easy estimate of the life of a material once the minimum creep rate is known. This ability of estimating the life of a material can be practically achieved by testing specimens at specified operating conditions until the minimum creep rate, which typically occurs well before the material's end-of-life, is reached and then, the test can be interrupted. This creep rate can then be used to predict the long-term creep life using the Monkman-Grant equation. However, one important disadvantage of using this relation to predict the creep life is that at operating conditions, it can still take tens of thousands of hours to reach the minimum creep rate and tests of this length are often not viable from the practical and the economical perspectives.

Therefore, although the Monkman-Grant relationship is applicable in some situations, there is still a disagreement about a few details such as the values of the constants used in this relationship and whether they are stress and/or temperature dependents and thus, more materials have to be tested and examined using this technique in order to generalise its use.

2.8. Review of the θ-projection methodology

The θ-projection method is one of the extrapolation methods which proved its applicability, in some situations, in predicting the creep life. It can be summarised in that creep curves under uniaxial constant stress are measured over a range of stresses and temperatures and their shapes are recorded. These shapes are then 'projected' to other stresses and temperatures at which full creep curves can be re-constructed. The required properties are then read off the constructed curves [1, 40]. Thus, the θ-projection concept, in its most general form, the 4-θ equation, describes the variation of creep strain, ε, with time, t, according to [41]:

$$\varepsilon = \theta_1 \left[1 - \exp\left(-\theta_2\, t\right)\right] - \theta_3 \left[1 - \exp\left(\theta_4\, t\right)\right] \tag{15}$$

where t and T are the time and temperature, respectively, θ_1 and θ_3 are scaling parameters defining the extent of the primary and tertiary stages with respect to strain, while θ_2 and θ_4 are rate parameters characterising the curvature of the primary and tertiary creep curves, respectively [42]. In this equation, the two terms on the right hand side describe the normal primary and tertiary components in which a deceleration in creep rate is observed during the primary stage whereas an acceleration is recorded during the tertiary stage [43, 44]. This method was extensively studied by Evans [41] who argued that this technique has an added advantage over the other traditional parametric procedures in that creep predictions are not only limited to the rupture time. However, it was found that the interpolation and/or the extrapolation of the θ-function, traditionally used by this method, was not really the best predictor of the long-term life as more accurate results were obtained using simpler functional forms. Moreover, this equation was quite poor in fitting the experimental creep curve at small strain values [41]. Deviations from the actual creep measurements were also found when this equation was used, particularly in the late tertiary stage, by Evans and Wilshire [44] who attributed these deviations to the intergranular cracks that present immediately prior to fracture.

Another study carried out by Evans [45] was in agreement with one done by Evans [41] in that the θ-projection method gave the poorest projections of creep properties at low strains.

Therefore, a modification to this equation has been suggested by Evans [45] in order to improve the fit of the experimental data at the very small strain values. This has been achieved by adding another two extra parameters to Eq. (2.20), which gave the (6-θ equation) as [45]:

$$\varepsilon = \theta_1 \left[1 - \exp\left(-\theta_2 \, t\right)\right] - \theta_3 \left[1 - \exp\left(\theta_4 \, t\right)\right] + \theta_5 \left[1 - \exp\left(-\theta_6 \, t\right)\right] \quad (16)$$

Now, in this equation, the first two right hand terms have the same physical meaning as in Eq. (15), whereas the third term describes the early primary creep behaviour that results from the initial sliding relaxation across grain boundaries [42]. According to Evans [41, 42, 45], this modified equation provided more precise results when it was used to fit experimental creep data, especially at the early stages of the primary creep. This was a result of the third term that has been added which took into account the effect of grain boundary relaxation during the primary creep that was completely neglected by Eq. (15).

In comparison to the previous parametric methods, the θ-projection method was considered to be more reliable and more accurate in estimating the long-term creep life and thus, it has been widely used and studied in an effort to prove its validity for a wider range of materials. However, further studies are still needed to assure that the errors encountered by the first proposed model of this equation are completely eliminated by the introduction of the modified version.

2.9. Review of the hyperbolic tangent methodology

This technique has been developed by Rolls-Royce plc in the 1990s for the purpose of creep lifing predictions. It implies that the highest stress that can be applied on a specified material at a certain creep temperature is the ultimate tensile strength of that material, σ_{TS}. The stress rupture behaviour is described by hyperbolic tangent curves over a wide range of temperatures, such that [46–48]:

$$\sigma = \sigma_{TS}/2 \left\{1 - \tanh\left[k \ln\left(t/t_i\right)\right]\right\} \quad (17)$$

where k and t_i are fitting parameters that can be obtained by regression analysis using the actual experimental data at each temperature. Once the values of k and t_i are obtained, they can be inserted into Eq. (17) to produce the stress rupture predictive curves. Alternatively, using the creep strain values, another hyperbolic function is used to predict the rupture behaviour, such that [46–48]:

$$\sigma = S_i \left\{1 + \tanh\left[S_L \ln\left(\varepsilon/\varepsilon_i\right)\right]\right\} \quad (18)$$

where in this equation, the (σ_{TS}/2) term of Eq. (18) has been eliminated and replaced by the parameter S_i whereas k, t and t_i have been replaced by S_L, ε and ε_i, respectively. Again, the values of these parameters can be obtained by regression analysis using the actual experimental data at each temperature. This method differs from the θ-projection method in that it does not try to fit the actual creep curves and then find an expression that relates the fitting constants with stress and temperature, but it represents the creep data at any temperature as a 3-D surface that combines stress, strain and time [47, 48]. This method provided a very good fit for the stress rupture and creep strain behaviour based on the time to fracture and creep strain measurements of many alloys. The only limitation is that inflexion points were found in these predictive curves with no theoretical explanation. Interestingly, in the stress rupture curves, these inflexion points took place at around $0.5\sigma_{TS}$ at each temperature as a result of changing the pattern of stress rupture behaviour, which might be expected above and below σ_y (or σ_{TS}). Moreover, in the strain-dependent rupture curves, this inflexion point was found at around ε_i which has a physical significance as the strain value at the minimum creep rate point of a creep curve [48].

2.10. Review of the minimum commitment (MC) methodology

This method was proposed by Manson and Ensign [49] in an effort to give a larger flexibility to the parametric analysis of creep data. In addition, it was invented in order to combine all the conflicting approaches into a single equation that will have a sufficient generality. This method is given by [49, 50]:

$$\log t + A\,P \log t + P = G \tag{19}$$

where t is the time, A is a constant dependent on the metallurgical stability of the alloy, P is a variable equal to: $R_1\,(T - T_{mid}) + R_2\,(1/T - 1/T_{mid})$, G is a variable equal to: $(B + C \log \sigma + D\,\sigma + E\,\sigma^2)$, and B, C, D, E, R_1 and R_2 are regression coefficients and T_{mid} is the mid-value of the temperature range for which the data are to be analysed. In this equation, it is apparent that there are seven constants that need to be determined by regression analysis. It was also found that the more unstable the material, the higher the negative value of A required to fit the data [51]. As the constant A defines the metallurgical stability of the material, a negative value means that the material has the tendency to precipitate embrittling phases whereas a zero value would mean that the material is stable [52]. Unfortunately, the use of any value of A other than zero led to non-linear multiple regressions [52].

Among those who studied this methodology was Jow-Lian Ding [53] who found that the results of the regression analyses indicated that the Minimum Commitment model fit the data slightly better than the Larson-Miller model. The reason was that this model has five independent variables whereas the Larson-Miller model has only two. This method was also studied thoroughly by Goldhoff [54] in his attempts to find the optimum value of A. In this regard, he found that when formulating a model using this technique, the resulting equations were always non-linear since the values of A and P were unknown. It was also found that when fitting the short-term data, there was, relatively, insensitivity to the value of A which is not true for the long-term creep data predictions.

In order to establish a confidence in the use and, alternatively, to reflect problems of this procedure, it should be applied to an existing set of data as well as much sparser data and there should be immediate research into the development of stability factors to enhance the effectiveness of this extrapolation procedure [54].

2.11. Review of the Goldhoff-Sherby (GS) methodology

This methodology pre-supposed the convergence of the iso-stress lines to the point $(1/T_a,\ t_a)$ located just below the region of the experimental data. The general equation of this technique is given by [55]:

$$P_{GS} = f(\sigma) = (\log t - \log t_a)/(1/T - 1/T_a) \quad (20)$$

where t_a and T_a are the time and temperature constants, respectively. For the purpose of examining this equation, it was used to analyse the results of the experiments carried out by Sobrinho and Bueno [56] on steels where it was found that the worst results were obtained when the Goldhoff-Sherby equation was used to fit the data in all cases. Therefore, due to the very narrow use of this methodology in creep data predictions in addition to the fact that only few studies were carried out to examine the validity of this technique, more research should be completed before generalising the use of this technique in predicting the creep properties for long-term purposes.

2.12. Review of the Soviet methodology

This method can be described by two models, namely: Soviet model (1) and (2), given by [57]:

$$\textbf{Soviet Model (1)}: \log t = a + b \log T + c \log \sigma + d/T + f\,\sigma/T \quad (21)$$

$$\textbf{Soviet Model (2)}: \log t = a + b \log T + c \log \sigma/T + d\,\sigma/T + f/T \quad (22)$$

where a, b, c, d and f are constants to be determined. In studying these models, some observations were presented by Evans [57] who stated that Soviet model (1) was highly effective in modelling the rupture times presented to it for estimation purposes, but it was totally inadequate for predicting data points not used in its estimation. However, this inability to generalise, or the tendency to overfit the interpolative data set, is a characteristic of all parametric techniques [57].

2.13. Review of the Wilshire equations method

By using this new methodology, the values of the minimum creep rate, $\acute{\varepsilon}_m$, and the time to fracture, t_f, recorded at different temperatures can be superimposed onto 'Master Curves' by simply normalising the applied stress through the ultimate tensile strength, σ_{TS}, measured at various creep temperatures [5, 6]. Superimposition can also be achieved using the yield strength, σ_y, but the data fit is usually poorer since the value of σ_y is more difficult to be measured precisely than σ_{TS} [58]. Therefore, by selecting σ_{TS} values for such purposes, property comparisons for different metals and alloys can be significantly simplified [5, 6]. Normalising the applied stress in the power law equation, $\acute{\varepsilon}_m = A\,\sigma^n \exp(-Qc/RT)$, and defining the minimum creep rate, $\acute{\varepsilon}_m$, as in the Monkman-Grant relationship, $\acute{\varepsilon}_m = M/t_f$, gives [5, 6]:

$$\acute{\varepsilon}_m = M/t_f = A^* (\sigma/\sigma_{TS})^n \exp(-Qc^*/RT) \quad (23)$$

where $A^* \neq A$ and $Qc^* \neq Qc$. In this case, Qc^* is determined from the temperature dependence of $\acute{\varepsilon}_m$ and/or t_f at constant (σ/σ_{TS}), in contrast to Qc which is normally calculated at constant σ. Although this equation still does not permit reliable extrapolation of the short-term measurements as a result of the unpredictable fall in n values as σ/σ_{TS} decreases, it reduces, at least, the scale and the number of the experimental tests undertaken to obtain long-term strength data, but not the maximum duration of these tests [5, 6].

The failure of the traditional procedures to give acceptable estimates of the 100,000 h strengths by the analysis of the 30,000 h data has frequently been attributed to different mechanisms of creep and/or creep fracture which become dominant in different stress and temperature regimes [5, 6]. If the dominant mechanism changes, measurements made at high stresses would not allow prediction of the low-stress behaviour. For this reason, the new methodology has been introduced to examine and assess whether the change in the failure characteristics after prolonged creep exposure prevents accurate predictions of the long-term rupture strengths by extrapolating the short-term creep measurements [5, 6]. In this regard, Wilshire and Scharning [5, 6] obtained very accurate estimation of the long-term creep-rupture strength using this technique, irrespective of the transition from transgranular to intergranular fracture, by extrapolating the short-term creep data.

This technique is mainly based on the data rationalisation achieved through Eq. (23), where it is possible to rationalise the minimum creep rate, $\acute{\varepsilon}_m$, and the time to fracture, t_f, measurements by normalising σ through σ_{TS}. Since σ_{TS} represents the maximum stress that can be applied on a material at a specific creep temperature, the data sets can be described over the entire stress range from (σ/σ_{TS} = 1) to (σ/σ_{TS} = 0). In addition, it is evident that ($\acute{\varepsilon}_m \rightarrow \infty$ and $t_f \rightarrow 0$) as ($\sigma/\sigma_{TS} \rightarrow 1$), whereas ($\acute{\varepsilon}_m \rightarrow 0$ and $t_f \rightarrow \infty$) when ($\sigma/\sigma_{TS} \rightarrow 0$). These essential criteria are met by replacing Eq. (23), so that the stress and temperature dependences of the creep lives are described by [5, 6, 58]:

$$\sigma/\sigma_{TS} = \exp\left(-k_1 \left[t_f \exp(-Qc^*/RT)\right]^u\right) \tag{24}$$

where the values of the coefficients k_1 and u can be easily evaluated from the plots of ln [t_f exp (− Qc^*/RT)] against ln [− ln (σ/σ_{TS})]. The slope of these plots represents the value of u whereas the intercept with the y-axis represents the value of ln (k_1) from which k_1 can be calculated. The value of Qc^* can be evaluated at constant σ/σ_{TS} by plotting ln (t_f) against 1/T where the slope of these plots represents the value of Qc^*/R from which Qc^* can be obtained. As with the representation of stress rupture properties through Eq. (24), the stress and temperature dependences of $\acute{\varepsilon}_m$ can be described using [5, 6, 58]:

$$\sigma/\sigma_{TS} = \exp\left(-k_2 \left[\acute{\varepsilon}_m \exp(Qc^*/RT)\right]^v\right) \tag{25}$$

where the values of the coefficients k_2 and v can be obtained from the plots of ln [$\acute{\varepsilon}_m$ exp(Qc^*/RT)] against ln [− ln(σ/σ_{TS})]. The slope of these plots represents the value of v whereas the intercept with the y-axis represents the value of ln (k_2) from which k_2 can be calculated. The value of Qc^* can be evaluated at constant σ/σ_{TS} by plotting ln ($\acute{\varepsilon}_m$) against 1/T where the slope of these plots represents the value of - Qc^*/R from which Qc^* can be obtained. In addition to Eqs. (24) and (25), the planned operational life for some components must take into account the times required to reach certain limiting strains, t_ε. As with t_f in Eq. (24) and $\acute{\varepsilon}_m$ in Eq. (25), the stress and temperature dependences of t_ε can be quantified as [5, 6]:

$$\sigma/\sigma_{TS} = \exp\left(-k_3 \left[t_\varepsilon \exp(-Qc^*/RT)\right]^w\right) \tag{26}$$

where the values of the coefficients k_3 and w can be calculated from the plots of ln [t_ε exp (− Qc^*/RT)] against ln [−ln (σ/σ_{TS})]. The slope of these plots represents the value of w whereas the intercept with the y-axis represents the value of ln (k_3) from which k_3 can be calculated. The value of Qc^* can be evaluated at constant σ/σ_{TS} by either plotting ln (t_f) and/or ln ($\acute{\varepsilon}_m$) against 1/T where the value of Qc^* can be obtained from the slope of these plots (the slope will be either Qc^*/R or $-Qc^*/R$, respectively). Studies by Wilshire and Scharning [5, 6, 59] revealed that using Eq. (24) allowed extrapolation of the short-term creep life measurements and accurately predicted the 100,000 h rupture strengths for several martensitic 9–12% chromium steels at different temperatures. Further studies by Wilshire and Scharning [5, 6] also showed that Eqs. (23)–(25) permitted effective rationalisation and extended extrapolation of the time to fracture, t_f, the minimum creep rate, $\acute{\varepsilon}_m$ and the time to certain strains, t_ε, data for 1Cr-1Mo-0.25 V steel, despite the tempering of the as received bainitic microstructure and the occurrence of a gradual transition from transgranular to intergranular fracture during creep exposure. In another study, Wilshire and Battenbough [58] proved that the stress and temperature dependences of $\acute{\varepsilon}_m$ and t_f were best described using Eqs. (24) and (25) when they used this technique on polycrystalline copper. Thus, using this new technique will certainly reduce the scale and duration of the test programmes currently undertaken to define the allowable creep strengths of power plants and aeroengine applications [59].

3. Analytical and modelling results

3.1. The Larson-Miller technique results

This technique has been investigated in order to find out whether the value of the constant, C_{LM}, used in its equation is actually a 'constant' or dependent on the test conditions. For this purpose, at constant stresses, log(t_f) was plotted against 1/T which gave straight lines of a slope equals to P_{LM} (the Larson-Miller parameter) and an intercept of - C_{LM} (the Larson-Miller Constant). The first observation that is in agreement with earlier studies [13, 14] was that even when these lines were extrapolated, they did not intersect at a certain point, which was assumed to represent the value of C_{LM}, as some studies [9] suggested. Besides, it is obvious from these plots that the value of C_{LM} is not constant (varied from ~ 14 to ~ 17). This analysis, therefore, suggests that the value of C_{LM} varies according to the test conditions, which agrees with previous studies [5, 6, 10, 14] and thus, disagrees with the assumption of the Larson-Miller technique [7]. However, as a first trial, an average value between 14 and 17 was used in order to obtain the stress rupture curves based on the Larson-Miller relation, but unfortunately, these curves did not fit the actual measurements accurately. The next attempt was to force all the creep data to collapse onto a single master curve by plotting the stress, σ, against the parameter P_{LM}, at randomly selected values of C_{LM}. The value of C_{LM} was considered only when it fitted the raw data perfectly based on the trial and error method. It was found that the best fit of the data was obtained when the value of C_{LM} was 20. From this plot, a relationship between the stress, time and temperature was obtained from which the stress-time predictive curves were constructed, **Figure 5**. The obtained curves were linear, equidistant and parallel. This implies that the relation between the stress and the time is, simply, linear which could lead, in return, to considerable errors as these curves did not fit the creep data accurately, especially at the higher stresses of each temperature, which agrees well with previous studies carried out on steels [12]. Actually, if fitting the creep data was that simple using a linear line,

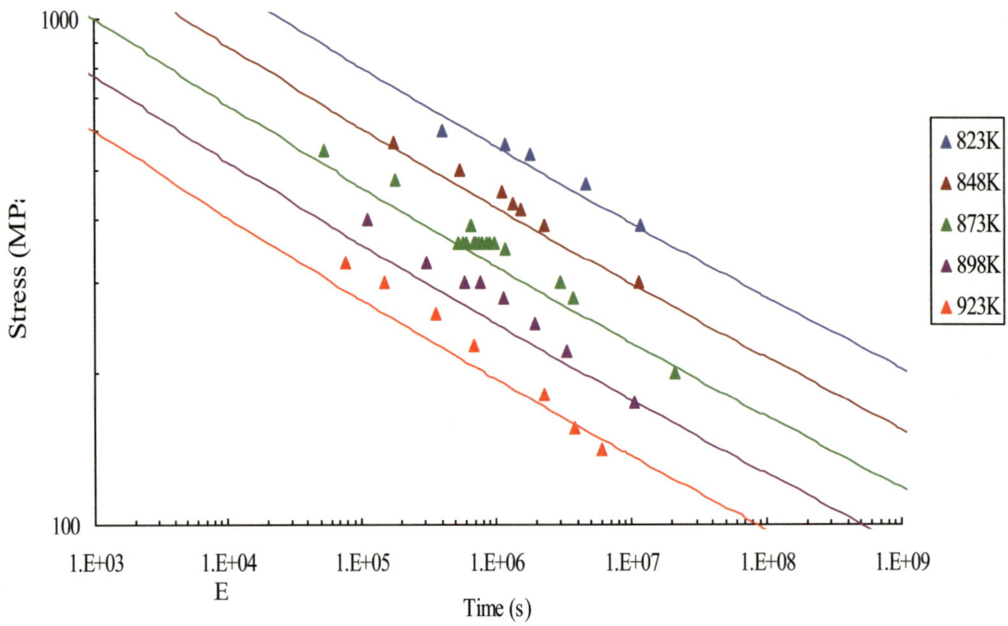

Figure 5. The Larson-Miller predictive curves.

there would not have been any need to develop complex relations to fit the data. But since the creep behaviour requires more complicated fitting equations to describe the actual creep behaviour, due to changes in creep mechanisms, linear relations will eventually lead to wrong estimations.

3.2. The Manson-Haferd technique results

As with all techniques, a relationship between the stress and the creep life at various temperatures is required. To start with the Manson-Haferd method, $\log(t_f)$ was plotted against T, at constant stresses, which gave straight linear lines of slope - P_{MH}, the Manson-Haferd parameter. When these lines were extrapolated, they did not meet at an intersection point of (T_a, t_a), as some studies [15] previously suggested. For this reason, another procedure was followed in order to calculate the values of these constants from the intersection point of the lines with the y and x-axes. The intercept of these linear lines represents the value of (P_{MH} T_a + log t_a) from which the value of T_a and t_a can be calculated, sequentially. The average calculated values of T_a and log t_a for Titanium IMI834 were ~ 1061 and 29.713, respectively, which differ from the values suggested by Manson and Haferd and agree with other literature studies [16, 17, 56]. These values were then inserted into the Manson-Haferd equation and plotted against the stress, σ, at constant temperatures from which a relation between the stress and the Manson-Haferd parameter was obtained. This plot disagrees with some studies [15] which assumed that plotting this parameter against the stress superimposes all the data points into a single master curve. However, the predictive stress-time curves were obtained and plotted along with the actual creep results, **Figure 6**. The curves showed a better capability of fitting the

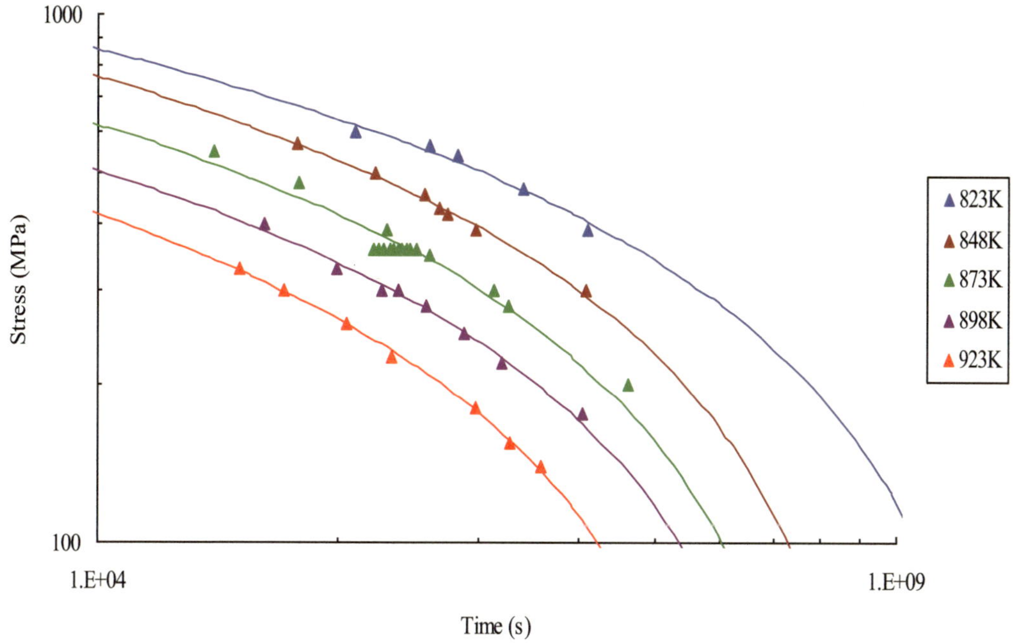

Figure 6. The Manson-Haferd predictive curves.

actual data points when compared with the Larson-Miller technique. This proves that the more complex the technique, the better its capability in predicting the creep properties.

3.3. The Orr-Sherby-Dorn technique results

The starting point of using this technique is similar to the Larson-Miller's analysis in which $\log(t_f)$ was plotted against 1/T, at constant stresses. These plots gave straight lines of a slope which represents the value of C_{OSD}, the Orr-Sherby-Dorn constant, and an intercept with the y-axis equals to - P_{OSD}, the Orr-Sherby-Dorn parameter. The first result that can be drawn from these plots is that the value of C_{OSD} is not constant as the slope was changing from ~ 16,244 to ~ 20,053 with changing the stress and temperature. This outcome disagrees with the assumption of Orr, Sherby and Dorn [20] who assumed that the value of C_{OSD} is constant. As with the Larson-Miller technique, the same method employed there was used here to force all the data points to collapse onto a master curve by plotting the stress, σ, against the Orr-Sherby-Dorn parameter, P_{OSD}, with randomly selected values of C_{OSD}. The best fit of data was obtained when the value of C_{OSD} was ~ 20,000. This is consistent with the fact that this value lies in the range between 16,244 and 20,053, that is, the values of the slopes of the constant stress lines previously discussed. From this master curve, a relationship between the stress, time and temperature can be obtained from which the predictive stress-time curves can be constructed, **Figure 7**, at all temperatures. The curves fitted the actual creep data quite well where the curvature of these curves improved the fit. When compared with the Larson-Miller curves, **Figure 5**, it showed much better fit of the data at all temperatures and stresses. However, the Manson-Haferd curves, **Figure 6**, showed better consistency of the predictive curves with the actual data than the Orr-Sherby-Dorn curves,

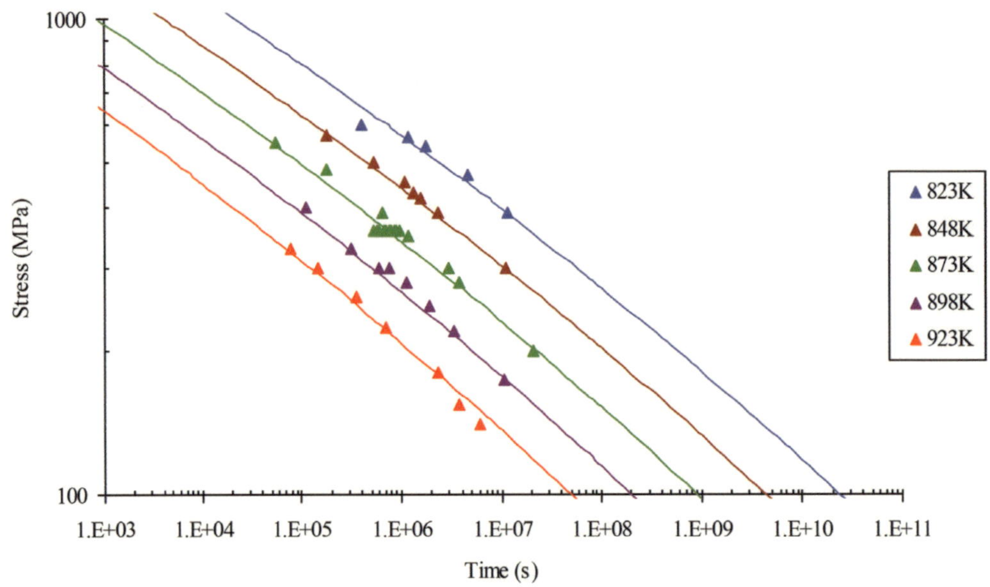

Figure 7. The Orr-Sherby-Dorn predictive curves.

Figure 7, as a higher degree of curvature was involved in the Manson-Haferd's curves as a result of the more complex function used in its equation.

3.4. The Manson-Succop technique results

The analysis using this technique started with plotting the values of log(t_f) against T, at constant stresses, which gave straight lines of slope equals to - C_{MS}, the Manson-Succop constant, and an intercept with the y-axis equals to P_{MS}, the Manson-Succop constant. These plots revealed that the slope, and hence the value of C_{MS}, varied between ~0.024 and 0.028 with varying the test conditions. This variation is relatively small but it could become more obvious if the tests conditions varied within a larger range of stresses and temperatures which might lead to a disagreement with the assumption of Manson and Succop [26] who confirmed that the value of C_{MS}, should be constant regardless of stress and temperature. However, an average value for C_{MS} was chosen, ~0.025, to superimpose all the data points onto a single curve by plotting the stress, σ, against the parameter, P_{MS}, from which a relation between the stress, time and temperature was obtained, **Figure 8**. This relation was then used to construct the stress-time curves on which the actual data points were projected, **Figure 8**. The stress-time curves were almost linear, equidistant and parallel (similar to the ones obtained using the Larson-Miller analysis). However, at the high temperatures (898 and 923 K), the fits were quite good in the high stress regime in comparison to the poor fits obtained in the low-stress regime. In contrast, the fits were quite good in the low-stress regime of the lower temperatures (823, 848 and 873 K), in comparison to the inferior fits obtained in high stress regime at these temperatures. Generally speaking, the fits were much better than those obtained from the Larson-Miller's analysis, but slightly less accurate than those obtained using the Manson-Haferd and Orr-Sherby-Dorn techniques.

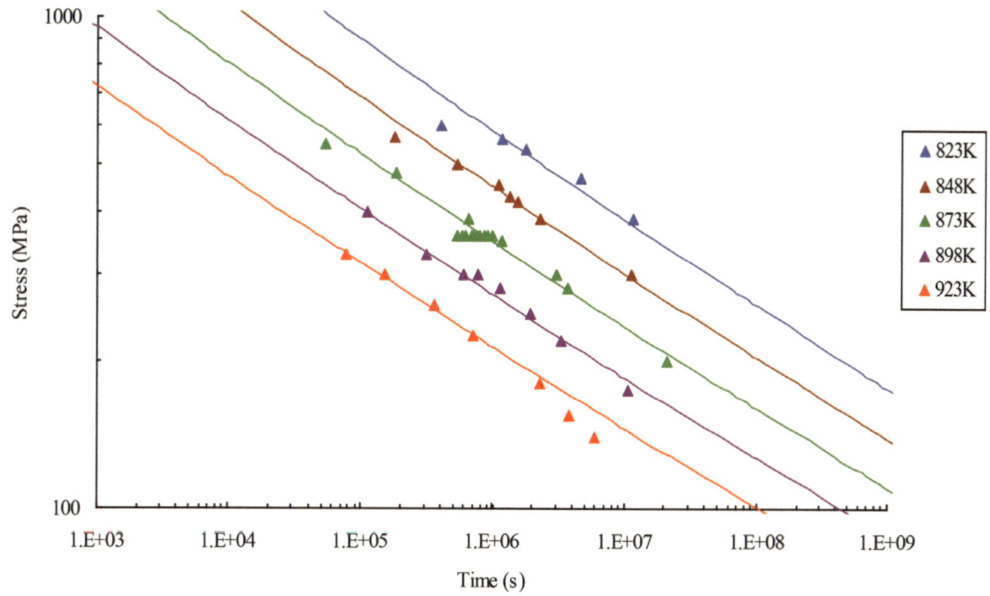

Figure 8. The Manson-Succop predictive curves.

3.5. The hyperbolic tangent technique results

For the purpose of finding the fitting parameters, plotting $\tanh^{-1}(1-2(\sigma/\sigma_{TS}))$ against $\ln(t_f)$, at constant temperatures, gave straight lines of a slope which represents the value of k and an intercept point with the y-axis equals to (k ln t_i). From these plots, the values of the constant k and t_i were calculated at each corresponding temperature. These values were then inserted into the hyperbolic tangent equation from which the predictive stress-time curves were obtained, **Figure 9**. These curves showed an impressive fit of the actual creep data as a result of the complex functions used in this technique and thus, the smooth curvature which improved the fit. It can also be observed that there is an inflexion point at around 50% σ_{TS}, at each corresponding temperature, which agrees with other studies [46–48] and implies that the creep mechanism is dependent on the applied stress level. Another observation is that at the intermediate temperatures, that is, 848 and 873 K, the curves slightly deviated from the actual creep data trend at the stresses between ~300 and 500 MPa. Even though, this technique can be considered as an easy and a straightforward method which directly relates the stress to the time and temperature without the need to superimpose the data onto a master curve to obtain the stress as a function of these two parameters, as with the previous techniques. Moreover, the predictions are much better and more reliable than all of the previously obtained results of the other methods, as can be seen from the constructed plots.

3.6. The Goldhoff-Sherby technique results

This technique is very similar to the Manson-Haferd methodology concerning the procedure of analysing the Titanium IMI834 data with the only difference that $\log(t_f)$ is plotted against the reciprocal of T at constant stresses where the slope of the lines represents the value of the Goldhoff-Sherby parameter, P_{GS}. Moreover, this plot provides the value of the constants log

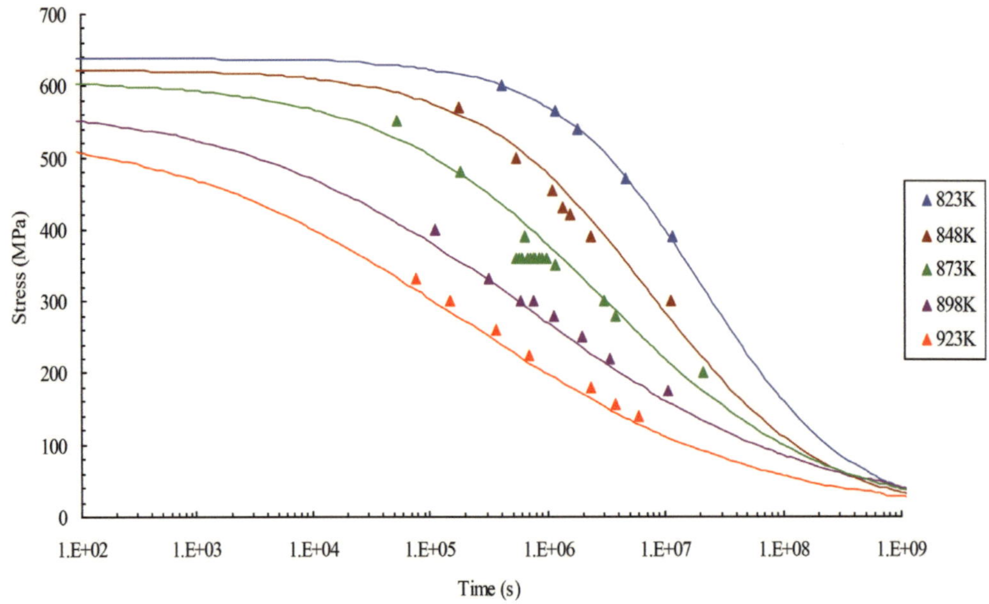

Figure 9. The hyperbolic tangent predictive curves.

t_a and $1/T_a$ from which a relationship between the stress and the Goldhoff-Sherby parameter, P_{GS}, can be obtained. For this purpose, an average value of log t_a and $1/T_a$ were taken as 15.824 and 0.0008, respectively. These relations between the stress and the parameter P_{GS} were then used in order to construct the stress rupture curves which showed a very good description of the actual creep results, **Figure 10**. The curves are very similar to those obtained by the Manson-Haferd technique which explains the similarity between these two methodologies in analysing the creep data. This again proves that the more complex the technique, the better its capability in predicting the long-term creep properties when compared to the simpler techniques.

3.7. The θ-projection technique results

Unlike the previously discussed models, this method was intended to fit the actual creep curves at various conditions and then express the fitting constants as functions of stress and temperature. The first version of this technique, the 4-θ was slightly able to fit the actual creep curves of Titanium IMI834. However, it did not give a very accurate description of the primary creep as many previous studies [41, 45] concluded, **Figure 11**. For this reason, the other version of this technique, the 6-θ, was used to fit the actual creep curves. Surprisingly, this equation provided a much better description of the primary creep behaviour which agreed very well with previous studies [41, 42, 45], **Figure 12**. This improvement in accurately fitting the primary creep confirms that the added two parameters, that is, θ_5 and θ_6, to the first version of this equation took into account the effect of grain boundary relaxation during the primary creep [42]. For both versions of the θ-method, the fitting procedure was possible by finding the values of the θ-parameters involved in their equation. The values of these parameters were

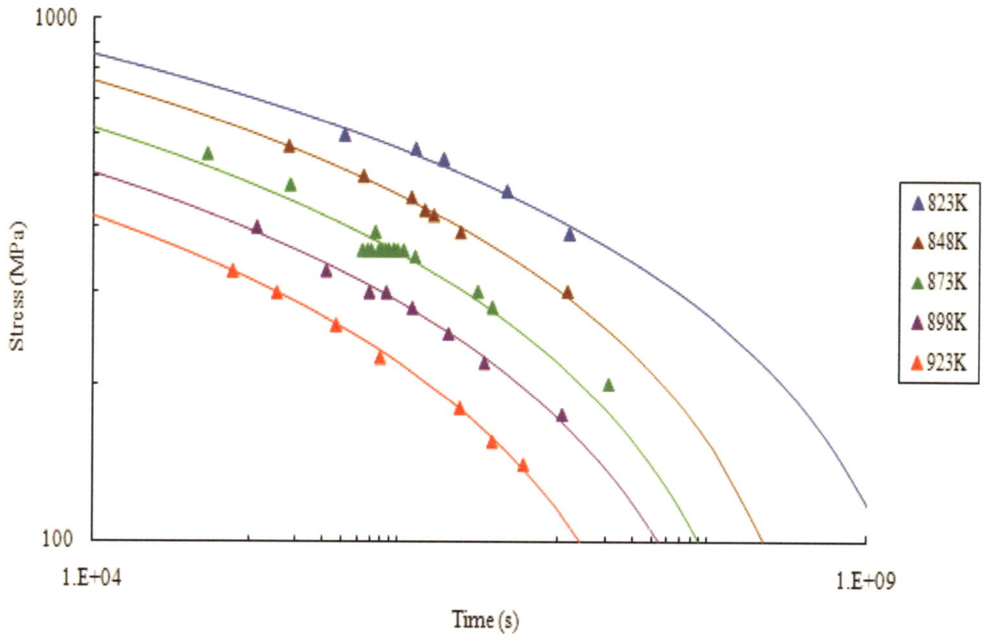

Figure 10. The Goldhoff-Sherby predictive curves.

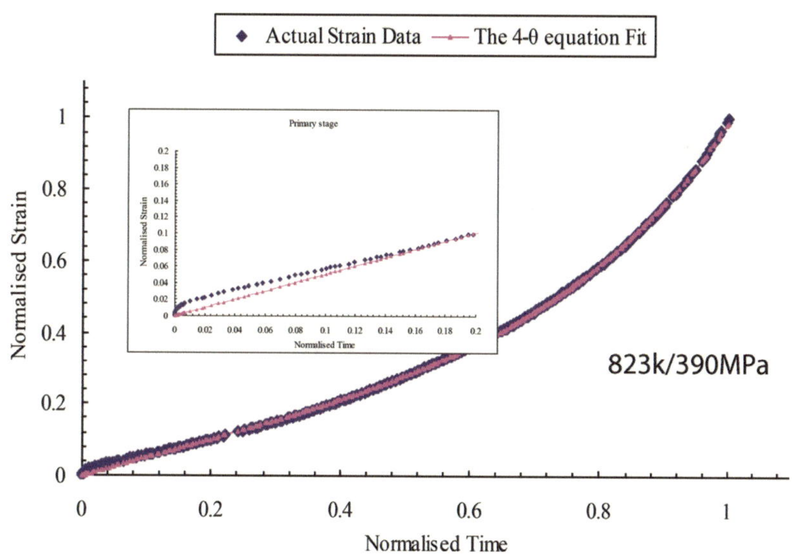

Figure 11. The fitting of the creep curve using the 4-θ method.

obtained by non-linear least square curve fitting routines (using SOLVER in Excel). Having obtained these parameters, many points and regions along the creep curve can then be defined, such as the primary and tertiary points, the minimum creep rate point, and the creep fracture, or the total ductility point. In these plots, the variation of each θ-term was plotted

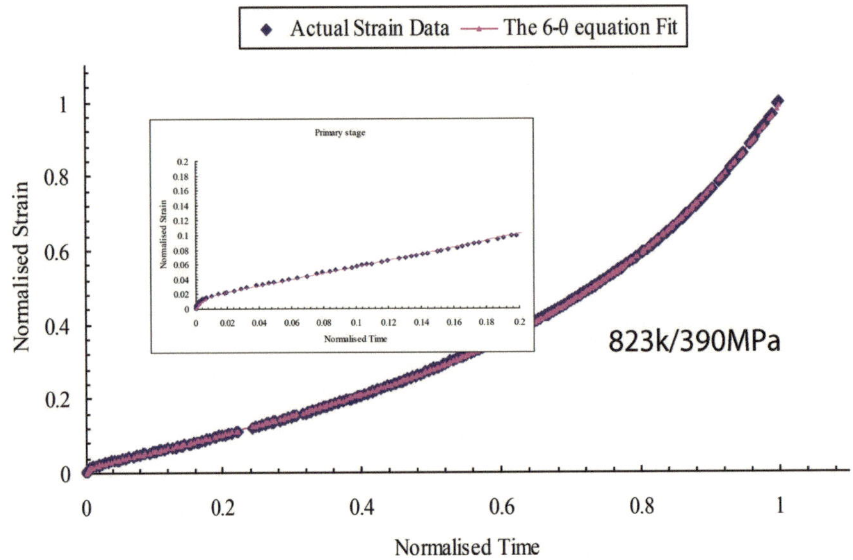

Figure 12. The fitting of the creep curve using the 4-θ method.

against stress at each individual temperature. The 4-θ results did not provide a systematic variation of the θ-parameters with stress for the primary creep region whereas the variation with the stress for the tertiary stage was slightly better, as it was found before [41]. This might be a result of the poor fit capability of this equation for the primary creep region. On the other hand, the 6-θ results provided a better description of all regions along the creep curve which is evident from the smooth and the linear variation with the stress. However, the trend of θ_3 and θ_5 was not purely linear, as they were, respectively, increasing/decreasing up to a certain stress level where they started to decrease/increase again at higher values of stress above that point. This unexpected change in the slope of these two parameters made it difficult to express them as a function of stress. If the trend of all parameters was completely linear, the values of these parameters could have been derived for any stress within the ranges studied experimentally. This means that this trend could have allowed interpolation of the data, although it might have also allowed reasonable extrapolation of creep properties. If the linear trends of the values of these parameters have been obtained, this means that they could have been expressed as functions of stresses and temperatures such that:

$$\theta = f\,(\sigma, \mathrm{T}) \tag{27}$$

which means that Eqs. (15) and (16) could have been re-written as:

$$\varepsilon = f\,(\mathrm{t}, \sigma, \mathrm{T}) \tag{28}$$

In conclusion, this method requires the availability of full creep curves prior to using it as a predictive tool. This technique can be considered as a 'fitting' technique rather than a 'predictive' model as the stress-time curves cannot be derived from its equation.

3.8. The Wilshire technique results

In order to start the analysis using this technique, it was essential to find the value of the apparent activation energy, Qc^*, the tensile strength, σ_{TS}, at the applied temperatures and the values of the fitting parameters (k_1, k_2, k_3, u, v and w). Unlike the calculations of Qc, described in the power law equation, at constant σ, the value of Qc^* was determined at constant σ/σ_{TS} using the power law principle. This was possible by either plotting $\ln(t_f)$ or $\ln(\acute{\varepsilon}_m)$ against 1/T at constant σ/σ_{TS} where the slope of these plots represents the value of Qc^*/R and - Qc^*/R, respectively. From the plot of $\ln(t_f)$ against 1/T, the value of Qc^* was ~305 kJ/mol whereas it was ~ 332 kJ/mol from the plots of $\ln(\acute{\varepsilon}_m)$ against 1/T. The difference in the value of Qc^* using either of these two procedures was not too large and thus, an overall average value of 320 kJ/mol was used to run the analysis. It can be seen that this overall value of Qc^* is not far away from the value of Qc (~327–344 kJ/mol) calculated at constant σ. The values of the constants k_1 and u were determined by plotting $\ln(-\ln \sigma/\sigma_{TS})$ against $\ln(t_f \exp(- Qc^*/RT))$ where the slope of these plots provided the value of u whereas the intercept is the value of ln k_1. However, it was observed that the linear trend of these plots deviated at a certain point that separated the data into two linear regimes, namely: the high- and the low-stress regimes. Based on this fact, different values of u and k1 were obtained from these two regimes. The predictive curves, **Figure 13**, showed a superb fit of the actual measurements in both the high and the low-stress regimes at all temperatures. It can be observed from these curves that there is a 'kink' point at which the trend of the creep data changed according to the stress level involved. This point exactly corresponds to the point found earlier in the plots of $\ln(-\ln \sigma/\sigma_{TS})$ against $\ln(t_f \exp(- Qc^*/RT))$ and this confirms that the dependence on stress level is more dominant than the temperature dependence, as the generated sigmoidal curve implied when

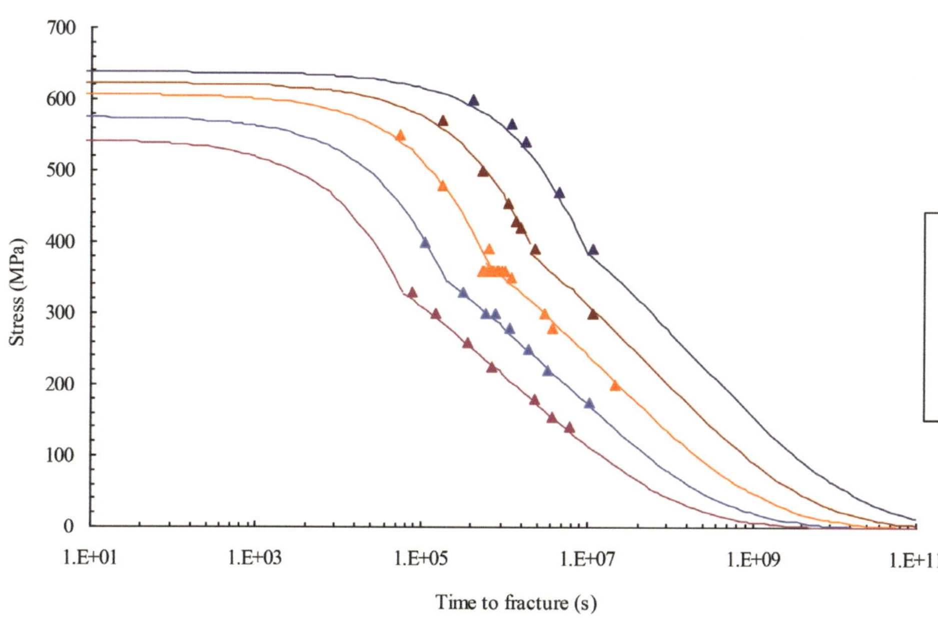

Figure 13. The Wilshire equations predictive curves.

the temperature dependence was eliminated. This predictability of the long-term creep behaviour using this equation proves that it is possible to extrapolate the short-term creep measurements at all test conditions.

Interestingly, the lines of the Wilshire 'kink' points and the yield stress regression line were linear, equidistant and parallel (slope ~ 0.6). Besides, the ratio of the stresses at the kink points was ~ 85% of the yield stress at each corresponding temperature. This implies that the inflexion points of the Wilshire curves are a result of the different deformation mechanisms above and below the material's yield point which play a key role in the creep behaviour. This physical explanation provides a possible reason for having two stress regimes and thus, the 'kink' in the predictive curves. It is worthwhile mentioning that the kink points were ~ 60% of the ultimate tensile strength, which is almost consistent with the hyperbolic tangent technique results, **Figure 9**, where the inflexion point of its curves was at ~50% of the ultimate tensile strength at each corresponding temperature.

It was found that the value of w and k_3 used in Eq. (26) is independent of stress and temperature at any selected strain level. This means that they can be expressed over a range of selected strains, such that:

$$w = f_1(\varepsilon) \tag{29}$$

and

$$k_3 = f_2(\varepsilon) \tag{30}$$

Inserting these two expressions into Eq. (26) gives:

$$\sigma/\sigma_{TS} = \exp\left(-f_2(\varepsilon)\,\left[t_\varepsilon \exp(-Qc^*/RT)\right]^{f_1(\varepsilon)}\right) \tag{31}$$

Rearranging this equation will provide an equation that relates the strain, ε, to stress, σ, and temperature, T, with time, t, such that:

$$\varepsilon = f\,(t, \sigma, T) \tag{32}$$

Obtaining Eq. (32) means that full creep curves at various stresses and temperatures can be re-produced based on the Wilshire equations technique. This was confirmed by the re-constructed creep curves obtained from the Titanium IMI834 data, **Figure 14**. These plots provided a full description of the creep curves at various conditions in addition to the very impressive description of the primary creep. The primary creep was described very well in mostly all cases of stresses and temperatures. The advantage of this capability can be summarised in that when the time required to reach a certain strain level is obtained from a creep curve, the stress-time curves for that strain level can be constructed based on this equation. Moreover, expressing w and k_3 as functions of strain can provide a description of the creep curves at any stress and temperature. Similarly, these equations present a way to define the end point of the creep curve. In other words, when the time to fracture is obtained from any creep curve, it can be used to construct the stress rupture curves based on this

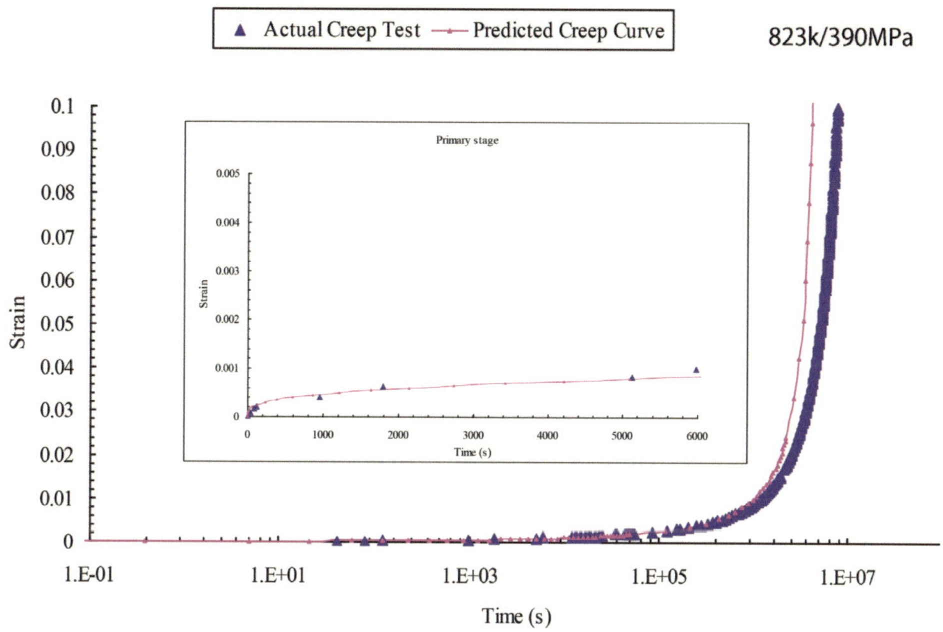

Figure 14. The re-produced creep curves using the Wilshire equations technique.

equation. In conclusion, in aerospace applications where the time to reach pre-defined strain levels is the main concern, typically ~1% strain level, then this technique provides an impressive description of the low strain levels required for such applications from the constructed creep curves.

4. Conclusions

- Using Titanium IMI834 data, it was revealed that the value of the stress exponent n and the activation energy Qc used in the power law equation are not constants which violates the original assumption of the power law which assumed that the value of these parameters is constant. This limits the use of this equation for long-term creep predictions. However, this equation can still be used as a mean to measure the value of the activation energy for different materials.
- When the Larson-Miller equation was examined using the Titanium IMI834, it was found that the value of the Larson-Miller constant CLM was actually not constant when the test conditions were altered. This disagrees with the assumption of Larson and Miller who assumed that the value of this 'constant' should be taken as 20 for all materials under all tests conditions. Moreover, the graphical method suggested by some scholars for obtaining the value of CLM was invalid for Titanium IMI834. Instead, an alternative procedure was used for determining the value of this constant. The stress rupture curves obtained from this equation were linear, equidistant and parallel. However, even with the

best chosen value of this constant, these curves did not fit the actual creep measurements accurately which led to overestimations of the long-term properties.

- The graphical method that was previously suggested by Manson and Haferd for determining the value of the constants Ta and ta was not applicable to Titanium IMI834. For this reason, another approach was used to find the values of these constants. Plotting the Manson-Haferd parameter PMH against the stress did not superimpose all the data points onto a single curve which violates the suggestions previously assumed by some researchers. The stress rupture curves showed a better consistency with the actual creep measurements when compared to the Larson and Miller curves. This capability of fitting the data points is a result of the more complex functions used in this equation in comparison to the Larson-Miller equation which suggests linear functions.

- The Orr-Sherby-Dorn equation was examined using the Titanium IMI834 creep data. The results showed that the constant used in their equation was not purely constant as it varied according to the applied test conditions. The stress rupture curves obtained fitted the actual creep data quite well and were more accurate than the Larson-Miller equation but less accurate than the Manson-Haferd relation.

- Applying the Titanium IMI834 data on the Manson-Succop equation revealed that the 'constant' used in their equation is actually not a constant and varied according to the test conditions which violates their assumption. Even when the best value of this 'constant' was used, the stress rupture curves were almost similar to those obtained from the Larson and Miller equation in that they were linear, equidistant and parallel. Despite this similarity, these curves fitted the actual measurements quite better than the Larson and Miller results. However, they were less accurate than the Manson-Haferd and the Orr-Sherby-Dorn equations.

- The hyperbolic tangent equation fitted the actual Titanium IMI834 creep measurements very well in comparison to all the previously used models. This equation is a straightforward and a more accurate procedure to be used for creep properties predictions. Interestingly, the inflexion points that can be observed in the stress rupture curves of this equation were found at ~50% of the ultimate tensile strength at each temperature. This means that the change in the long-term creep behaviour corresponds to the change in the applied stress level.

- The results of the 6-θ equation described the primary creep of Titanium IMI834 much more accurately when compared to the 4-θ equation at all test conditions. These two equations require full creep curves to be available in advance before they can be used in any application which makes them as 'fitting' equations rather than 'predictive' techniques. However, it was difficult to express the θ-parameters used in these equations as functions of stress which made it impossible to re-produce full creep curves based on these equations.

- The Wilshire equations showed a superb capability in fitting the actual measurements of Titanium IMI834. This was proved using the three forms of the Wilshire technique which accurately predicted the stress rupture, the minimum creep rate and the time to predefined strain values, respectively.

- In the Wilshire predictive curves, it was observed that there are inflexion, or 'kink', points at all temperatures. Investigations confirmed that these inflexion points took place at ~75% of the yield stress value at each corresponding temperature which split each curve into a high- and a low-stress regime. These points were also ~60% of the ultimate tensile stress which is almost consistent with the hyperbolic tangent equation results. This physical explanation implies that different deformation mechanisms are involved at each of these stress regimes.
- Full creep curves were re-constructed based on the Wilshire technique. This was possible by expressing the constants used in the Wilshire equation that predicts the time to reach certain strain levels as functions of strain, which was found to be impossible with the θ-technique. The re-constructed creep curves showed a very good description of the creep behaviour of Titanium IMI834 at all stresses and temperatures. The primary creep was also described very accurately, especially at the lower stress levels. This ability of re-producing the creep curves will, in return, save the time and cost required to carry out creep tests that might last for very long durations of time, especially at the lower stress levels.

Author details

Zakaria Abdallah*, Karen Perkins and Cris Arnold

*Address all correspondence to: Z.A.M.Abdallah@Swansea.ac.uk

Institute of Structural Materials, Swansea University, UK

References

[1] Wilshire B, Evans R. Introduction to Creep. London: The Institute of Materials; 1993

[2] Penny R, Marriott D. Design for Creep. 2nd ed. London: Chapman & Hall; 1995

[3] Cane B, Aplin P. Creep life assessment methods. The Journal of Strain Analysis for Engineering Design. Cane B & Aplin P, 1994. Creep life assessment methods. The Journal of Strain Analysis for Engineering Design. 1994;**29**(3):225-232

[4] Brown S, Evans R, Wilshire B. Creep strain and creep life prediction for the cast nickel-based Superalloy IN-100. Materials Science and Engineering. 1986;**84**:147-156

[5] Wilshire B, Scharning P. A new methodology for analysis of creep and creep fracture data for 9-12% chromium steels. International Materials Reviews. 2008;**53**(2):91-104

[6] Wilshire B, Scharning P. Prediction of long-term creep data for forged 1Cr-1Mo-0.25V steel. Materials Science and Technology. 2008;**24**(1):1-9

[7] Larson F, Miller J. A time-temperature relationship for rupture and creep stresses. Trans. ASME. 1952;**74**:223-249

[8] Monkman F, Grant N. An empirical relationship between rupture life and minimum creep rate in creep rupture tests. S.L. Proceedings ASTM. 1956;**56**:593-620

[9] Grote K, Antonsson E. Springer Handbook of Mechanical Engineering: Mechanical Properties. New York: Springer Science & Business Media; 2009

[10] Krivenyuk V, Mamuzic I. Correlation of creep-rupture data for complex alloys at elevated temperatures. Metalurgija. 2007;**46**(2):79-85

[11] Gilbert J, Long Z, Ningileri S. Application of Time-Temperature-Stress Parameters to High Temperature Performance of Aluminium Alloys. The Minerals, Metals & Materials Society, London, UK; 2007, 2007

[12] Cipolla L, Gabrel J. New Creep Rupture Assessment of Grade 91. University of Cambridge, Department of Materials Science and Metallurgy, Phase Transformations & Complex Properties Research Group, Cambridge, UK; 2005, 2005. Available at: https://www.phase-trans.msm.cam.ac.uk/2005/LINK/162.pdf

[13] Larke E, Inglis N. A critical examination of some methods of analysing and extrapolating stress-rupture data. s.l. Proceedings of the Institution of Mechanical Engineers. 1963

[14] Murry G. Extrapolation of the results of creep tests by means of parametric formulae. s.l. Proceedings of the Institution of Mechanical Engineers. 1963

[15] Manson S, Haferd A. A linear time-temperature relation for extrapolation of creep and stress-rupture data, s.l. In: NACA Technical Note 2890. 1953

[16] Murray J, Truman R. The high temperature properties of Cr-Ni-Nb and Cr-Ni-Mo austenitic steels. s.l. Proceedings of the Institution of Mechanical Engineers. 1963

[17] Bueno L, Sordi V, Marino L. Constant load creep data in air and vacuum on 2.25Cr-1Mo steel from 600°C to 700°C. Materials Research. 2005;**8**(4):401-408

[18] Sobrinho J, Bueno L. Correlation between creep and hot tensile behaviour for 2.25 Cr-1Mo steel from 500°C to 700°C. Part 2: An assessment according to different parameterization methodologies. Revista Matéria. 2012;**17**(3):1098-1108

[19] Pink E. Physical significance and reliability of Larson-Miller and Manson-Haferd parameters. Materials Science and Technology. 1994;**10**(4):340-346

[20] Orr R, Sherby O, Dorn J. Correlation of rupture data for metals at elevated temperatures. Trans. ASM. 1954;**46**

[21] Carreker R. Plastic flow of platinum wires. Journal of Applied Physics. 1950;**21**

[22] Mullendore A, Dhosi JW, Grant N. Study of parameter techniques for the extrapolation of creep rupture properties, in Conference Proceedings 1963. S.L. Proceedings of the Institution of Mechanical Engineers. 1963

[23] Garofalo F, Smith G, Royle B. Validity of time compensated temperature parameters for correlating creep and creep rupture data. Trans. Amer. Soc. Mech. Engs. 1956

[24] Allen N. The Extrapolation of Creep Tests, A Review of Recent Opinion. Institute of Metals, London, UK; 1960

[25] Brozzo P. A method for the extrapolation of creep and stress-rupture data of complex alloys. s.l. Proceedings of the Institution of Mechanical Engineers. 1963

[26] Manson S, Succop G. Stress-Rupture Properties of Inconel 700 and Correlation on the Basis of Several Time-Temperature Parameters. ASTM; 1956 Special Technical Publication (No. 174)

[27] Zharkova N, Botvina L. Estimate of the life of a material under creep conditions in the phase transition theory. Doklady Physics. 2003;**4**(7). (translated from Doklady Akademii Nauk, 2003. Volume: 391 (No. 3): p. 334-336. Original Russian Text Copyright)

[28] Manson S, Brown W. Time-temperature-stress relaxations for the correlation and extrapolation of stress-rupture data. s.l. Proceedings of the ASTM. 1953

[29] Viswanathan R. Damage Mechanisms and Life Assessment of High-Temperature Components. 2nd ed. Ohio: ASM International; 1993

[30] Manson S. Design consideration for long life at elevated temperatures. s.l. Proceedings of the Institution of Mechanical Engineers. 1963

[31] Lin C, Chu D. Creep rupture of lead-free Sn-3.5Ag and Sn-3.5Ag-0.5Cu solders. Journal of Materials Science: Materials in Electronics. 2005;**16**:355-365

[32] Davies P, Wilshire B. An Interpretation of the Relationship Between Creep and Fracture. s.l. The Iron and Steel Institute; 1960

[33] Baldan A, Kaftelen H. Comparative creep damage assessments using the various models. Journal of Materials Science. 2004;**39**(13):81-87

[34] Dobes F, Milicka K. The relation between minimum creep rate and time to fracture. Metal Science. 1976;**10**(11):382-384

[35] Dlouhy A, Kucharova K, Orlova A. Long-term creep and creep rupture characteristics of TiAl-base intermetallics, structural materials: Properties, microstructure and processing. Materials Science and Engineering A. 2009;**510/511**:350-355

[36] Davanas K, Solomon A. Theory of Intergranular creep cavity nucleation, growth and interaction. Acta Metallurgica. 1990;**38**(10):1905-1916

[37] Baldan A. Effects of carbides and cavitation on the Monkman-Grant ductility of a nickel-base superalloy. Journal of Materials Science Letters. 1992;**11**(19):1315-1318

[38] Menon M et al. Creep and stress rupture behaviour of an advanced silicon nitride: Part III, stress rupture and the Monkman-Grant relationship. Journal of the American Ceramic Society. 1994;**5**:77

[39] Evans M. A generalised Monkman-Grant relation for creep life prediction: An application to 1CrMoV rotor steel. Journal of Materials Science. 2006;**12**:41

[40] Tancret F, Sourmail T, Yescas M, Evans R. Design of a creep resistant nickel-base superalloy for power plant applications: Part 3 (experimental results). Materials Science and Technology. 2003;**19**

[41] Evans M. Sensitivity of the theta projection technique to the functional form of the theta interpolation/extrapolation function. Journal of Materials Science. 2002;**14**:37

[42] Evans R, Scharning P. The theta projection method applied to small strain creep of commercial aluminum alloy. Materials Science and Technology. 2001;**17**(5):31-45

[43] Evans R, Little E, Preston J, Wilshire B. 1993. Rationalisation of the creep behaviour of oxide-dispersion-strengthened alloys. s.l.. Fifth International Conference on Creep and Fracture of Engineering Materials and Structures

[44] Evans R, Wilshire B. The Role of Grain Boundary Cavities During Tertiary Creep. s.l. UK: Department of Metallurgy & Materials Technology, Swansea University; 2001. pp. 303-314

[45] Evans R. The theta projection method and low creep ductility materials. Materials Science and Technology. 2000;**16**(1):48-65

[46] Williams S. An Automatic Technique for the Analysis of Stress Rupture Data—Report MFR30017, s.l. Derby, UK: Rolls-Royce Plc; 1993

[47] Williams S. 1999. The implementation of creep data in component FE analysis. s.l. Compass 1999 Proceedings of the 1st International Conference on Component Optimisation, University of Wales, Swansea, United Kingdom

[48] Williams S, Bache M, Wilshire B. Recent developments in the analysis of high temperature creep and creep fracture behaviour. Materials Science and Technology. 2009

[49] Manson S, Ensign C. A Specialised Model for Analysis of Creep-Rupture Data by the Minimum Commitment, Station-Function Approach. s.l.: NASA TM-X-52999; 1971

[50] White W, May I, Silviera T. Design parameters for high temperature creep and the minimum-commitment method. Journal of Materials for Energy Systems. 1980;**2**(2):51-59

[51] Manson S, Ensign C. Interpolation and extrapolation of creep rupture data by the minimum commitment method, Part 1: Focal-point convergence. s.l. NASA TM-78881; 1978

[52] Park J, Manson S. A new approach for the characterisation of creep rupture properties for a newly fabricated material. s.l. The Sixth International Conference on Creep and Fatigue Design and Life Assessment at High Temperatures; 1996

[53] Ding J, Kenneth C, Brinkman C. A comparative study of existing and newly proposed models for creep deformation and life prediction of Si_3N_4. In: Life Prediction Methodologies and Data for Ceramic Materials. s.l.: s.n.; 1994. pp. 62-83

[54] Goldhoff R. Towards the standardisation of time-temperature parameter usage in elevated temperature data analysis. Journal of Testing and Evaluation, JTEVA 2. 1974;**5**:387-424

[55] Goldhoff R, Hahn G. Correlation and Extrapolation of Creep-Rupture Data of Several Steels and Superalloys Using Time-Temperature Parameters. Cleveland: American Society for Metals; 1968. ASM Publication D-8-100

[56] Sobrinho J, Bueno L. Correlation Between Creep and Hot Tensile Behaviour for 2.25 Cr-1Mo Steel from 500°C to 700°C. Part 2: An Assessment According to Different Parameterization Methodologies. s.l.: s.n; 2005

[57] Evans M. Method for improving parametric creep rupture life of 2.25Cr-1Mo steel using artificial neural networks. Materials Science and Technology. 1999;**15**:647-658

[58] Wilshire B, Battenbough A. Creep and creep fracture of polycrystalline copper. Materials Science and Engineering A. 2007;**443**:65-78

[59] Wilshire B, Scharning P. Long-term creep life prediction for a high chromium steel. Scripta Materialia. 2007;**56**:645-724

Chapter 8

Small Punch Creep

Robert J. Lancaster and Spencer P. Jeffs

Additional information is available at the end of the chapter

http://dx.doi.org/10.5772/intechopen.70375

Abstract

A thorough characterisation of the creep properties of any modern alloy designed for a structural application can be an expensive and timely process. As such, significant effort is now being placed in identifying suitable alternative characterisation techniques. The small punch creep (SPC) test is now widely regarded as an effective tool for ranking and establishing the creep properties of a number of critical structural materials from numerous industrial sectors. Over recent years, the SPC test has become an attractive miniaturised mechanical test method ideally suited for situations where only a limited quantity of material is available for qualification testing. Typically, the method requires only a modest amount of material and can provide key mechanical property information for highly localised regions of critical components. As such, SP creep testing offers a feasible option of determining the creep properties of novel alloy variants still at the experimental stage and the residual life of service-exposed material.

Keywords: small punch, creep, fractography, empirical correlations, numerical correlations

1. Introduction

Small punch (SP) test methods were initially developed in the 1980s by the nuclear industry, then known as the miniaturised disc bend test, where it was used to estimate the mechanical properties and residual life of irritated materials [1–3]. It was then in the 1990s that it was proposed the technique could be utilised to establish elevated temperature creep properties of materials [4], from which it has received significant interest in terms of its potential advantages and applications and is now more commonly known as the Small Punch Creep (SPC) test. **Figure 1** demonstrates the continuously growing research and interest in SPC testing by the number of documents listed searching the expression 'small punch creep' in Scopus since 1970 [5].

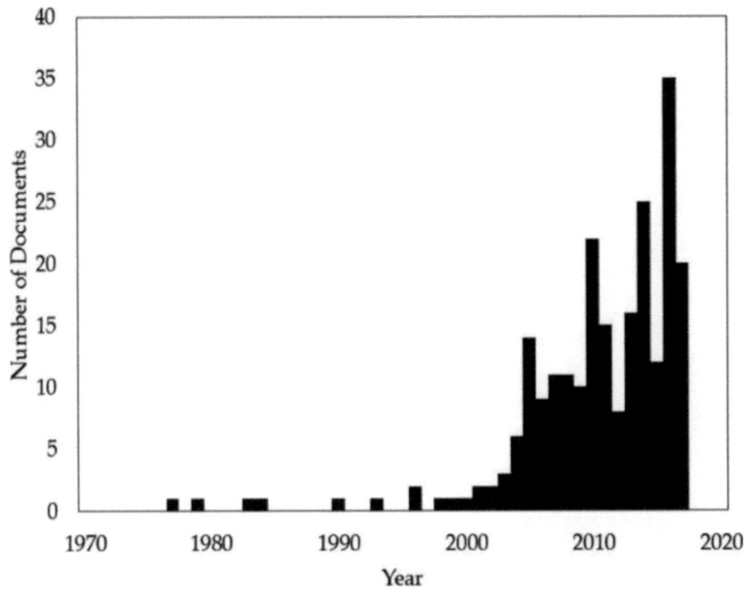

Figure 1. Number of documents listed in Scopus each year where the expression 'small punch creep' is included in the title, abstract or the keywords in the engineering, materials science and energy subject areas [5].

The key feature of the SPC test in comparison to conventional mechanical test methods derives from its use of small disc specimens, the dimensions of which have been seen to be between 3 and 10 mm in diameter and 0.2–0.5 mm in thickness depending on the test being performed [6].

The use of small disc specimens breeds a range of advantages and applications for the SPC method. As was the original intent of its development the technique has found success in remnant life assessment of in-service components such as pressure vessels and boiler pipes [7–9]. Disc specimens for determining remnant life are generally extracted from in-service components by means of scoop sampling [10], a schematic for which is shown in **Figure 2** [11]. This is then typically followed by Electrical Discharging Machining (EDM) and a grinding/polishing process to the desired thickness, thus generating disc samples ready for testing. The different stages of this process for sample preparation are shown in **Figure 3**. Other means of extracting SP discs include EDM directly from components or structures without the necessity of the scoop sampling step or by turning down thread ends from uniaxial specimens on a CNC lathe. In cases where sampling does not compromise the structural integrity of a component, the technique may be considered a Non-Destructive Test (NDT) [12].

Besides its initial development for remnant life assessment, SPC has been proven to be effective at determining localised material behaviours over a range of material systems, processes and engineering sectors. These include:

- In power plants, welding processes are used for the integration of complex pipe systems resulting in the necessity to understand the evolution of properties throughout the base material, weld material and heat-affected zone (HAZ) [13]. As a result, the SPC test has been implemented to assess the creep properties of weldments at high temperatures over a range of materials [13–16].

- In additive manufacturing (AM) processes, such as direct laser deposition (DLD) which is of large interest to the aerospace sector, SPC has enabled mechanical characterisation that may not be possible with conventional test methods due to geometry restrictions. Therefore, SPC has offered the possibility of providing effective ranking of different process variables and orientations on component representative geometries [17]. This principle of ranking through SPC can then be transferred across into early stage alloy development where only limited material stock may be available.
- Changes in mechanical behaviour such as establishing the effect of anisotropy in textured materials [18, 19], identifying grain size sensitivity along with revealing the influence of chemical compositions [20] have been unveiled through the SPC test.

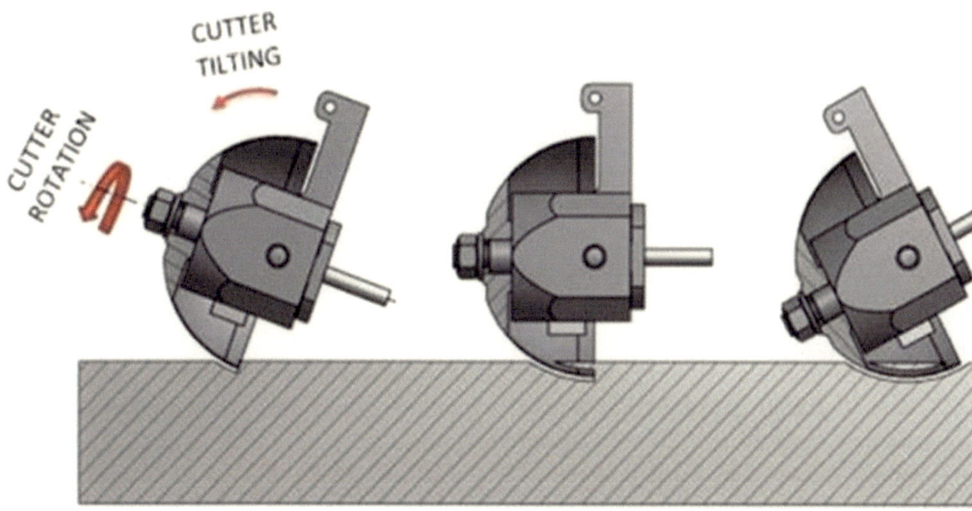

Figure 2. Schematic of scoop sampling process [11].

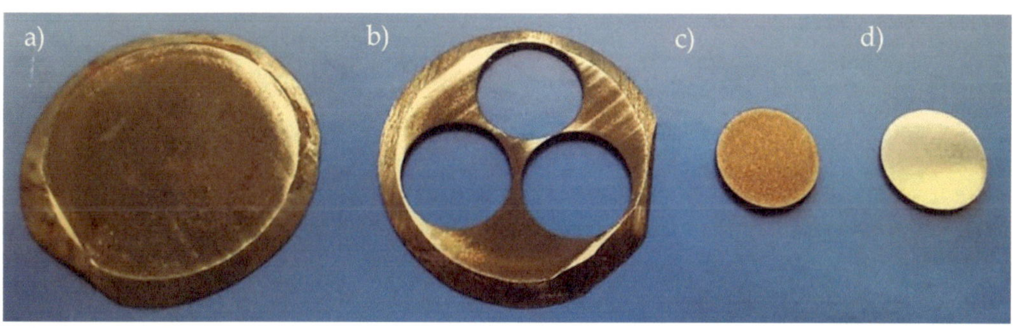

Figure 3. (a) Scoop sample, (b) scoop sample after EDM, (c) SP samples obtained through EDM and (d) polished SP specimen ready for testing.

- Finally, the technique has long been adopted as a means of determining equivalent or correlated uniaxial properties over a wide range of material systems including nickel based single crystal material ®CMSX-4 [21, 22], γ-titanium aluminide [23], aluminium alloys [24] and stainless steel [25].

Overall the technique has been adopted for a variety of research and development purposes to further highlight the range of potential benefits of using miniature disc specimens to derive creep properties of materials. Currently, SPC is undergoing a stringent procedure to progress from a working code of practice [26] into a European standard [27]. This has been supported by the creation of a specialist group of worldwide researchers who are collaborating towards standardising the procedure with a targeted publication date of 2019. This effort is supplemented by round robin testing across different institutions to establish the reliability and repeatability of the technique and confirm its place as an accepted mechanical testing methodology.

2. Experimental procedures

In a small punch creep (SPC) test, a miniature disc specimen with typical dimensions of 8–9.5 mm in diameter and 500 μm in thickness is subjected to a constant load that is applied to the material through a punch indenter. As the load is applied to the disc, the material undergoes a biaxial form of deformation prior to the onset of final rupture.

2.1. Test frame assembly

The SPC test is usually performed in a dead-weight test frame assembly as displayed in **Figure 4**. In this arrangement, the miniature disc is positioned into the recess of a lower die, which locates the specimen in the centre of the rig. The disc is then fixed securely via a circumferential clamping load to prevent any residual flexing of the material during test. However, over clamping should be avoided to reduce any plastic deformation which may influence the final test result. Hand tight clamping is usually suffice to ensure repeatability in results. Loading is then applied through the central axis of the frame via an upper load pan which exerts a force onto the specimen via a punch indenter.

2.1.1. Punch indenter

The punch tip typically consists of a hemi-spherical end with a diameter ranging from 2 to 2.5 mm, but an alternative assembly can also be used such as a flat tip indenter and a 2–2.5 mm diameter ceramic ball. Typical ball materials include Si_3N_4. In this instance, the ball should be replaced after every test. The geometries of the two punch configurations are displayed in **Figure 5**. Depending on the material that is to be tested, the punch material can vary from high strength steel for testing relatively soft and ductile materials such as copper, to high temperature ceramic materials including zirconium oxide and aluminium oxide for testing creep resistant materials such as single crystal alloys. For testing at elevated temperatures, oxidation and wear of the punch is a concern and may require refurbishment after every test.

http://dx.doi.org/10.5772/intechopen.70375

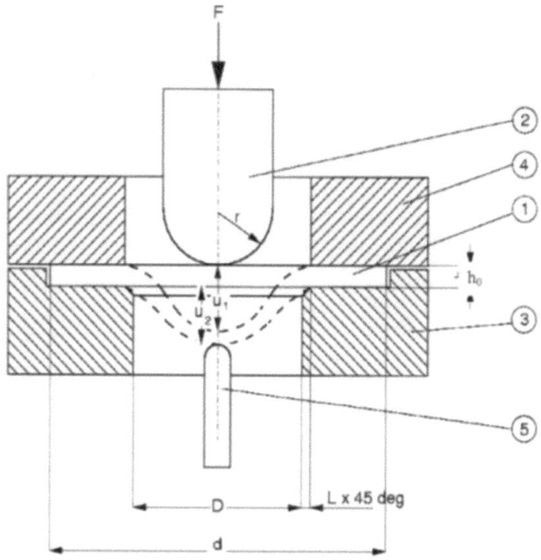

1. SP disc test piece
2. Hemispherical ended punch
3. Lower Die
4. Upper Die
5. Transducer quartz rod

u_1, u_2 = displacement
h_0 = 0.5mm
d = 8-9.5mm
r = 1-1.25mm
D = 4mm

Figure 4. Small Punch Creep Test Set-up [26].

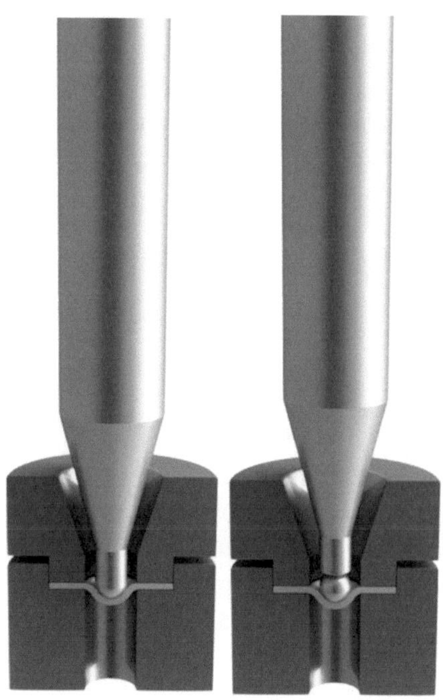

Figure 5. Schematic diagrams of the punch geometries for (a) one piece punch and (b) ball-punch configuration.

2.1.2. Measurement of deformation

Upon exertion of the testing load, disc deformation is monitored and recorded from two locations via linear variable displacement transducers (LVDTs). One LVDT is usually positioned in a location to detect the movement of the load pan and records the displacement on the top surface of the disc. The other LVDT is located within the lower die—pull rod assembly and measures the deflection on the underside surface of the specimen via a quartz rod. This enables the deformation on both surfaces to be closely monitored throughout a test and the resultant SPC curve usually displays an average of the top and bottom recordings.

2.1.3. Heating instrumentation and measurement

In a SPC test, heat is usually applied via a digitally controlled single zone furnace or induction heating and is constantly monitored throughout the test by one or two thermocouples (either type K or N depending on the desired temperature). One thermocouple should be placed in contact with the underside of the disc, with an optional second thermocouple located in a drilled hole in the upper die, close to the top surface of the specimen.

For high temperature SP testing, an inert gas environment is usually recommended to avoid oxidation of the disc specimen.

2.2. Specimen preparation

For SPC testing, disc specimens are usually removed from a larger piece by extracting cylindrical rods of the test material through EDM. The rods are turned down to the desired diameter (8–9.5 mm) and sectioned into slices approximately 800 μm thick. The material is then ground on both faces with progressively finer grit papers until a 500 ± 5 μm thickness is achieved with a 1200 grit finish. The use of smaller test pieces (diameter = 3 mm, thickness = 250 μm) is also permissible to allow the use of TEM specimens. In either specimen geometry, a representative volume must be contained in the thickness of the disc to obtain macroscopic material properties. This would typically include at least five grains in the cross-sectional thickness.

Prior to testing, the disc thickness shall be measured at four locations around the perimeter at 90° intervals to ensure uniformity.

3. Results

The result of a SPC test is characterised by a time-displacement/deflection curve with recognisable stages of deformation, notably primary, secondary and tertiary creep. This behaviour is comparable to that expected from a conventional uniaxial creep test for a given material. However, in reality the actual material deformation is considerably different. In a SPC test, the period of primary creep represents the elastic bending period of disc deformation. This leads to the onset of plastic deformation and membrane stretching under a biaxial stress condition, coinciding with an extended phase of secondary creep. The deformation then continues as the disc starts to experience considerable necking due to thinning, before accelerating to failure, akin to tertiary creep. This contrasts with the behaviour observed in a

http://dx.doi.org/10.5772/intechopen.70375

traditional uniaxial creep arrangement where the test specimen simply elongates under an applied tensile load with time.

Figure 6a displays a series of typical SPC curves from tests performed on the single crystal superalloy CMSX-4, at 950°C under a variety of applied loads. The same results can also be represented in terms of the minimum displacement rate with time, as shown in **Figure 6b**. In a similar manner to uniaxial creep data, the results can also be plotted in terms of time to rupture, t_f, an example of which is given in **Figure 7**. As the graph shows, SPC testing demonstrates a sensitivity to temperature and loading, as would be expected in conventional creep test results.

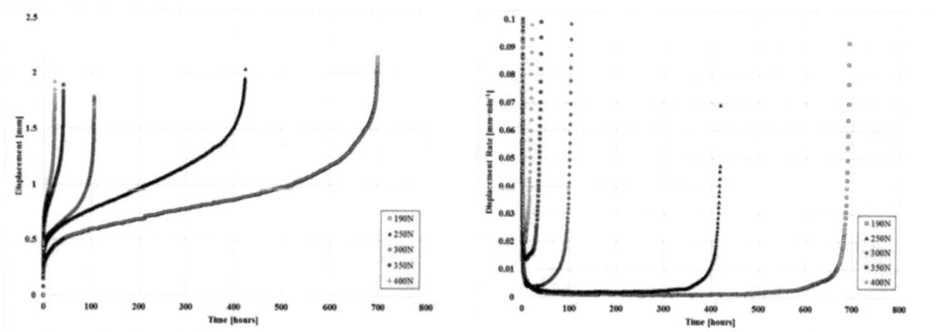

a b

Figure 6. (a) SP creep curves and (b) minimum displacement rates for CMSX-4 tested at 950°C under a range of applied loads [21].

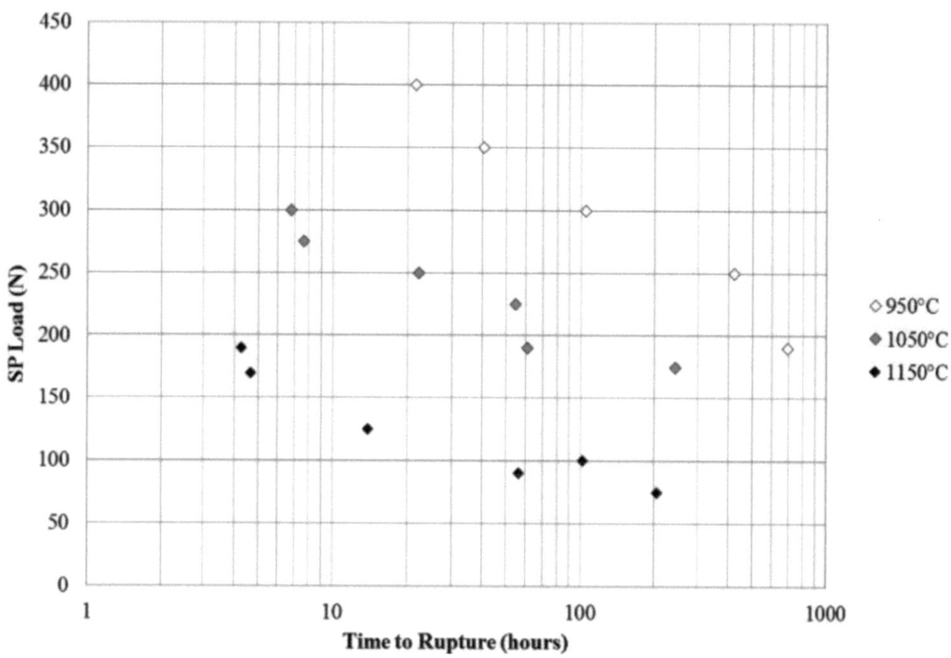

Figure 7. Load vs Time to Rupture for SPC tests on CMSX-4 at 950°C, 1050°C and 1150°C [21].

3.1. Fractography

In a SP creep test, the manner of failure usually resembles one of two dominant modes of fracture; ductile or brittle. In a ductile failure, the fracture usually initiates around the periphery of the punch head in the vicinity of tensile membrane stretching, propagating along the circumference indicating that the maximum stress is found away from the central region of the disc. **Figure 8** illustrates a typical example of a ductile fracture for a ruptured high nitrogenic ferritic steel specimen, tested at 725°C under a load of 140N [28]. The fracture morphology displays a cap-like form of deformation whilst severe necking can be seen around the central region of the disc. The higher magnification image, given in **Figure 8b**, shows a dominant transgranular type behaviour with ductile dimples covering the surface.

In contrast, SP testing has also been employed to characterise the creep properties of more inherently brittle type materials, such as γTiAl [23] or Mg alloys [24].

Figure 9 presents the typical fracture morphology for a brittle small punch disc. In this example, a SP creep test was performed at 650°C on a 14Cr ODS steel, a ferritic/martensitic

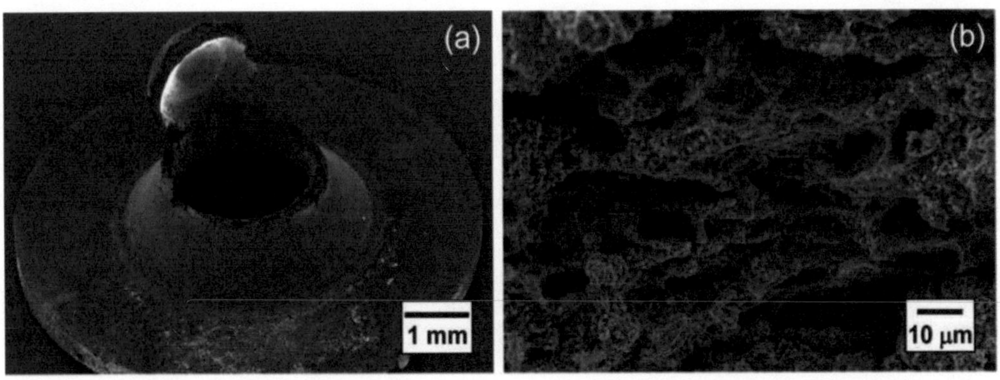

Figure 8. (a) A typical SP creep ductile fracture surface for a test on HN9L at 725°C at 140N and (b) fracture surface depicting ductile transgranular mode of fracture [28].

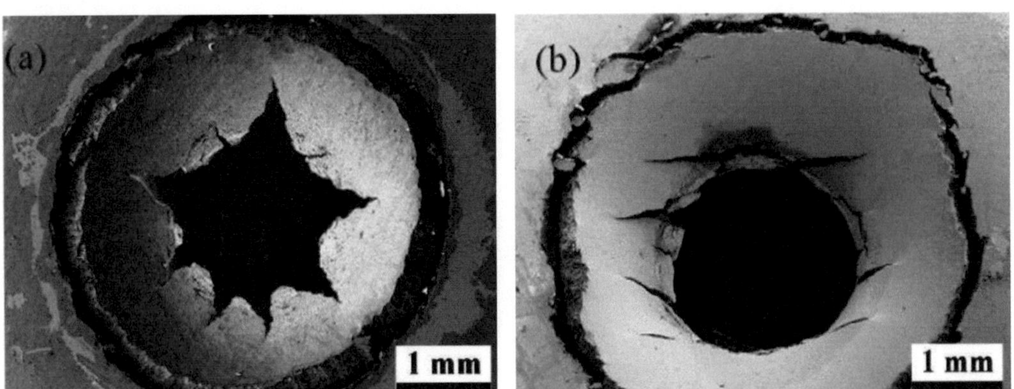

Figure 9. SEM micrographs of SP creep specimens for a 14Cr ODS Steel with a brittle fracture in (a) the transverse orientation and (b) the longitudinal orientation [18].

oxide dispersion strengthened material that exhibits a highly anisotropic microstructure and mechanical response [18].

Figure 9 shows the fracture behaviour in the two specimen orientations denoted as transverse (a) and longitudinal (b). The fracture morphology in the transverse specimen clearly shows radial type cracking emanating from the central point of the disc, in direct contrast to the behaviour seen in a more ductile material. However, in the longitudinal specimen, a series of parallel intergranular cracks can be seen which are aligned to the extrusion process of the material fabrication.

4. Data interpretation and correlations to uniaxial creep results

Previous research [13, 23] has demonstrated that the shape of the time displacement/deflection curves from SP creep tests are qualitatively similar and representative of the shape typically observed from uniaxial testing approaches. Now, over recent years, many researchers and scientists have devoted significant effort in establishing a series of correlative means capable of comparing the results obtained from SP creep testing to data gathered from more traditional creep testing methodologies. These include approaches to obtain equivalent stress values rather than load, strain values rather than deflection/displacement and a minimum creep rate rather than minimum displacement rate.

4.1. Deflection – Chakrabarty membrane stretch model

In 1970, Chakrabarty [29] proposed a theory for a material stretching over a hemispherical punch head, now widely known as the membrane stretch model. This theory has provided the basis for much of the understanding of the relationship between central deflection and strain for a small punch creep test [30]. Assumptions include a rigid punch that is well lubricated so friction is considered negligible, rigid-plastic material due to the large strains and deformation occurring through membrane stresses alone.

Based upon Chakrabarty's model, Li and Šturm [31] devised the following third order polynomial [Eq. (1)] for the equivalent strain at the contact boundary, ε. In this equation, central deflection is given as δ, the receiving hole diameter is 4 mm and the punch radius is 1 mm, as recommended in the European Code of Practice (CoP) for small punch testing [26]. The strain at the contact boundary is considered since necking and failure in SPC are commonly located away from the specimen centre [32].

$$\varepsilon = 0.204656\delta + 0.12026\delta^2 + 0.00950\delta^3 \quad (1)$$

When considering other SPC setups, specifically a punch diameter of 2.5 mm, the constant parameters within the polynomial equation change accordingly [31]. Not only has Chakrabarty's model provided an insight into the strain-deflection relationship, but has further been implemented to help determine the relationship between stress and load with SPC, as discussed further in the k_{SP} method.

Nevertheless, the Chakrabarty model has its limitations in that it is only really suitable for large deformations which is not always the case with brittle materials where specimens may crack upon loading or in the early phase of the test [23], or fractures that reveal a star type pattern as illustrated earlier [18]. In addition, any strain hardening effects must be considered rather than an exponential hardening law only as is the case here [32].

4.2. Empirical approaches

4.2.1. The k_{SP} method

The most widely recognised technique for empirical correlation of SP creep and uniaxial results is the SP creep correlation factor, k_{SP}. In this approach, the SP load can be correlated to a uniaxial creep stress to compare the two sets of data, using the following equation [Eq. (2)]:

$$\frac{F}{\sigma} = 3.33 k_{SP} R^{-0.2} r^{1.2} h_0 \tag{2}$$

where F is the SP load, σ is the applied stress in a uniaxial creep test, R is the radius of the receiving hole, r is the radius of the punch indenter and h_0 is the thickness of the disc specimen.

This a $\frac{F}{\sigma} = 3.33 k_{SP} R^{-0.2} r^{1.2} h_0$ approach has been derived from Chakrabarty's membrane stretching theory but remains empirical in nature due to the inclusion of a material constant, represented here as the k_{SP} value. In situations where no uniaxial creep results are available, this value is usually referred to as 1. However, in such instances the method is less reliable and can only serve as a rough approximation of the corrected creep stress.

The k_{SP} method has been successfully applied to several material systems, notably CMSX-4 [21], P91 [33] and P92 [13] steels. **Figure 10** demonstrates the correlation obtained from the technique for P92 steel, both as a base metal and as weld material [13], and for the lightweight aluminium alloy AlSi9Cu3 [24]. In both examples, the results show good agreement between

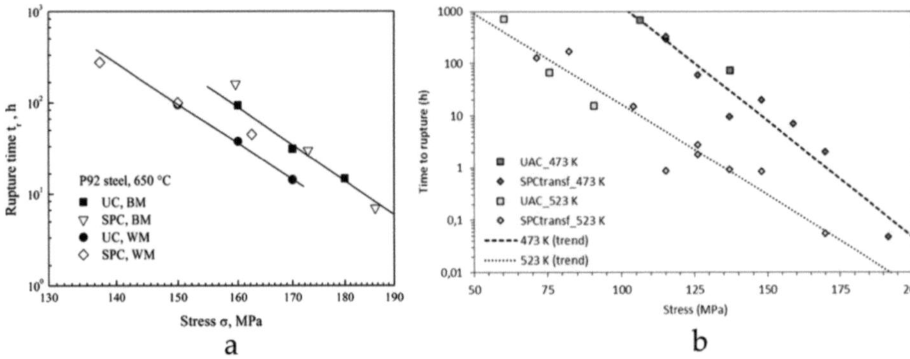

Figure 10. Comparison of uniaxial creep and SPC data using the k_{SP} method for a. P92 steel at 600°C [13] and b. AlSi9Cu3 at 473 and 523K [24].

http://dx.doi.org/10.5772/intechopen.70375

uniaxial and SPC results but the k_{SP} value was found to change for the base metal (1.035) and weld material (0.977) in P92. This difference is attributed to the higher yield strength observed in the base material, which leads to a change in the manner of deformation.

Conversely, the k_{SP} method has been found to be limited to materials with proven ductility and the correlations for more inherently brittle alloys are not as impressive. **Figure 11** presents the relationship for γTiAl 45-2-2 (XD) [23] at elevated temperatures and a clear breakdown in the correlation is observed. This is related to the use of the k_{SP} value as a ductility factor, and the catastrophic manner in which γTiAl 45-2-2 (XD) was found to crack on the onset of loading during a SPC experiment.

4.2.2. Monkman-Grant

The Monkman-Grant model is a well recognised predictive approach for uniaxial creep data that can also be observed in SPC. The technique typically considers the relationship between t_f and the minimum strain rate, $\dot{\varepsilon}$, but can also be modified to account for SPC when considering the minimum deflection rate $\dot{\delta}_M$, in the form of:

$$\log t_f + m \log \dot{\delta}_M = C_S \quad (3)$$

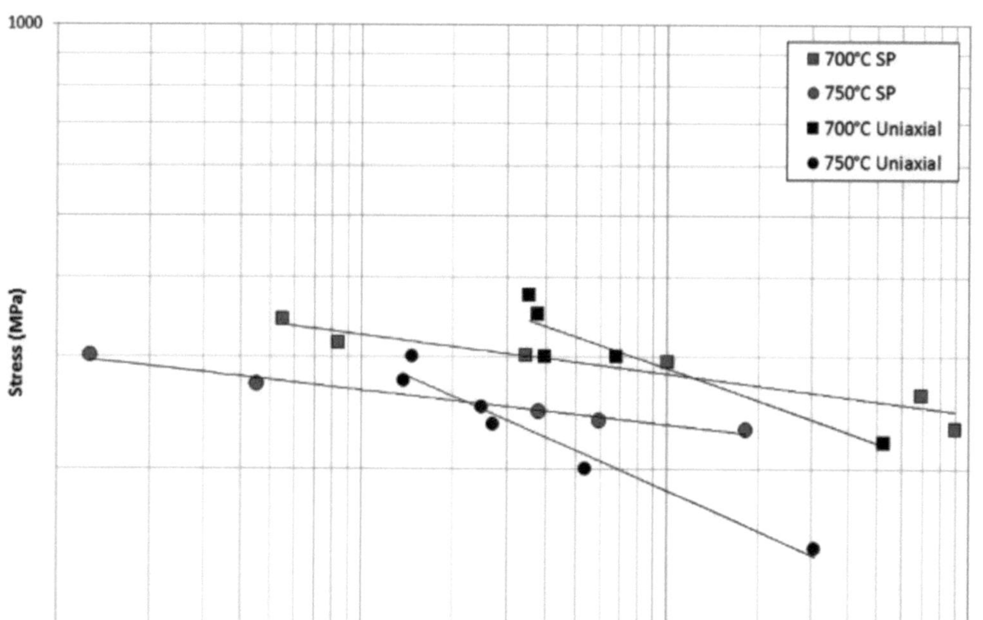

Figure 11. k_{SP} correlation for uniaxial and converted SPC data for γTiAl 45-2-2 (XD) at 700 and 750°C [23].

where C and m are material constants typically defined by linear regression. If the rupture times are equal across the two test types, a correlation may be derived [34] which sanctions a direct comparison of the two rates, $\dot{\varepsilon}$ and $\dot{\delta}_M$, through the following equation:

$$\dot{\varepsilon} = 10^{C - C_{s/m}} \dot{\delta}_M \tag{4}$$

Dobes and Milicka [35] used such a relationship to correlate the minimum creep rate from a uniaxial creep test to the minimum deflection rate from a SPC experiment. Dymáček [36] applied this methodology to P92 NT steel and found parallel trends in the two data sets, demonstrating the applicability of the Monkman-Grant relationship to SPC data (**Figure 12**).

4.2.3. Larson-Miller

The Larson-Miller parameter (LMP) is an established means of predicting the lifetime of materials subjected to creep damage by using a correlative based approach that incorporates the Arrhenius equation. The LMP correlates the relationship between temperature (T), stress and t_f using the following equation:

$$LMP = T(C + \log t_f) \tag{5}$$

where C is a material constant.

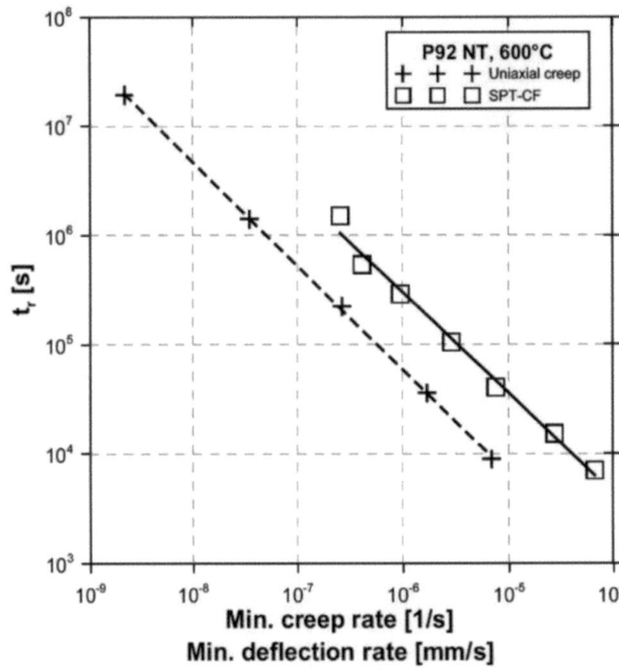

Figure 12. Monkman-Grant relationship for SPC and uniaxial creep results on P92 NT steel at 600°C [36].

A series of researchers [24, 28] have applied the LMP to interpret SPC data, but to also establish a relationship between SPC and uniaxial creep results. Andrés et al. [24] obtained a LMP design curve from SPC results on the magnesium alloy AZ31 through linear regression, once the equivalent uniaxial creep stress was defined. In this research, C was assumed as a constant value to guarantee matching LMP values in both test approaches. The results are illustrated in **Figure 13** which correlates the two data sets and shows good agreement, confirming the suitability of the approach.

4.2.4. Theta projection

The theta-projection approach has found extensive use in the life prediction of uniaxial creep data. The method, developed by Evans and Wilshire [37], has attracted widespread attention [38] and is capable of obtaining an accurate creep deformation curve with a true representation of the three dominant stages of creep. This method can be simplified to accommodate SPC tests via the following expression:

$$\varepsilon = At + B\left(e^{\alpha t} - 1\right) \tag{6}$$

where t is the creep time, A and B are parameters related to the hardening and weakening of strain and α is the constant of creep deformation rate, making the equation more applicable to constant load deformation.

Zheng et al. [39] used this approach to extrapolate the SPC deformation curves of service-exposed Cr5Mo Steel at 550°C, as presented in **Figure 14**, and choose 20% of the creep life as

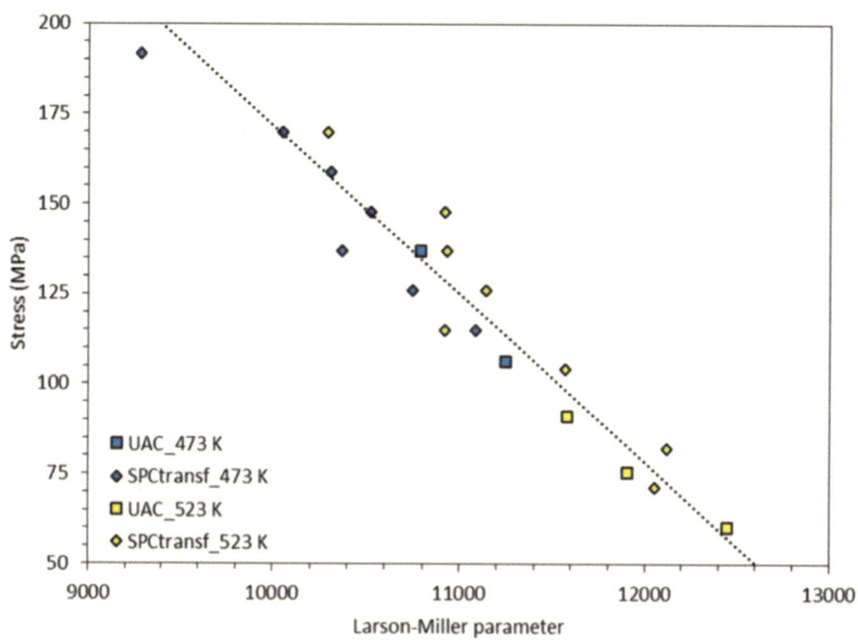

Figure 13. Correlation of SPC equivalent stress to uniaxial creep against LMP for AZ31 alloy [24].

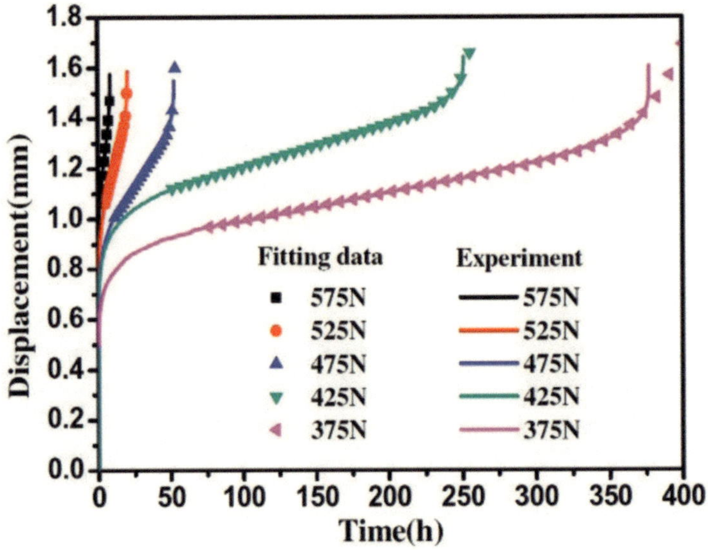

Figure 14. SPC time-deflection fitting curves using the theta-projection method on service-exposed Cr5Mo Steel at 550°C [39].

the starting point for the prediction. As **Figure 14** illustrates, this approach shows good agreement with the generated experimental data

4.2.5. Norton (1/2)

The most important and useful properties from the SPC test are t_f and $\dot{\delta}_M$. In an identical manner to traditional uniaxial creep test approaches, both properties in a SPC test are heavily dependent on the applied test temperature and load. These parameters directly influence the controlling creep mechanism which can vary from diffusional creep to dislocation creep, with dislocation creep more prevalent at higher and intermediary applied stress levels and at temperatures high enough to activate creep (0.4 Tm).

The relationship between $\dot{\delta}_M$ and the applied force F of a SPC test can be described by a power law in a similar manner to the well established Norton law for uniaxial creep approaches, where:

$$\dot{\delta}_M = AF^n \tag{7}$$

Here, A is a proportionality constant. Gülçimen and Hähner [14] used such an approach to determine and compare the local creep properties of a P91 weldment across three distinct zones, namely the base metal, weld metal, and the heat-affected zone. Yang et al [40] also used a similar approach to establish the relationship between $\dot{\varepsilon}$ and $\dot{\delta}_M$, allowing for an approximate conversion of the small punch deflection to the equivalent creep strain in Incoloy880H. **Figure 15** shows the Norton relations for the individual zones in P91 and in Incoloy880H.

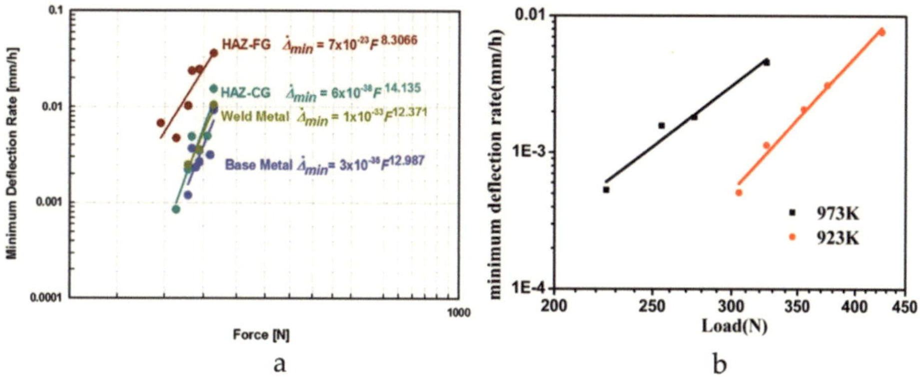

Figure 15. Norton laws for a. base metal, weld metal, HAZ-FG and HAZ-CG in P91 steel [14] and b. Incoloy880H [40].

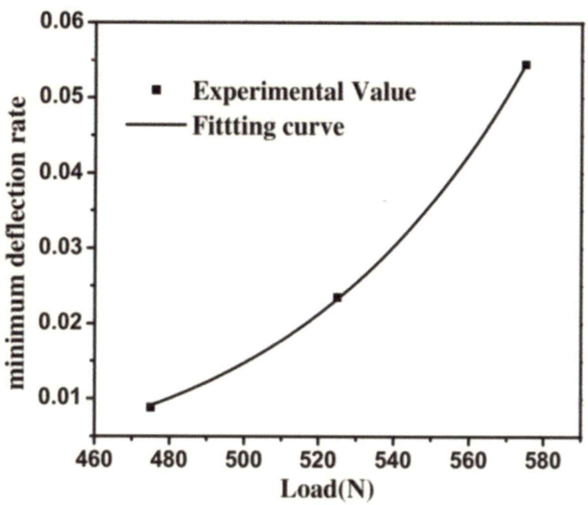

Figure 16. Relationship between minimum deflection rate and load for Cr5Mo steel [39].

Zheng et al. [39] used the Norton relationship to determine the correlation between the stable creep rate and the applied load. This was achieved by using a curve fitting procedure, where the material's $\dot{\delta}_M$ can be predicted from the value of the applied load. The result of this curve fit is displayed in **Figure 16** for Cr5Mo steel. The authors found the prediction represents a 15% error compared to the actual value and can be used as an accurate prediction of the remaining life of the material.

4.2.6. Wilshire equations (1)

The Wilshire equations have been well established over the last few decades and offer the potential to predict long term creep life based on relatively short term creep data [41–43]. The principle is based upon the assumptions that $t_f \rightarrow 0$ as $\sigma \rightarrow \sigma_{TS}$ while $t_f \rightarrow \infty$ as $\sigma \rightarrow 0$, with applied stress, σ, related to t_f through the following equation:

$$(\sigma/\sigma_{TS}) = \exp\{-k_1 [t_f \exp(-Q_c * /RT)]^u\} \quad (8)$$

where σ_{TS} is the ultimate tensile strength in MPa, Qc^* is the apparent activation energy for creep in J mol^{-1} and R is the gas constant (8.314 J mol^{-1} K^{-1}). Qc^*, k_1 and u parameters are then derivable from a comprehensive data set, although it is important to consider an appropriate explanation for activation energy magnitudes.

In the case of SPC data, the Wilshire equations have been exploited for different means and approaches. Jeffs et al. [22] compared the creep life predictions as determined by the Wilshire equations for SPC results, correlated via the k_{SP} and uniaxial data. This relationship is displayed in **Figure 17**, with a good agreement between each test type seen.

In a similar nature, the Wilshire equations may be modified such that normalisation is carried out using loads rather than stresses [44] as shown in Eq. (8) below:

$$(F/F_m) = \exp\{-k_1 [t_f \exp(-Q_c * /RT)]^u\} \quad (9)$$

Here, F is the applied SPC test load and F_m the ultimate load through a constant displacement rate SP test. Using such an approach could provide an effective ranking method of material types exclusively through SPC. Alternatively, Holmström et al. [15] has utilised the Wilshire equations to determine a strength correlation factor between weld material variants.

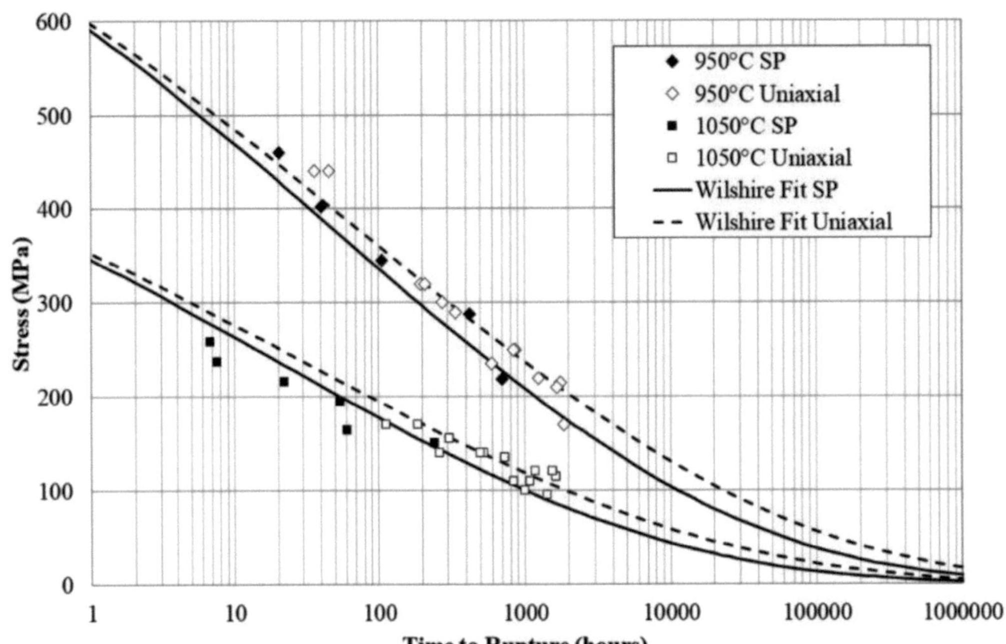

Figure 17. Wilshire equation predictions of σ vs. t_f for uniaxial creep and SPC test types for CMSX-4 [22].

4.3. Numerical approaches

4.3.1. Finite element analysis (FEA)

Several authors have used FEA to predict the SP creep behaviour in different material systems. Evans [45] analysed the SP deformation of 0.5Cr0.5Mo0.25V ferritic steel and assumed deformation was viscoplastic with a constant volume, with deformation rates governed by the creep properties of the material. They used a phenomological constitutive relationship for the creep properties of the internal variable type, where the variables used include hardening, softening and continuum damage with an additional creep failure criterion also adopted. They constructed a model using uniaxial creep data and verified it by comparing the creep curves and t_f from actual SPC tests to those produced through the FEA model, as given in **Figure 18**. As can be seen, the results show good correlation with the FEA model.

Lancaster et al. [23] used a constitutive creep model based on the theta-projection method first developed by Evans [45], which was implemented into Abaqus using a creep derived subroutine. The authors used such an approach to relate creep rate to the accumulation of internal material state variables, which has been previously shown in Ref. [47] to predict creep behaviour better than other strain or time based hardening approaches, for situations where a transient stress state is present, such as in a SPC test. In this research, an axisymmetric model was used to predict the material deformation in a SPC test on γTiAl 45-2-2 (XD) with model geometries replicating those that are used in a SPC experiment, with all features modelled as rigid bodies and surfaces. The model showed that the Von-Mises stress was found to be high across the thickness of the disc and as such, the rate of creep deformation was predicted to be high. However, the maximum principle stress adjacent to the disc was found to be low, but due to a combination of high radial and hoop stresses, the maximum principle stress on the underside of the specimen was high in the central region. As specimen deformation increases with displacement, the stress field is then found to evolve and the peak stress in the disc slowly decreases away from the centre of the disc, leading to the onset of thinning and steady state deformation. An additional damage model dependent on maximum principle stress was incorporated to capture the rupture characteristics of the material and was able to predict rupture to a high level of accuracy when compared to experimental results [23].

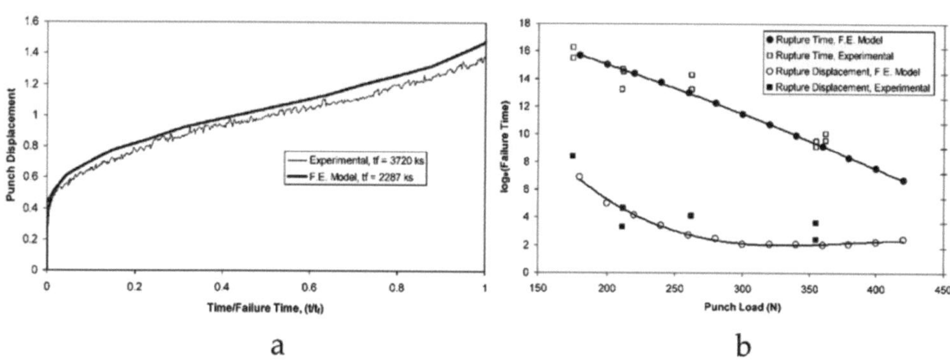

Figure 18. Experimental and modelled SPC results at 848K on 0.5Cr0.5Mo0.25V ferritic steel a. time vs punch displacement for a test performed under 211N and b. failure time vs punch displacement for all simulated and experimental tests [46].

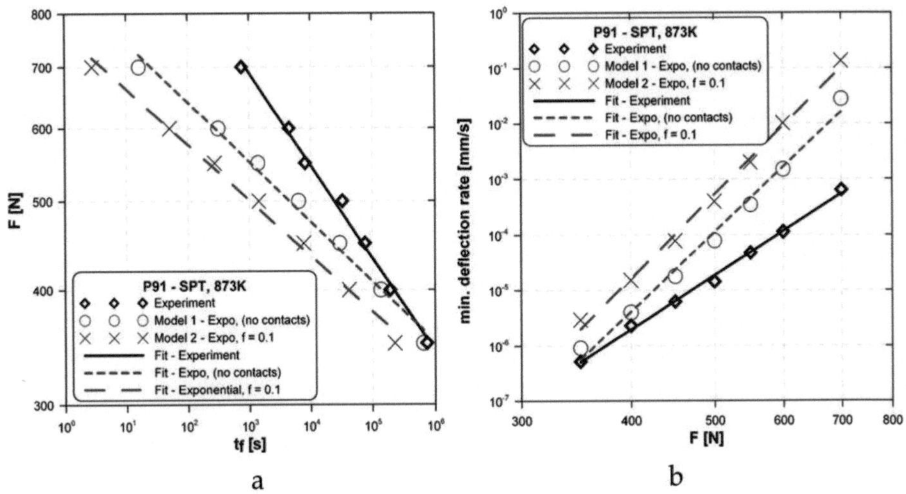

Figure 19. Comparison of results generated from alternative FEA models and experimental SPC tests for P91 steel a. F vs. t_f and b. $\dot{\delta}_M$ vs. F [48].

Dymáček and Milička [48] utilised two basic creep constitutive models, namely Norton power-law ('Model 1') and the exponential relationship ('Model 2'), to run their FEA simulations. Model 1 considered only the simple boundary conditions as supports and assumed the pressure of the load without actually using the specific puncher shape. However, for Model 2, they included contact elements whilst accounting for the surface frictional contact between contacting surfaces. In either case, a simplified 2D axisymmetric model was adopted and the numerical findings from each model were compared with real experimental results, as displayed in **Figure 19**.

The authors found that Model 1 was somewhat limited due to the absence of a friction parameter which was deemed necessary from the research. However, Model 1 was capable of predicting more accurate results under high forces and in general, the values of t_f and $\dot{\delta}_M$ showed good agreement with the experimental data in both models.

Other notable contributions are provided by Nakata et al. [49] for their work on F82H ferritic/martensitic steel; Zhao et al. [13] who simulated the SPC behaviour of P92 steel from several locations around a welded joint; Kobayashi et al. [50] who used FEA to support research into the SPC deformation of aluminium alloys from the duralumin series and Jeffs et al. [22] who adopted ABAQUS to simulate the SPC deformation curves of the single crystal superalloy, CMSX-4.

Author details

Robert J. Lancaster* and Spencer P. Jeffs

*Address all correspondence to: r.j.lancaster@swansea.ac.uk

Swansea University, Swansea, UK

References

[1] Manahan MP, Argon AS, Harling OK. The development of a miniaturizsed disk bend test for determination of post irradiation mechanical properties. Journal of Nuclear Materials. 1981;**104**:1545-50

[2] Manahan MP. A new post-irradiation mechanical behavior test–The miniaturized disk bend test. Nuclear Technology. 1983;**63**(2):295-315

[3] Corwin WR, Lucas GE. The Use of Small-Scale Specimens for Testing Irradiated Material: A Symposium Sponsored by ASTM Committee E-10 on Nuclear Technology and Applications, Albuquerque, N.M., 23 Sept. 1983. [Internet]. ASTM; 1986 [cited 2017 Jun 8]. Available from: https://inis.iaea.org/search/search.aspx?orig_q=RN:18065732

[4] Parker JD, James JD. Disc-bend creep deformation behaviour of 1/2Cr1/2Mo1/4V low alloy steel. In: 5th International Conference: Creep and Fracture of Engineering Materials and Structures, Swansea, Wales: CRC Press, USA; 1993. pp. 651-660

[5] Available from: www.scopus.com. Elsevier B.V

[6] Hyde TH, Sun W, Williams JA. Requirements for and use of miniature test specimens to provide mechanical and creep properties of materials: A review. International Materials Review. 2007;**52**(4):213-255

[7] Kusumoto J, Kanaya A, Kobayashi T. Residual life assessment of 2.25Cr-1Mo boiler pipe by small punch creep test. In: ASME 2006 Pressure Vessels and Piping/ICPVT-11 Conference; Vancouver, BC: ASME, USA; 2006. pp. 637-642

[8] Ule B, Šuštar T, Dobeš F, Milička K, Bicego V, Tettamanti S, et al. Small punch test method assessment for the determination of the residual creep life of service exposed components: Outcomes from an interlaboratory exercise. Nuclear Engineering and Design. 1999;**192**(1):1-11

[9] Izaki T, Kobayashi T, Kusumoto J, Kanaya A. A creep life assessment method for boiler pipes using small punch creep test. International Journal of Pressure Vessels and Piping, [Internet]. Elsevier Ltd; 2009;**86**(9):637-642. Available from: http://dx.doi.org/10.1016/j.ijpvp.2009.04.005

[10] Rolls-Royce Power Engineering plc. Scoop sampling – Extraction of material samples for examination and analysis [Internet]. Commissioning and In-service Support; 2010. [Online]. Available from: http://www.rolls-royce.com/~/media/Files/R/Rolls-Royce/documents/customers/nuclear/scoop-sampling-tcm92-50918.pdf

[11] Omacht D, Kubanek Z, Dolezal R. Development of testing machines and equipment for small punch testing, proposals for improvement of CWA 15627. Key Engineering Materials [Internet]. 2017;**734**:237-248. Available from: http://www.scientific.net/KEM.734.237

[12] Kumar K, Madhusoodanan K, Rupani BB. Miniature specimen technique as an NDT tool for estimation of service life of operating pressure equipment. International Conference of Pressure Vessels and Piping, OPE-2006-Chennai, Chennai, India: BARC Newsletter, India; 2006. p. 92-102

[13] Zhao L, Jing H, Xu L, Han Y, Xiu J, Qiao Y. Evaluating of creep property of distinct zones in P92 steel welded joint by small punch creep test. Materials and Design [Internet]. 2013;**47**:677-686 Available from: http://dx.doi.org/10.1016/j.matdes.2012.12.032

[14] Gülçimen B, Hähner P. Determination of creep properties of a P91 weldment by small punch testing and a new evaluation approach. Materials Science and Engineering A [Internet]. Elsevier; 2013;**588**:125-131. Available from: http://dx.doi.org/10.1016/j.msea.2013.09.029

[15] Holmström S, Auerkari P, Hurst R, Blagoeva D. Using small punch test data to determine creep strain and strength reduction properties for heat affected zones. Materials Science and Technology [Internet]. 2014;**30**(1):63-66. Available from: http://www.maneyonline.com/doi/abs/10.1179/1743284713Y.0000000311

[16] Kim B, Lim B. Local creep evaluation of P92 steel weldment by small punch creep test. Acta Mechanica Solida Sinica [Internet]. The Chinese Society of Theoretical and Applied Mechanics; 2008;**21**(4):312-317. Available from: http://dx.doi.org/10.1007/s10338-008-0836-z

[17] Jeffs S, Lancaster R, Davies S. Effect of build orientation and post processing of a direct laser deposited nickel superalloy as determined by the small punch creep test. Key Engineering Materials. 2017;**734**:128-136

[18] Bruchhausen M, Turba K, De Haan F, Hähner P, Austin T, De Carlan Y. Characterization of a 14Cr ODS steel by means of small punch and uniaxial testing with regard to creep and fatigue at elevated temperatures. Journal of Nuclear Materials [Internet]. Elsevier B.V.; 2014;**444**(1–3):283-291. Available from: http://dx.doi.org/10.1016/j.jnucmat.2013.09.059

[19] Turba K, Hurst RC, Hähner P. Anisotropic mechanical properties of the MA956 ODS steel characterized by the small punch testing technique. Journal of Nuclear Materials [Internet]. Elsevier B.V.; 2012 Sep;**428**(1–3):76-81. Available from: http://linkinghub.elsevier.com/retrieve/pii/S0022311511008191

[20] Mathew MD, Ganesh Kumar J, Ganesan V, Laha K. Small punch creep studies for optimization of nitrogen content in 316LN SS for enhanced creep resistance. Metallurgical and Materials Transactions A, Physical Metallurgy and Materials Science. 2014;**45**(2):731-737

[21] Jeffs SP, Lancaster RJ. Elevated temperature creep deformation of a single crystal superalloy through the small punch creep method. Materials Science and Engineering A. 2015 Feb;**626**:330-337

[22] Jeffs SP, Lancaster RJ, Garcia TE. Creep lifing methodologies applied to a single crystal superalloy by use of small scale test techniques. Materials Science and Engineering A. 2015;**636**:529-535

[23] Lancaster RJ, Harrison WJ, Norton G. An analysis of small punch creep behaviour in the γ titanium aluminide Ti-45Al-2Mn-2Nb. Materials Science and Engineering A. 2015 Feb;**626**:263-274

[24] Andrés D, Lacalle R, Álvarez JA. Creep property evaluation of light alloys by means of the small punch test: Creep master curves. Materials and Design [Internet]. Elsevier Ltd; 2016;**96**:122-130. Available from: http://dx.doi.org/10.1016/j.matdes.2016.02.023

[25] Kumar JG, Laha K. Small punch creep deformation and rupture behavior of 316L (N) stainless steel. Materials Science and Engineering A [Internet]. Elsevier; 2015;**641**:315-322. Available from: http://dx.doi.org/10.1016/j.msea.2015.06.053

[26] CEN Workshop Agreement CWA 15267. European Code of Practise: Small Punch Test Method for Metallic Materials. CEN Members National Standards Body; 2007. p. 1-69

[27] Matocha K, Hurst R. Small punch testing–The transition from a code of practice to a European testing standard. Key Engineering Materials [Internet]. 2017;**734**:3-22. Available from: http://www.scientific.net/KEM.734.3

[28] Naveena, Komazaki S. Evaluation of creep rupture strength of high nitrogen ferritic heat-resistant steels using small punch creep testing technique. Materials Science and Engineering: A [Internet]. Elsevier; 2016;**676**:100-108. Available from: http://dx.doi.org/10.1016/j.msea.2016.08.102

[29] Chakrabarty J. A theory of stretch forming over hemispherical punch heads. International Journal of Mechanical Sciences. 1970;**12**:315-325

[30] Yang Z, Wen Wang Z. Relationship between strain and central deflection in small punch creep specimens. International Journal of Pressure Vessels and Piping. 2003;**80**(6):397-404

[31] Li Y, Šturm R. Determination of creep properties from small punch test. In: ASME 2008 Pressure Vessels and Piping Conference. Chicago, IL: ASME, USA; 2008. pp. 739-750

[32] Rouse JP, Cortellino F, Sun W, Hyde TH, Shingledecker J. Small punch creep testing: review on modelling and data interpretation. Materials Science and Technology [Internet]. 2013 Nov;**29**(11):1328-1345. Available from: http://www.maneyonline.com/doi/abs/10.1179/1743284713Y.0000000278

[33] Blagoeva DT, Hurst RC. Application of the CEN (European Committee for Standardization) small punch creep testing code of practice to a representative repair welded P91 pipe. Materials Science and Engineering A. 2009;**510–511**(C):219-223

[34] Monkman FC, Grant NJ. An empirical relationship between rupture life and minimum creep rate in creep-rupture tests. Proceedings of ASTM International. 1956;**56**:593-620

[35] Dobeš F and Milička K. Application of creep small punch testing in assessment of creep lifetime. Materials Science and Engineering: A. 2009 Jun;**510–511**:440-443

[36] Dymáček P. Recent developments in small punch testing: Applications at elevated temperatures. Theoretical and Applied Fracture Mechanics. 2016;**86**:25-33

[37] Evans RW, Wilshire B. Creep of Metals and Alloys. London, UK: The Institue of Metals; 1985. 197-243 p.

[38] Law M, Payten W, Snowden K. Modelling creep of pressure vessels with thermal gradients using theta projection data. International Journal of Pressure Vessels and Piping. 2002;**79**(12):847-851

[39] Zheng Y, Yang S, Ling X. Creep life prediction of small punch creep testing specimens for service-exposed Cr5Mo using the theta-projection method. Engineering Failure Analysis

[Internet]. Elsevier Ltd; 2017;**72**:58-66. Available from: http://dx.doi.org/10.1016/j.engfailanal.2016.11.009

[40] Yang S, Ling X, Zheng Y. Creep behaviors evaluation of Incoloy800H by small punch creep test. Materials Science and Engineering A [Internet]. Elsevier; 2017;**685**(Nov 2016):1-6. Available from: http://dx.doi.org/10.1016/j.msea.2016.12.092

[41] Whittaker MT, Evans M, Wilshire B. Long-term creep data prediction for type 316H stainless steel. Materials Science and Engineering A [Internet]. Elsevier B.V.; 2012;**552**: 145-150. Available from: http://linkinghub.elsevier.com/retrieve/pii/S0921509312007150

[42] Wilshire B, Battenbough AJ. Creep and creep fracture of polycrystalline copper. Materials Science and Engineering A [Internet]. 2007 Jan;**443**(1–2):156-166. Available from: http://linkinghub.elsevier.com/retrieve/pii/S0921509306019460

[43] Wilshire B, Scharning PJ. Long-term creep life prediction for a high chromium steel. Scripta Materialia [Internet]. 2007 Apr;**56**(8):701-704. Available from: http://linkinghub.elsevier.com/retrieve/pii/S1359646207000231

[44] Holmström S, Hähner P, Hurst RC, Bruchhausen M, Fischer B, Lapetite J-M, et al. Small punch creep testing for material characterization and life time prediction. Materials for Advanced Power Engineering. Juelich, Germany: FZ. Juelich, Germany; 2014. pp. 627-635

[45] Evans RW. A constitutive model for the high-temperature creep of particle-hardened alloys based on the θ projection method. Proc R Soc A Math Phys Eng Sci. 2000;**456**:835-68

[46] Evans RW, Evans M. Numerical modelling of small disc creep test. Materials Science and Technology. 2006;**22**(June):1155-1162

[47] Harrison WJ, Whittaker MT, Deen C. Creep behaviour of Waspaloy under non-constant stress and temperature. Materials Research Innovations. 2013;**17**(5):323-326

[48] Dymáček P, Milička K. Creep small-punch testing and its numerical simulations. Materials Science and Engineering A [Internet]. 2009 Jun;**510–511**:444-449. Available from: http://linkinghub.elsevier.com/retrieve/pii/S0921509308014780

[49] Nakata T, Komazaki S, Kohno Y, Tanigawa H. Development of a small punch testing method to evaluate the creep property of high Cr ferritic steel: Part II–Stress analysis of small punch test specimen by finite element method. Materials Science and Engineering A [Internet]. Elsevier; 2016;**666**:80-87. Available from: http://dx.doi.org/10.1016/j.msea.2016.03.102

[50] Kobayashi K, Kajihara I, Koyama H, Stratford GC. Deformation and fracture mode during small punch creep tests. Journal of Solid Mechanics and Materials Engineering. 2010;**4**(1):75-86

Chapter 9

Thermomechanical Time-Dependent Deformation and Fracturing of Brittle Rocks

Tao Xu and Guang-lei Zhou

Additional information is available at the end of the chapter

http://dx.doi.org/10.5772/intechopen.72326

Abstract

A thermomechanical numerical model is proposed to describe the time-dependent brittle deformation of brittle rocks under different constant temperatures and confining pressures. The mesoscale model accounts for material heterogeneity and local material degradation, and the model introduces the concept of a mesoscopic renormalization to capture the cooperative interaction between microcracks in the transition from distributed to localized damage. The thermophysical parameters for the model were determined based on creep experiments of granite at temperatures of 23, 50, and 90°C. The numerical simulations agree well with the experimental data. We then explore the influence of temperature, differential stress, confining pressure, and sample homogeneity on brittle creep in granite using the same parameters. The simulated results show that the creep strain rate increases with an increase in temperature and differential stress and time to failure decreases, while creep strain rate decreases with an increase in confining pressure and sample homogeneity, and therefore time to failure increases. The proposed model is of great help to control and optimize rock engineering in granite.

Keywords: time-dependent deformation, temperature, creep strain rate, damage evolution

1. Introduction

Disposal of radioactive waste is currently considered as the preferred option to store and isolate this waste from biosphere over hundreds of thousands of years for waste management worldwide. In the geological disposal, the temperature in the near-field rock mass will increase due to the heat generated by the radioactive waste and in turn affect the time-dependent behavior of the host rock and consequently change the overall long-term performance of the disposal. Therefore, a good understanding of the time-dependent behavior of the host rock at elevated temperatures is of great significance to evaluate the stability of nuclear waste disposal.

Extensive effort has been made to investigate the time-dependent behavior of rocks like sedimentary rock and rock salt [1–4] and hard rocks like granite [5–10] and basalt [11]. Granite is a widely recognized potential host rock for the disposal of radioactive waste due to its low permeability and high strength [12]. Moreover, a good understanding of time-dependent deformation in granite will also assist in our understanding of crustal deformation and natural hazards since granite is a major constituent of the continental crust. The distribution of stress and permeability evolution in granite is of great importance when we analyze natural hazards and rock engineering projects. Despite the importance of granite in these high-temperature environments, very few studies of time-dependent brittle creep on granite have been made at elevated temperature [2, 8, 9, 13]. During a brittle creep experiment of rock at a constant differential stress for an extended period of time, the strain against time first decelerates before accelerating as macroscopic sample failure is approached [14]. The onset of the acceleration to failure in brittle creep experiments has been ascribed as the result of the sample reaching a microcrack density at which microcracks can interact and coalesce, sometimes referred to as the critical damage threshold [2, 11, 13, 15]. Time-dependent brittle deformation is often attributed to a mechanism of subcritical crack growth, namely stress corrosion cracking [14, 16]. Stress corrosion describes fluid-rock reactions that occur preferentially between a chemically active pore fluid and strained atomic bonds at the crack tips and therefore is sensitive to environmental factors such as stress, temperature, and pore fluid reactivity [16]. Temperature influences the crack growth rate through the Arrhenius temperature dependence of crack growth rate and because temperature affects the stress dependency of the rate of crack growth. Experimental tests have shown that speed of crack growth during experiments increases with an increase in temperature [14, 17, 18].

Due to the aforementioned importance of understanding the long-term mechanical behavior of rock for disposal of radioactive waste, here we present a two-dimensional constitutive creep model to describe the time-dependent deformation of granite at different temperatures and confining pressures. We first formulate the coupled thermomechanical time-dependent model. We then validate the model and determine the required thermo-physico-mechanical parameters with published experimental data [10, 19]. Finally, the results of brittle creep simulations at different temperatures, differential stresses, confining pressures, and sample homogeneities were presented and discussed.

2. Description of time-dependent constitutive model

In this chapter, a quantitative model is proposed to describe the coupled heat transfer and rock failure problems associated with rock exposed to elevated temperatures. For a model used to investigate time-dependent creep deformation at high temperature, the coupled effect of the medium deformation and heat transfer must be important. Three components including a heat transfer description, a stress description, and a failure description must be accounted for. The descriptions of heat transfer, stress, and failure in the model are, respectively, presented.

2.1. Material heterogeneity description

It is important to introduce material heterogeneity to reveal a collective macroscopic behavior different from that of the individual elements. The mechanical parameters such as uniaxial compressive strength and Young's modulus of the mesoscopic material elements are assumed to be homogeneous and isotropic and are randomly assigned with a Weibull statistical distribution (Eq. (1)) [19] to reflect the material heterogeneity at a mesoscale:

$$f(u) = \frac{\chi}{u_0}\left(\frac{u}{u_0}\right)^{\chi-1} \exp\left[-\left(\frac{u}{u_0}\right)^{\chi}\right] \tag{1}$$

in which u is the scale parameter of an individual element at a mesoscale and the scale parameter u_0 is related to the average element parameter. The shape parameter χ indicates material homogeneity and denotes as a homogeneity index. In order to get a set of values of uniaxial compressive strength and Young's modulus following the Weibull statistic distribution, a suit of random numbers ω_i ranging from 0 to 1 are generated according to the Monte Carlo method, and then the uniaxial compressive strength and Young's modulus of each element can be obtained from Eq. (2):

$$u_i = u_0 \ln\left(\frac{1}{1-\omega_i}\right)^{\frac{1}{\chi}} \tag{2}$$

in which u_i is the uniaxial compressive strength and Young's modulus of the element i.

Figure 1 shows two numerical standard rock specimens with a homogeneity index χ of 5. The uniaxial compressive strength and the Young's modulus are randomly determined according to the Weibull distribution. The different colors in **Figure 1** correspond to different values of element strength and Young's modulus. Therefore, the homogeneity index is an important controlling parameter for the macroscopic mechanical response of a numerical sample [20].

2.2. Heat conduction description

It is assumed that the conductive flux is saturated at a value comparable to the enthalpy per unit volume. The heat conduction equation in a medium is known as Fourier's law:

$$q_i = -k_{ij}\,\mathrm{grad}(T) \tag{3}$$

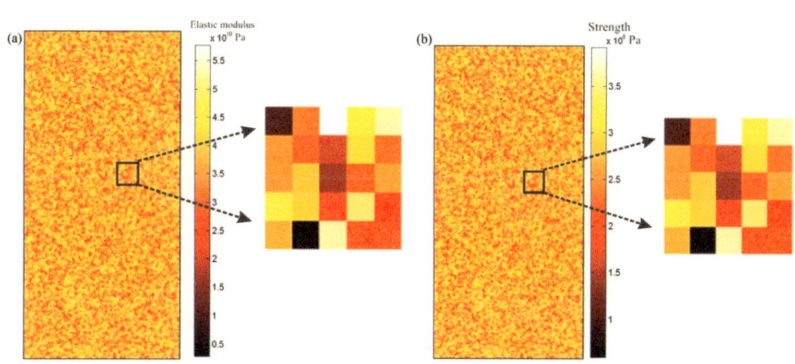

Figure 1. Numerical specimens with a geometry of 100 mm in length and 50 mm in width at a homogeneity index χ of 5.

in which q_i is the rate of heat conduction in W/m^2, k_{ij} is the thermal conductivity tensor of the medium in W/(m °C), and grad(T) is the temperature gradient in °C/m. The negative sign indicates that heat is transferred in the direction of decreasing temperature.

It is assumed that thermal equilibrium between the phases and heat conduction as the dominant mechanism of heat transfer and the energy balance equation in an anisotropic medium can be expressed as

$$\frac{\partial}{\partial x_i}\left(k_{ij}\frac{\partial T}{\partial x_j}\right) + q = \rho c\frac{\partial T}{\partial t} \tag{4}$$

in which q denotes the heat source generated in the medium in W/m^3, ρ is the bulk density of medium in kg/m^3, c is the specific heat of the medium in J/(kg °C), and t is the time. The energy balance equation for an isotropic material can be reduced:

$$k\nabla^2 T + q = \rho c\frac{\partial T}{\partial t} \tag{5}$$

2.3. Stress description

Assuming the total strain for a stressed medium is made up of elastic, creep, and thermal components. The total strain can be decomposed:

$$\varepsilon = \varepsilon_e + \varepsilon_c + \varepsilon_T \tag{6}$$

in which the subscripts e, c, and T refer to the elastic strain, creep strain, and thermal strain, respectively. The elastic strain ε_e can be expressed in a tensor form for elastic medium:

$$\varepsilon_e = \varepsilon_{ij} = \frac{1}{2G}S_{ij} + \frac{1}{3}\varepsilon_{kk}\delta_{ij} \tag{7}$$

in which G is the shear modulus, ε_{ij} is the tensor form of the elastic strain ε_e, S_{ij} is the stress deviator tensor of the elastic stress components σ_{ij}, $S_{ij} = \sigma_{ij} - \frac{1}{3}\sigma_{kk}\delta_{ij}$, and δ_{ij} is the Kronecker delta. The Kronecker delta equals zero when $i \neq j$ and one when $i = j$. The thermal strain ε_T due to the variation of temperature can be expressed:

$$\varepsilon_T = \alpha\Delta T\delta_{ij} \tag{8}$$

in which α is the coefficient of linear thermal expansion and ΔT is the temperature change. The creep strain ε_c is a function of the stress σ, temperature T, and time t, i.e., $\varepsilon_c = F(\sigma, T, t)$. It is customary to assume that the effects are separable and can be written as

$$\varepsilon_c = f(\sigma)g(T)h(t) \tag{9}$$

where $f(\sigma)$, $g(T)$, and $h(t)$ are the functions related to stress, temperature, and time, respectively.

For creep problems, a Norton-Bailey equation [4] known as a constitutive law of the creep strain rate was adopted to characterize time-dependent creep deformation based on the approach of the equation of state theory:

$$\varepsilon_c = A\sigma^n t^m \exp\left(-\frac{U}{RT}\right) \tag{10}$$

where A, m, and n are constants dependent on temperature. The constant n usually denotes stress component of greater than one, m is usually a fraction, U is the creep activation energy determined empirically as proportional to the slope of a plot of log ε_c as a function of $1/T$ at constant σ, R is the universal gas constant, and T is the absolute temperature in Kelvin.

Equation (10) can also be expressed in a strain rate form because the strain rate is of great interest for our study:

$$\dot{\varepsilon}_c = mA\sigma^n t^{m-1} \exp\left(-\frac{U}{RT}\right) \tag{11}$$

Creep flow rule under multiaxial stress conditions can be expressed in tensor form:

$$\dot{\varepsilon}_{ij}^c = \frac{3}{2}\frac{d\bar{\varepsilon}_c}{dt}\frac{S_{ij}}{\sigma_e} \tag{12}$$

in which σ_e is the effective stress, $\sigma_e = \sqrt{\frac{3}{2}S_{ij}S_{ij}}$, $d\bar{\varepsilon}_c$ is the effective creep strain, and $d\bar{\varepsilon}_c = \sqrt{\frac{2}{3}d\varepsilon_{ij}^c d\varepsilon_{ij}^c}$. Now, substitute Eq. (12) into Eq. (11); the creep strain rate is also extended to the multiaxial stress case:

$$\dot{\varepsilon}_{ij}^c = \frac{3}{2}AmS_{ij}\sigma_e^{n-1}t^{m-1} \exp\left(-\frac{U}{RT}\right) \tag{13}$$

where $\dot{\varepsilon}_{ij}^c$ is the creep strain rate, S_{ij} is the deviatoric part of σ_{ij}, and σ_e is effective stress defined as

$$\sigma_e = \left(\frac{1}{\sqrt{2}}\right)\sqrt{(\sigma_{11}-\sigma_{22})^2 + (\sigma_{33}-\sigma_{22})^2 + (\sigma_{11}-\sigma_{33})^2 + 6\left(\sigma_{12}^2+\sigma_{23}^2+\sigma_{13}^2\right)} \tag{14}$$

This creep model can describe the decelerating creep commonly seen in laboratory [14], but it fails to capture the acceleration in strain rate in the approach to macroscopic sample failure. Thus, a damage evolution law for the accelerating creep of rock is incorporated at this stage.

The static stress equilibrium equation can be expressed in tensor form:

$$\sigma_{ij,j} + f_i = 0 \tag{15}$$

in which σ_{ij} is the total stress tensor in MPa and f_i is the body force per unit volume in MPa. The continuity equation can be expressed in terms of the displacement gradient:

$$\varepsilon_{ij} = \frac{1}{2}\left(u_{i,j} + u_{j,i}\right) \tag{16}$$

in which ε_{ij} is the small strain tensor and u is the displacement of the medium. The constitutive equation for elastic isotropic medium can be expressed as

$$\sigma_{ij} = \lambda\delta_{ij}\varepsilon_{ii} + 2G\varepsilon_{ij} - \beta\delta_{ij}\Delta T \tag{17}$$

in which G is shear modulus, λ is Lamé's constant, β is the coefficient of thermal stress in 1/°C, and $\beta = (3\lambda + 2G)\alpha$.

Thus, the final governing equation of heat transfer in a medium can be written in displacement form:

$$(\lambda + G)\cdot u_{j,ji} + Gu_{i,jj} + f_i - \beta\delta_{ij}\Delta T = 0 \tag{18}$$

2.4. Damage evolution description

An elastic damage constitutive law is incorporated in the model to describe its stress-strain relationship. The elastic modulus for an isotropic elastic medium is expressed:

$$E = E_0(1 - D) \tag{19}$$

where E and E_0 are the Young's moduli of the damaged and the undamaged material, respectively, and D is the isotropic damage variable. The constitutive relation can be rewritten in a uniaxial state of stress:

$$\sigma_{11} = E_0(1 - D)\varepsilon_{11} \tag{20}$$

According to damage mechanics theory, the Young's modulus of an element will degrade according to Eq. (19) when the stress on the element exceeds a damage threshold. The stress-strain curve of the element is considered linear elastic with a constant residual strength until the given damage threshold is reached. The elastic damage constitutive law of each element under uniaxial stress condition is shown in **Figure 2**. The maximum tensile stress criterion and modified Mohr-Coulomb criterion are chosen as the damage thresholds to determine whether any elements are damaged in tension or shear, respectively:

$$\begin{aligned} \sigma_3 &= -f_{t0} \\ \sigma_1 - \sigma_3\frac{1 + sin\varphi}{1 - sin\varphi} &= f_{c0} \end{aligned} \tag{21}$$

where σ_1 is the maximum principal stresses, σ_3 is the minimum principal stresses, f_{c0} and f_{t0} are the uniaxial compressive and tensile strength, respectively, and φ is the internal friction angle of the element. The tensile strain criterion is always preferential because the uniaxial tensile strength is far below uniaxial compressive strength of rock material.

The damage variable D according to the constitutive law as shown in **Figure 2** can be described.

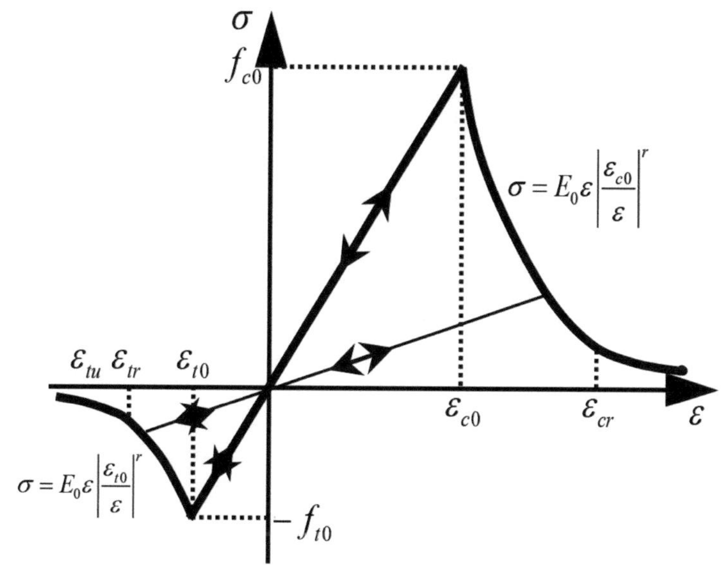

Figure 2. Elastic damage constitutive law for element under uniaxial compression and tension.

$$D = \begin{cases} 0 & F_1 < 0 \quad and \quad F_2 < 0 \\ 1 - \left|\dfrac{\varepsilon_{t0}}{\varepsilon_1}\right|^r & F_1 = 0 \quad and \quad dF_1 > 0 \\ 1 - \left|\dfrac{\varepsilon_{c0}}{\varepsilon_3}\right|^r & F_2 = 0 \quad and \quad dF_2 > 0 \end{cases} \tag{22}$$

where ε_1 is the maximum principal strain and ε_3 is the minimum principal strain. ε_{t0} is the maximum principal strain in tension, ε_{c0} is the maximum principal strain in compression, and r is a constitutive coefficient with a value of 2. The element behaves elastically during loading and unloading when the major tensile or compressive strain of the element is less than the maximum principal strain in tension or compression, respectively. The element will be damaged by Eq. (22) when either of these strain thresholds is exceeded.

3. Model validation

3.1. Constant strain rate experiments

In this section, the proposed model was calibrated with experimental data [10, 18, 21] to determine appropriate input parameters. Mechanical data from uniaxial and triaxial constant displacement rate experiments at room temperature were used to obtain the physico-mechanical input parameters at the mesoscale. The geometry of the granite samples in laboratory tests was 100 mm in length and 50 mm in width. The compressive strength of air-dried Beishan granite under confining pressures of 0, 1, and 5 MPa is 165.2, 174, and 216.8 MPa, respectively. The Young's modulus and Poisson's ratio of air-dried Beishan granite in uniaxial compression are 43 GPa and 0.25, respectively. The randomly generated numerical samples were discretized into 20,000 square elements. The size of the modeled sample was kept the same for all numerical simulations in the present paper. A suite of unconfined and confined compressive tests on the modeled rock samples were performed at constant room temperature and constant confining pressures of 0, 1, and 5 MPa. The physico-mechanical input parameters of the individual elements at a mesoscale used in the simulations were listed in **Table 1**.

The numerical and experimental stress-strain curves for the granite simulations are plotted in **Figure 3**. It can be seen from **Figure 3** that the simulated stress-strain curves are in good agreement with the experimental stress-strain curves. It needs to be noted that the nonlinear behavior at the early beginning of the experimental stress-strain curves is a result of the closure of preexisting compliant microcracks. This nonlinearity is not replicated in the present model because the stress-strain behavior of an element is considered linear elastic until the given damage threshold is attained. We highlight that the model input parameters were the same for each of the simulations in **Figure 3**, adding confidence that the model is capable of accurately capturing the short-term mechanical behavior of Beishan granite under different confining pressures.

3.2. Single-step brittle creep experiments

The numerical model was validated via a successful attempt to replicate published uniaxial and triaxial experimental data for Beishan granite. We will now run a suite of conventional brittle creep simulations on Beishan granite under different constant temperatures of 23, 50, and 90°C in order to find the required thermo-physico-mechanical properties of the granite. These thermophysical parameters were determined from the experimental data, and material

Input parameters	Granite
Homogeneity index	5
Mean elastic modulus (GPa)	43
Mean uniaxial compressive strength (MPa)	350
Poisson's ratio	0.25
Frictional angle (°)	30
Ratio of uniaxial compressive to tensile strength	10
Axial stress (MPa)	150
Specific heat capacity (J/(kg K))	900
Coefficient of linear thermal expansion (1/K)	4.6×10^{-6}
Thermal conductivity (W/(m K))	3.48
A	6.8×10^{-11}
n	1.75
m	0.39
U	3000

Table 1. Thermo-physico-mechanical parameters of the numerical model.

constants A, m, and n for the numerical model are listed in **Table 1**. It is noted that the same physico-mechanical input parameters listed in **Table 1** were used for the simulations. The temperatures of 23, 50, and 90°C represent room temperature, the in situ rock temperature, and the maximum temperature on the canister surface, respectively.

The numerically simulated creep curves together with the experimental creep curves are plotted in **Figure 4**. It can be seen that the simulated creep curves agree well with the experimental curves. The numerical creep curves clearly capture the phenomenology of brittle creep: the strain rate first decelerates followed by an acceleration in strain rate prior to final macroscopic failure.

Figure 5 presents the snapshots of the damage evolution of the numerical specimens deformed at constant temperatures of 23, 50, and 90°C. **Figure 5** shows when and where damage and

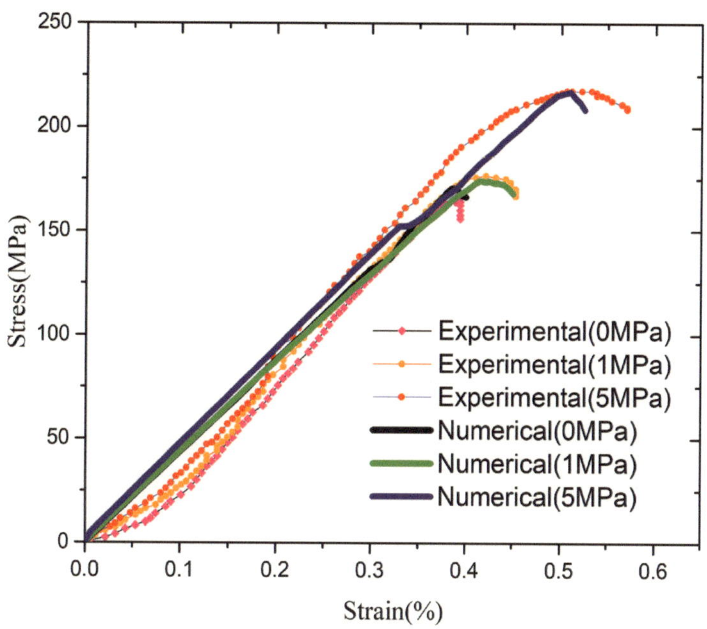

Figure 3. Comparison between numerical and experimental stress-strain curves for granite.

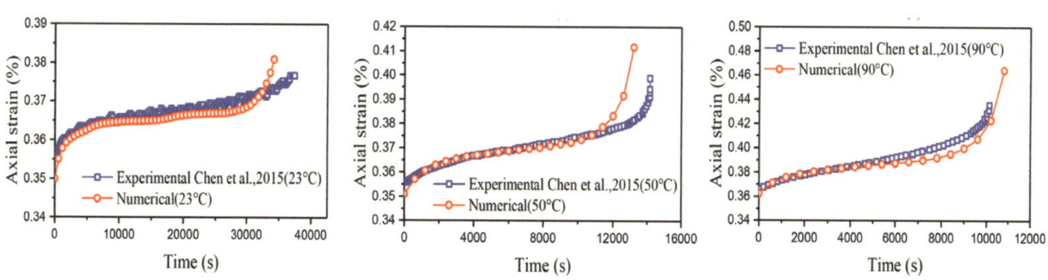

Figure 4. Comparisons between numerical and experimental creep curves under uniaxial conditions (σ_x = 0 MPa) at constant temperatures of 23, 50, and 90°C.

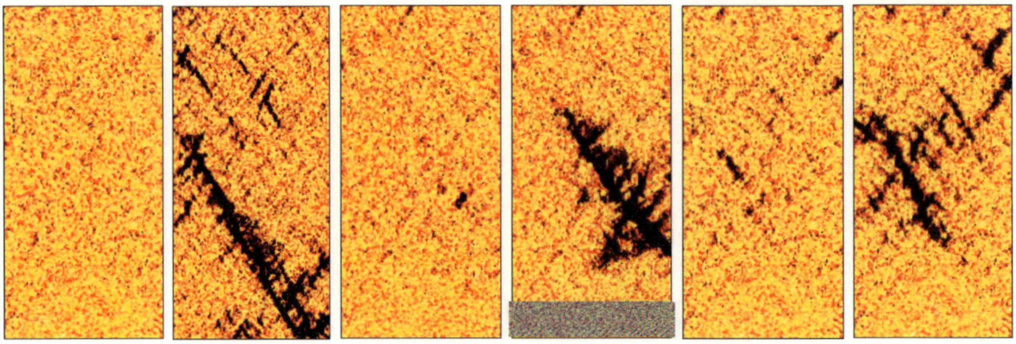

Figure 5. Snapshots of the simulated failure process during the brittle creep.

failure occur in the numerical specimen. The value of Young's modulus is randomly distributed since the numerical rock samples are heterogeneous. Therefore, the elements with a low Young's modulus act as nucleation sites for damage. As time goes on, these damaged elements grow and form localized damage zones. The localized damaged zones alter the stress field in their surrounding region, and these alterations further trigger the dynamic extension of the damage zone. Eventually, a thoroughgoing macroscopic fracture forms that signals the macroscopic failure of the sample.

3.3. Multistep brittle creep experiments

Chen et al. [18] also performed a series of multistep creep experiments to investigate the influence of temperature and stress on the time-dependent behavior of Beishan granite. Multistep creep tests were performed at confining pressures of 0, 1, and 5 MPa and at temperatures of 23°C (room temperature) and 90°C (the maximum temperature on the canister surface according to the current disposal conceptual design in China), respectively. The stress applied on the sample during the experiments was increased stepwise to predetermined percentages of the average peak stress (20, 40, 60, and 80%). The sample was kept at each level of stress for 1 week. We also performed numerical multistep creep simulations under the same conditions to further validate our model. It is noted that the numerical multistep creep simulations were all performed using the determined physico-mechanical and thermophysical parameters listed in **Table 1**. During the simulations, the rock samples are fixed in the vertical direction but can move freely in the horizontal direction.

The numerically simulated multistep creep curves together with the experimental multistep creep curves are plotted in **Figure 6**. It can be seen from **Figure 6** that the simulated multistep creep curves are in good agreement with the experimental curves. In detail, the model captures the influence of both confining pressure and temperature on the mechanical behavior, and the strain at the different stress steps and the time to failure are very similar between the experiment and the model (**Figure 6**).

We therefore conclude that, based on the above validations, our model can be used to investigate time-dependent creep of low-porosity granite at different temperatures. We will now use

http://dx.doi.org/10.5772/intechopen.72326

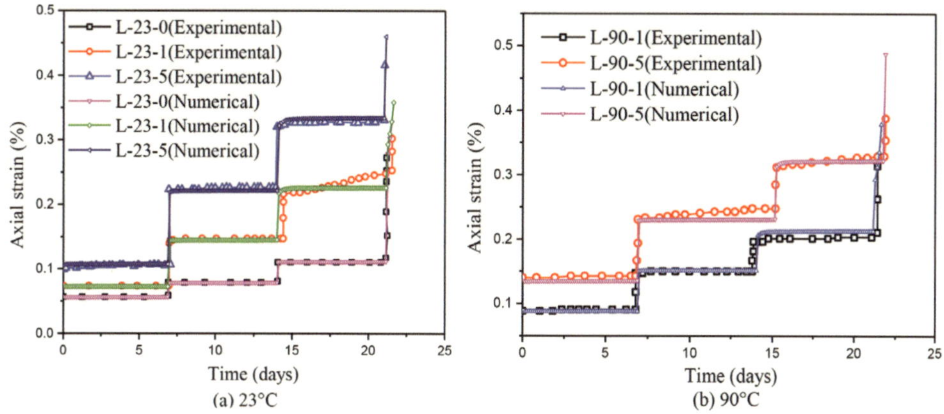

Figure 6. Multistep creep curves for simulations performed under different confining pressures (σ_x = 0, 1, and 5 MPa) and temperatures (23 and 90°C).

our model to further explore the influence of temperature, differential stress, confining pressure, and sample heterogeneity on brittle creep in low-porosity granite.

4. Numerical simulations

4.1. Model setup

In this section, the influence of temperature, differential stress, confining pressure, and sample heterogeneity on brittle creep of granite was investigated with the proposed model. The numerical samples as shown in **Figure 7** are the same as the samples modeled in the validation described above. The modeled samples were discretized into 20,000 square elements. We applied various axial stresses of 140, 145, 150, 155, and 160 MPa and various constant temperatures 23, 40, 50, 75, and 90°C on numerical samples with a inhomogeneity index 4, 5, and 6, respectively, to study the influence of temperature, differential stress, and sample heterogeneity on brittle creep of granite. The loading conditions are also shown in **Figure 7**. The relevant model parameters in the simulations are the same as the parameters listed in **Table 1**.

4.2. Effect of temperature on brittle creep

In addition to the validations above, two additional simulations of brittle creep were conducted under uniaxial compressive stress of 150 MPa at constant temperatures of 40 and 75°C. The creep curves and creep strain rate curves for various constant temperatures are shown in **Figure 8**, and the curves clearly show the accelerating-decelerating phenomenology of brittle creep observed in laboratory tests. We can see that there is a clear temperature effect on brittle creep in granite from these simulations.

It is noted that the creep strain rate strongly depends on temperature as observed in brittle creep experiments. The evolution of creep strain rate at various constant temperatures is

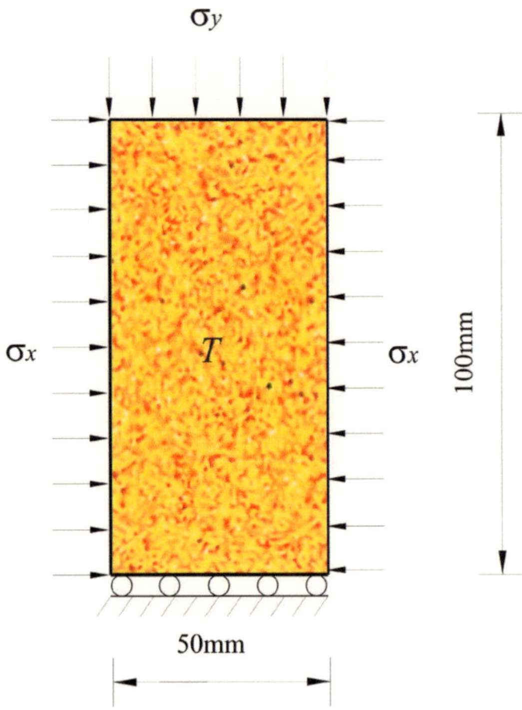

Figure 7. Numerical model and loading conditions for the simulated creep experiments.

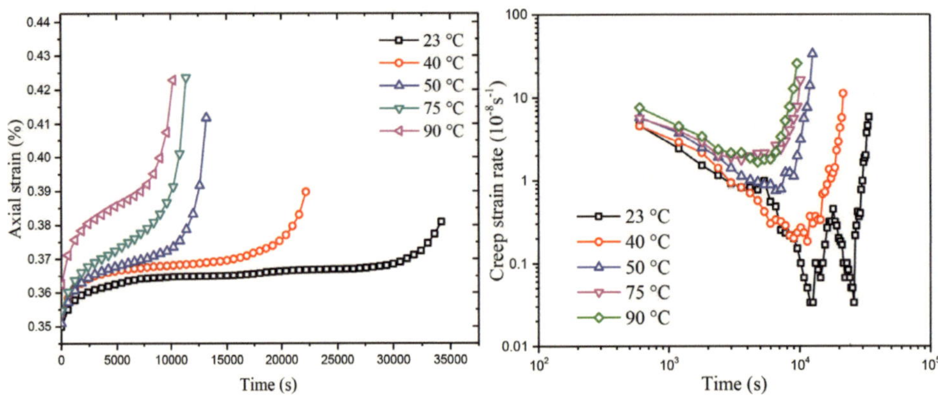

Figure 8. Creep curves and creep strain rate curves for simulations performed under uniaxial conditions at constant temperatures.

shown in **Figure 8**. The strain rate first decreases, reaches the minimum creep strain rate, and finally increases as the sample approaches macroscopic failure. The simulations show that there are several orders of magnitude difference in the minimum creep strain rate between a lower temperature (23°C) and a higher temperature (90°C). A large increase in strain rate at a higher temperature results in a large decrease in the time to failure (**Figure 9**) as observed in laboratory experiments on granite [8].

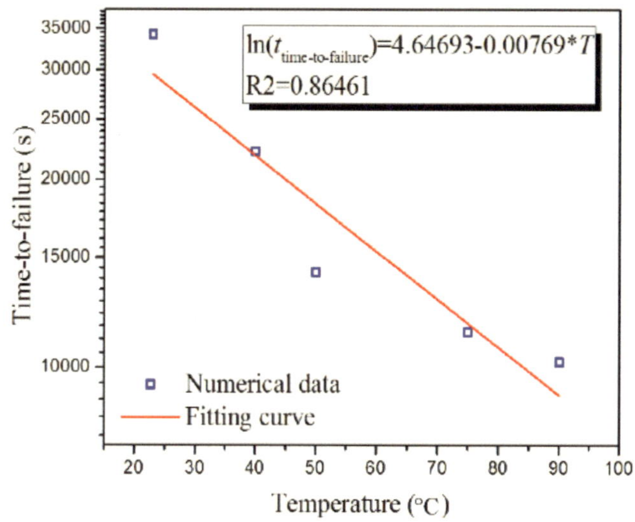

Figure 9. Time-to-failure vs. temperature curve for simulations performed under uniaxial conditions at constant temperatures of 23, 40, 50, 75, and 90°C.

4.3. Effect of differential stress on brittle creep

Here, the results of single-step brittle creep experiments under different constant-applied differential stresses were presented to study the effect of differential stress on brittle creep in granite. Uniaxial creep tests at a constant temperature of 50°C and constant axial stresses of 140, 145, 150, 155, and 160 MPa are numerically performed, and the numerically obtained creep curves for the five simulations are shown in **Figure 10**. Likewise, **Figure 10** clearly shows the accelerating-decelerating phenomenology of brittle creep seen in laboratory tests. The simulations indicate that the differential stress has a great influence on brittle creep in granite, as observed in brittle creep experiments [14]. First, the creep strain rate is higher when the differential stress is higher. A change in differential stress from 140 to 160 MPa leads to an increase in the minimum strain rate by over an order of magnitude. Similar observations have

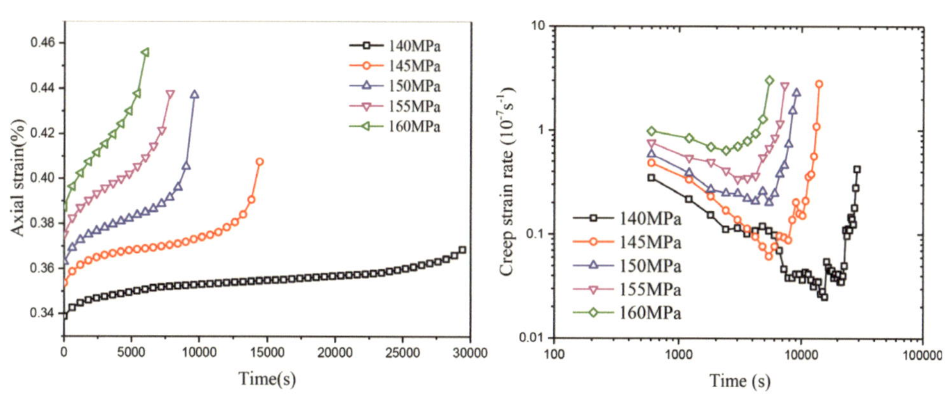

Figure 10. Creep curves and creep strain rate curves for simulations performed under uniaxial conditions at constant differential stresses between 140 and 160 MPa.

been experimentally obtained for many rock types [14]. As a result of the higher strain rate at higher differential stress, the time-to-failure is reduced as differential stress is increased (**Figure 11**).

Moreover, brittle creep tests under different constant confining pressures of 0, 2, and 10 MPa, but the same constant temperature and applied axial stress of 50°C and 150 MPa, respectively, were also performed, and the brittle creep curves were presented in **Figure 12**. The simulations capture the decelerating-accelerating phenomenology in laboratory experiments and reveal that confining pressure plays a great influence on brittle creep in granite. The creep strain rate reduces with an increase in the confining pressure (**Figure 12**): an increase in confining

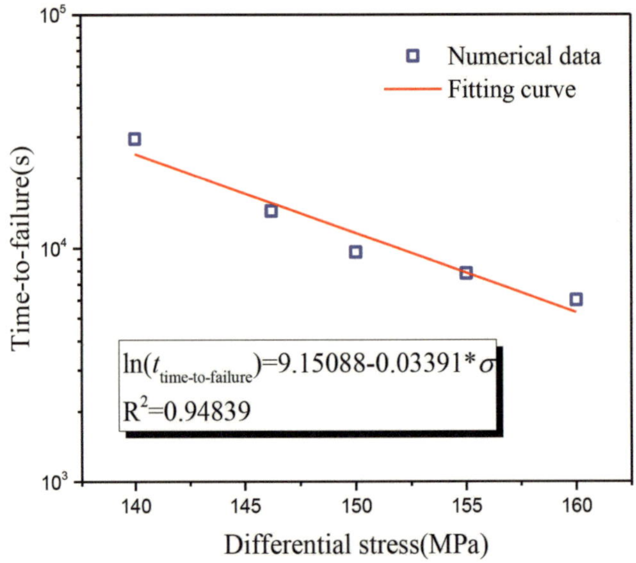

Figure 11. Time-to-failure vs. temperature curve for simulations performed under uniaxial conditions at constant differential stresses between 140 and 160 MPa.

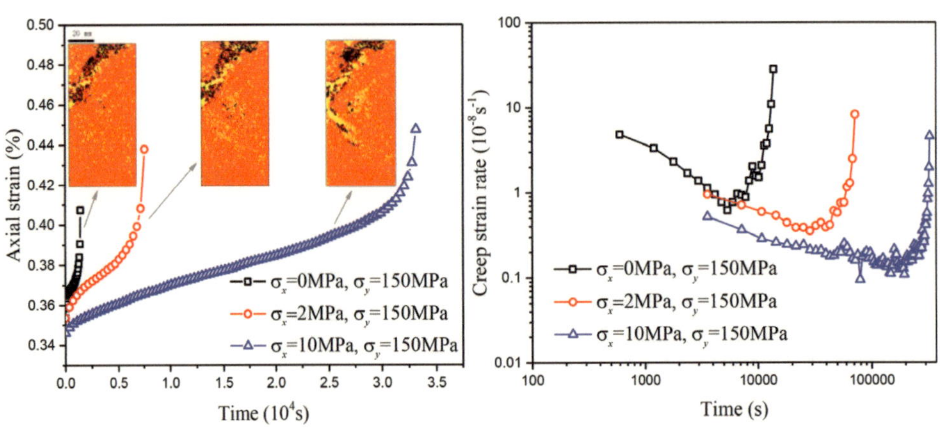

Figure 12. Creep and creep strain rate curves for simulations performed under a constant differential stress of 150 MPa and confining pressures of 0, 2, and 10 MPa.

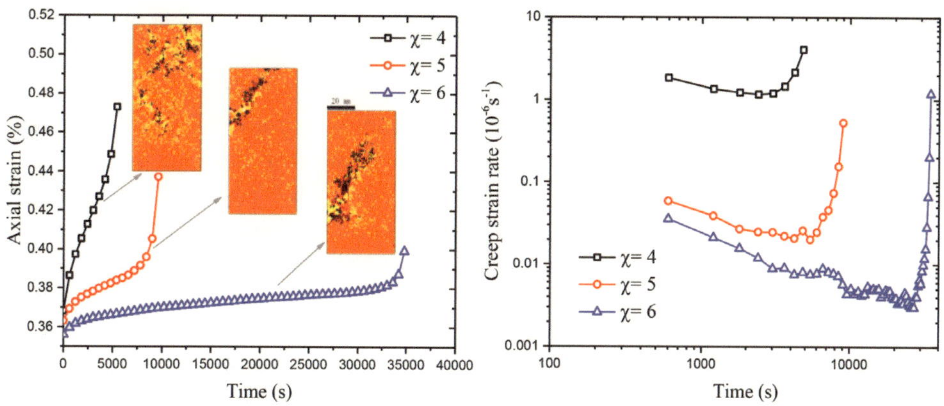

Figure 13. Creep curves and creep strain rate curves for simulations performed on granite samples characterized by a different homogeneity index.

pressure from 0 to 10 MPa leads to a reduction in the minimum strain rate by about an order of magnitude; the results are consistent with those from brittle creep experiments at different confining pressures. **Figure 12** also shows snapshots of each of the failed samples. It can be seen that more localized shear damage occurred in rock samples at higher confining pressures.

4.4. Effect of material heterogeneity on brittle creep

As we know, rock is heterogeneous, and thus we use a statistical Weibull distribution to reproduce material heterogeneity in rock. A set of simulations were performed to study the effect of material heterogeneity on brittle creep in granite with different homogeneity indices but at a temperature of 50°C, confining pressure of 0 MPa, and constant-applied axial stress of 150 MPa. The mean Young's modulus and mean UCS of the elements were kept the same (43 GPa and 350 MPa, respectively), and only the homogeneity indices were changed in these simulations. As described above, a larger homogeneity index implies that the elements within the sample are closer to the mean Young's modulus and the mean UCS. Therefore, a sample with a larger homogeneity index indicates that there are fewer low-strength elements in rock sample, and it will become stronger and more brittle.

The simulated creep curves and the evolution of creep strain rate presented in **Figure 13** show that an increase in material homogeneity leads to a decrease in creep strain rate and an increase in time to failure. For example, the minimum creep strain rate decreases by about an order of magnitude, and the time to failure increases from about 111,600 to 358,200 s when the homogeneity index increases from 4 to 6. On the contrary, a decrease in material homogeneity results in an increase in the creep strain rate and a decrease in the time to failure.

5. Concluding remarks

A numerical time-dependent thermomechanical model was proposed to simulate brittle creep in granite under different loading conditions and different temperatures. The mechanical parameters such as uniaxial compressive strength and Young's modulus of the mesoscopic material elements assumed to be homogeneous and isotropic are randomly assigned with a Weibull statistic distribution to reflect the material heterogeneity at a mesoscale. It is noted that the model can well capture the cooperative interaction between microcracks in the transition from distributed to localized damage. The model is validated against published experimental data, and then conventional brittle creep experiments at various constant temperatures, applied differential stresses, confining pressures, and sample homogeneities were simulated. The simulations accurately capture the short- and long-term mechanical behavior of the experimental brittle creep tests using unique thermo-physico-mechanical properties. The simulations further show that an increase in temperature and differential stress leads to an increase in the creep strain rate and therefore a decrease in time to failure and an increase in confining pressure and sample homogeneity leads to a decrease in creep strain rate and an increase in time to failure. Thus, the model proposed in the present paper will help the management and optimization of rock engineering projects in granite.

Acknowledgements

The supports provided by the National Basic Research Program (973) of China (Grant no. 2014CB047100), Natural Science Foundation of China (Grant nos. 41,672,301 and 51,474,051), and Fundamental Research Funds for the Central Universities of China (N150102002) are highly acknowledged.

Author details

Tao Xu* and Guang-lei Zhou

*Address all correspondence to: neuxutao@gmail.com

Center for Rock Instability and Seismicity Research, Northeastern University, Shenyang, China

References

[1] Carter NL, Hansen FD. Creep of rocksalt. Tectonophysics. 1983;**92**(4):275-333

[2] Heap MJ et al. Time-dependent brittle creep in Darley Dale sandstone. Journal of Geophysical Research. 2009;**114**(B7):1-22

[3] Brantut N et al. Mechanisms of time-dependent deformation in porous limestone. Journal of Geophysical Research Solid Earth. 2014;**119**(7):5444-5436

[4] Heap MJ, Baud P, Meredith PG. Influence of temperature on brittle creep in sandstones. Geophysical Research Letters. 2009;**36**(19):L19305

[5] Lin QX et al. Time-dependent strength degradation of granite. International Journal of Rock Mechanics and Mining Sciences. 2009;**46**(7):1103-1114

[6] Lockner D. Room temperature creep in saturated granite. Journal of Geophysical Research. 1993;**98**(B1):475-487

[7] Kranz RL. The effects of confining pressure and stress difference on static fatigue of granite. Journal of Geophysical Research. 1980;**85**(B4):1854-1866

[8] Kranz RL, Harris WJ, Carter NL. Static fatigue of granite at 200°C. Geophysical Research Letters. 1982;**9**(1):1-4

[9] Wang C et al. Experimental characterisation of thermo-mechanical coupling properties of Beishan granite. European Journal of Environmental and Civil Engineering. 2015;**19**(sup1): s29-s42

[10] Chen L et al. Experimental investigation on the creep behaviour of Beishan granite under different temperature and stress conditions. European Journal of Environmental and Civil Engineering. 2015;**19**(sup1):s43-s53

[11] Heap MJ et al. Brittle creep in basalt and its application to time-dependent volcano deformation. Earth and Planetary Science Letters. 2011;**307**:71-82

[12] Wang J et al. Deep geological disposal of high-level radioactive wastes in China. Chinese Journal of Rock Mechanics and Engineering. 2006;**4**:000

[13] Kranz RL, Scholz CH. Critical dilatant volume of rocks at the onset of tertiary creep. Journal of Geophysical Research. 1977;**82**(30):4893-4898

[14] Brantut N et al. Time-dependent cracking and brittle creep in crustal rocks: A review. Journal of Structural Geology. 2013;**52**(1):17-43

[15] Baud P, Meredith PG. Damage accumulation during triaxial creep of darley dale sandstone from pore volumometry and acoustic emission. International Journal of Rock Mechanics and Mining Sciences. 1997; **34**(3–4):24.e1–24.24.e10

[16] Atkinson BK. Subcritical crack growth in geological materials. Journal of Geophysical Research. 1984;**89**(B6):4077-4114

[17] Meredith PG, Atkinson BK. Stress corrosion and acoustic emission during tensile crack propagation in Whin sill dolerite and other basic rocks. Geophysical Journal International. 1983;**75**:1-21

[18] Chen L et al. Effects of temperature and stress on the time-dependent behavior of Beishan granite. International Journal of Rock Mechanics and Mining Sciences. 2017;**93**:316-323

[19] Xu T et al. The influence of temperature on time-dependent deformation and failure in granite: A mesoscale modeling approach. Rock Mechanics & Rock Engineering. 2017;**50**(9): 2345-2364

[20] Xu T et al. The influence of temperature on time-dependent deformation and failure in granite: A mesoscale modeling Approach. Rock Mechanics & Rock Engineering. 2017:1-20

[21] Chen L et al. Study of acoustic emission characteristics of beishan deep granite under different stress conditions. Chinese Journal of Rock Mechanics and Engineering. 2012;**31**(S2): 3618-2624

Chapter 10

Review on Creep Analysis and Solved Problems

Vahid Monfared

Additional information is available at the end of the chapter

http://dx.doi.org/10.5772/intechopen.71184

Abstract

This chapter presents a useful literature reviews and applied solved problems that focus on the creep phenomenon and behavior of it in the solids. Various insights and available studies are reviewed and investigated regarding the creep behavior analysis in three categories such as analytical, numerical and experimental methods. In addition, novel and recent findings are presented in this chapter such as predicting and obtaining the viscosity of the solids at high temperatures using steady state creep phenomenon (i.e., introducing a simulation and analogy between creeping solids and viscous fluids).

Keywords: creep, analytical, numerical, experimental methods, solved problems

1. Introduction

Creep is a slow, continuous deformation of a material under constant stress and temperature. On the other hand, creep phenomenon in solids under high stress and temperature is one of the important topics in the scientific societies, and therefore, the creep analysis become more significant in various industries. Therefore, study on the process of the creep phenomena is essential and significant for engineering applications concerning high temperature and high stress. For instance, spaceships, turbine blades and discs are commonly under the creep effects. Additionally, predicting the creep behavior is very important for designing the advanced reinforced/nonreinforced materials. Consequently, a logical analysis of the creep behavior and its mechanisms for these materials is crucial. Creep phenomenon may be happened in short fiber composites or nanocomposites because of any elevated temperatures and applied loads which can be dangerous for composites and structures. Therefore, a thorough knowledge of creep characteristics and deformation mechanisms of reinforced and nonreinforced materials is required to utilize these composites in high stress and high temperature applications. In recent years, extensive investigations have been conducted to predict the steady state creep behaviors of the materials (see **Figure 1**).

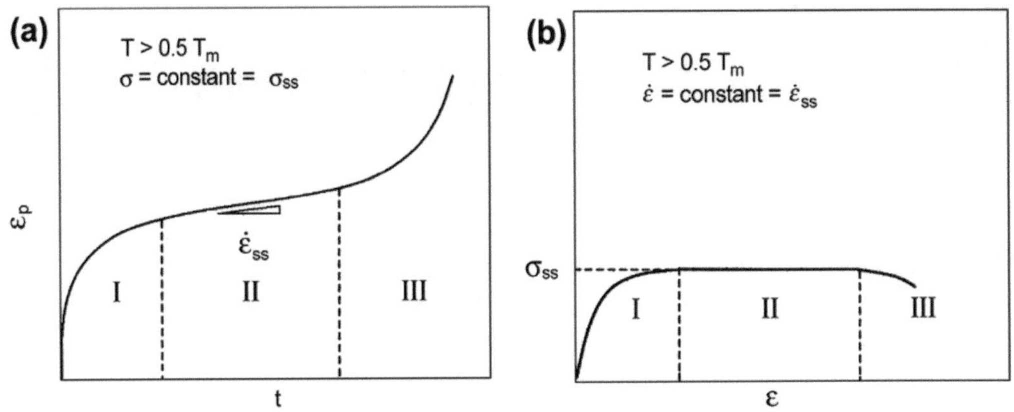

Figure 1. (a) Creep $\varepsilon - t$ diagram (b) creep $\sigma - \varepsilon$ diagram [105].

These advanced materials are generally used in various industries such as aircraft and aerospace industries and other engineering applications because of numerous advantages. One of the hazardous phenomena is creep occurrence which may occur in these materials. Consequently, the creep must be accurately studied and analyzed to prevent unsafe and undesired events. Several investigations have been performed to predict the creep behaviors of the advanced reinforced/nonreinforced materials. In this chapter, in addition to presenting some insights, the related previous studies are reviewed in three categories such as analytical, numerical and experimental methods.

2. Analytical methods

In reinforced materials, various theoretical studies are based on the shear-lag theory [1–13]. For more illustrations, 1D shear-lag theory was initially proposed by Cox [1], which is a strong model for the stress transfer analysis of the unidirectional fibrous composites. Recently, Mondali et al. [12] introduced a theoretical approach to predict the second-stage creep behavior of the short fiber composites by using shear-lag model and imaginary fiber method. The creep behavior of the creeping matrix was predicted using a creep exponential law. Moreover, various investigators have studied fibrous composites by employing imaginary fiber method and technique (fictitious fiber) elastically [14–18].

For instance, one of the useful and valuable researches was done by Weng and Sun [14]. They employed the fictitious fiber technique and method to study the effects of the fiber length on elastic moduli of randomly oriented chopped-fiber composites theoretically. The method and technique is analogous to the imaginary fiber technique to analyze the short fiber composites. The mentioned solution method based on fictitious fiber technique [14] is similar to Hsueh's elastic solution [15], known as the imaginary fiber technique. The scientific application of the fictitious and imaginary fiber method is very significant for analyzing the matrix located at the top of the fiber. In addition, this model may assist to the analysis of a full and complete fibrous composite.

Stress transfer from the matrix to the embedded fiber in fiber-reinforced composites has been studied elastically while the loading direction is parallel to the fiber axis and the fiber is bonded to the matrix. Stress transfer occurs both at the interface along the fiber length and at the ends of the fiber. Nevertheless, the boundary condition at the bonded ends is unclear, and different suppositions were made analytically to achieve solutions for this stress transfer problem. To this end, a novel insight of supposing imaginary fibers in the fibrous composite was introduced. In comparison to the prior theoretical solutions, this method [15] bears more physical meaning and is in better agreement with numerical (FEM) and experimental results [15].

In addition, many investigations were carried out concerning the creep of the fibrous composites and applications by different methods theoretically [19–24]. Some investigators have employed different methods in place of the shear-lag model to solve the problems like Eshelby's thought experiment with a formulation based on the Schapery model [19] and variational approach [21]. Also, Monfared et al. [22] proposed the second-stage creep strain rate of the short fiber composites without utilizing the shear-lag theory. They determined various key unknowns in the steady state creep of the short fiber composites using mapping function and dimensionless parameter techniques.

Furthermore, novel theoretical insight and formulation have been proposed for studying the second-stage creep in fibrous composites utilizing complex variable method, at which, both the matrix and fiber creep at low stresses and temperatures. To study the creeping fiber, a plane stress model has been employed. Significant novelties of the research work of Monfared et al. [23] are determination of the displacement rates with suitable boundary conditions in the creeping fibers and also utilizing the complex variable method in the creep analyzing. It is important that the model and method may be helpful to investigate the creep behavior in polymeric matrix composites. Furthermore, another major application of the method is to theoretically analyze the creep or elastic behavior of carbon nanotube polymer composites. Here, as an example, a result presented in Ref. [23] is introduced as the following:

$$\dot{u}_1(z,\bar{z}) = \varphi(z) + \Psi(z - \lambda\bar{z}) \tag{1}$$

$$\dot{u}_2(z,\bar{z}) = \mu\left[\oint_c 2\aleph\left(\frac{5z-\lambda\bar{z}}{4}\right)dz - \oint_c \lambda\aleph\left(\frac{5z-\lambda\bar{z}}{4}\right)d\bar{z}\right] + \rho\left(\frac{z}{2}+\frac{5}{6}\bar{z}\right) \tag{2}$$

λ is equal to $\frac{5}{3}$ obtained by solving the partial equations, μ is equal to $\frac{1}{3}$ and c is closed finite region in steady state creep. Using the residual theorems in complex variable analysis, gives

$$\oint_c \aleph(z,\bar{z})d(z,\bar{z}) = 2\pi i Res\, \aleph(z,\bar{z})\Big|_{z,\bar{z}=z_0} \tag{3}$$

In above equation, *Res* indicates the residual value of $\aleph(z,\bar{z})$. Moreover, two analytical functions ρ *and* $\aleph$ are introduced through Cauchy-Goursat theorem. Therefore, $\dot{u}_1$, $\dot{u}_2$, U and $\overline{U}$ may be obtained utilizing complex variable in the second-stage creep of the fibrous composites. All the mentioned theoretical and analytical functions like $\varphi(z)$ can be supposed by a polynomial function of degree n with its unknown coefficients. The unknowns may be obtained employing the appropriate boundary conditions. As the major benefit of the method and formulation of [23], Eq. (3) will be able to analyze and solve the improper integrals [23]. Moreover, displacement rates $\dot{w}_1$ and $\dot{w}_2$ are obtained utilizing different boundary conditions in the crept matrix, distinctive geometric relations and same procedures. That is, $\dot{w}_1 = \beth(z) + \Lambda(z - \overline{z}) = \Upsilon(z - \overline{z})$ and $\dot{w}_2 = \left[\oint_c 2\mathcal{L}(z - \overline{z})dz - \oint_c \mathrm{T}\mathcal{L}(z - \overline{z})d\overline{z}\right]\mathcal{B} + \mho(z + \overline{z}) = \mathcal{G}(\overline{z} + z)$. Using the boundary conditions, displacement rates $\dot{u}_1, \dot{u}_2$ and $\dot{w}_1, \dot{w}_2$ are coupled simultaneously. In above equations, (u_1, u_2) and (w_1, w_2) are displacement fields in the fiber and matrix, respectively. Also, subscripts 1 and 2 are related to x_1 and x_2 directions [23].

3. Numerical methods

Numerical attempts have been made for analyzing the creep problems by various investigators. Numerical methods have helped the researchers for analyzing the problems [25–33]. The purpose of the paper [30] is to obtain the general creep mechanisms for the fibrous composites (MMCs) subjected to axial stress states and to build a relation between the macroscopic stable creep behavior and the material micro-geometric parameters numerically. The unit cell models have been employed to compute the macroscopic creep behavior with different micro-geometric parameters of the fibers on different loading directions. The influence of the geometric parameters of the fibers and loading directions on the macroscopic creep behavior was determined previously. The matrix and fiber interface were considered by a third layer, matrix and fiber interlayer, in the unit cells with different creep properties and thickness. Based on the numerical-obtained results of the unit cell models, a statistic model was introduced for the plane randomly distributed-fiber MMCs. The fiber breakage was taken into account in the statistic model because it starts early in the creep life experimentally. With the distribution of the geometric parameters of the fibers, the obtained results of the statistic model agree well with the experiments. With the statistic model, the influence of the geometric parameters and the breakage of the fibers in addition to the properties and thickness of the interlayer on the macroscopic steady state creep rate were discussed [30].

FEM investigation for steady state creep behavior of the creeping metal matrix composite was done with supposing the fiber-matrix debonding parameter in the modeling by Mondali et al. [31]. Accuracy of simulation and being time consuming are several major difficulties of FEM generally. In addition, one of the significant sections of reference [33] is to predict the limited creep debonding at the interface in the second-stage creep of the fibrous composites subjected to axial stress. At this point, a key formulation, presented in Ref. [33], is generally introduced as the following:

The average axial stress of fiber is semi-analytically obtained as a function of interfacial shear stress which is presented as follows:

$$\overline{\sigma}^f_{zz} = \Gamma \times \tau_i \times \left(\mathbb{Z} l^\delta D^{-\delta} - \mathbb{C} z^\delta D^{-\delta}\right) \tag{4}$$

where parameters Γ, $\mathbb{Z}$ and $\mathbb{C}$ are constant coefficients and also τ_i and $\overline{\sigma}^f_{zz}$ are the interfacial shear stress and the average axial stress of the elastic fiber, respectively. In addition, the parameters of

$l^\delta D^{-\delta}$ and $z^\delta D^{-\delta}$ are dimensionless parameters, which are employed to determine the quasi shear lag formulation, that is, Eq. (4). Recent formulation may be obtained using linear combination of the elastic fiber diameter ($D = 2a$), axial position (z), dimensionless parameters (lD^{-1}, zD^{-1}) and the fiber length (l) considering semi-theoretical approaches. After determining the unknown coefficients, QSL formulation will be determined as below:

$$\overline{\sigma}^f_{zz} \cong \tau_i|^{r=a}_{z=l} \times \left(\frac{54l - 35z}{10D}\right) \tag{5}$$

where $\tau_i|^{r=a}_{z=l}$ is the maximum value of shear stress along the fiber length at the interface (i.e., at $z = l, r = a$). Note that the formulation of interfacial shear stress, i.e. τ_i, is proposed by Monfared et al. [22], which has been used for determining average axial stress in the fiber $\overline{\sigma}^f_{zz}$ as follows [22]:

$$\tau_i(z) = \frac{\eta}{\sqrt{3}} \ln\left(1 + \frac{\sqrt{3}C}{\eta} z\right) \tag{6}$$

In which,

$$C = \exp\left(\frac{\sqrt{3}}{\eta}\tau_i\right)\frac{d\tau_i}{dz} \tag{7}$$

Other relation for obtaining a shear stress at the interface has the following form [22]:

$$\tau_i = \tau^{\mathrm{m}}_{\mathrm{rz}}\big|_{r=a} = \underset{(\mathrm{r,z})\to(\mu,\aleph)}{limit}\left[\frac{\eta \ln\left(\frac{\dot{\varepsilon}_{eq}}{\lambda}\right)\dot{\gamma}_{\mathrm{rz}}}{3\dot{\varepsilon}_{eq}}\right]_{r=a} \tag{8}$$

In which,

$$\dot{\gamma}_{rz} = \frac{\partial \dot{u}}{\partial z} + \frac{\partial \dot{w}}{\partial r} = 2\dot{\varepsilon}_{rz} \tag{9}$$

where $\mu = [\mathrm{a,b}]$, $\aleph = (0, l]$, which imply $a \le \mu \le b$ and $0 < \aleph \le l$. Also, $\dot{\gamma}_{\mathrm{rz}}$ is the shear strain rate in the direction indicated by its subscript and $\dot{u}$ and $\dot{w}$ are, respectively, the radial and axial displacement rates.

On the other hand, the average axial stress in the fiber is approximately linear in elastic state, that is, $\overline{\sigma}^f_{zz} \cong g_1 z + g_2$. Note that the unknown coefficients g_1 and g_2 are obtained employing mentioned calculations and engineering estimations. In general, average axial stress in the fiber is calculated using maximum shear stress at the interface and geometric factors. As mentioned earlier, a weight coefficient is considered for any parameter such as g_1 and g_2 with parameter z employing numerical methods, weighted calculus (computer coding) and meta-calculus with considering nonlinear regression and neural network approaches [33].

4. Experimental methods

Unlike several difficulties of the experimental methods such as complexity of them (they may be time-consuming, expensive and intricate), they are useful to predict the steady state creep behaviors [34–49]. For instance, the second-stage creep behavior of the composite "SiC/Al6061" was experimentally and analytically analyzed by Morimoto et al. [36], in which happening of the non-aligned fibers and creation of the microcracks in the creeping composite are some difficulties during the experimental process. For more illustration, the creep behavior of the TiC-particulate-reinforced Ti alloy composite was studied at temperatures from 500 to 650°C and applied stresses from 230 to 430 MPa [37].

The steady state creep behavior of poly (vinylidene fluoride) (PVDF)/multiwall carbon nanotubes nanocomposites was investigated at various stress levels and temperatures experimentally. To fine-tune the ability to transfer stress from matrix to carbon nanotubes, bud-branched nanotubes were fabricated. The PVDF showed improved creep resistance by adding carbon nanotubes. However, bud-branched nanotubes showed a modified stress- and temperature-dependent creep resistance in comparison with carbon nanotubes. Also, at low stress levels and low temperatures, bud-branched nanotubes showed better improvement of the creep resistance than that of virgin carbon nanotubes, whereas at high stress levels and high temperatures, the virgin carbon nanotubes presented better creep resistance than that of bud-branched nanotubes. Differential scanning calorimetry (DSC), WAXD and Fourier transform infrared spectroscopy (FTIR) were used to characterize the crystalline structures, and dynamic mechanical properties were characterized by dynamic mechanical analysis (DMA) testing. The Burger's model and the Findley power law were utilized to model the creep behavior, and both were found to well describe the creep behavior of PVDF and its nanocomposites. The relationship between the structures and properties was analyzed based on the parameters of the modeling. The improved creep resistance for PVDF by adding the nanotubes would be beneficial for its application in thermoset composite welding technology [47].

The high-temperature creep behaviors of 7075 and 2124 aluminum alloys have been analyzed by the constant-stress uniaxial tensile creep experiments. In addition, constitutive models for describing the high-temperature creep behaviors of the studied aluminum alloys were established based on the continuum damage mechanics (CDM). Initially, the continuum damage mechanics (CDM) models were simplified to explain the primary and secondary creep of 7075 aluminum alloy. Then, because the effects of the applied stress on the creep damage during tertiary stage were not entirely considered in the original CDM model, the modified constitutive models were presented to predict the creep-rupture behavior of 2124 aluminum alloy. A stress exponent D, which can preferably reveal the effects of the applied stress on the stationary creep damage, was presented in the modified model. The results of this investigation showed that a good agreement between the measured and predicted results was determined, which confirms that the established creep constitutive models can give an precise and exact approximation of the high-temperature creep behaviors for 7075 and 2124 aluminum alloys [48]. Briefly, the high-temperature creep behaviors of 7075 and 2124 Al alloys were investigated; a stress exponent was proposed to preferably reveal the effects of the stress on creep damage and the established CDM models correlate well with the experimental results [48].

In addition, the second-stage creep and recovery behaviors of polystyrene composites filled with 2D chemically diminished graphene oxide sheets have been investigated, in which a series of stationary creep and recovery characterization tests have been carried out for thermoplastic composites filled with three different types of carbon nanoadditives (including CB, CNT and graphene). Furthermore, the thermal and dynamic mechanical properties of the composites have been studied and analyzed to find and understand the relative formation mechanisms generally [49].

5. Complementary methods

Complementary methods may be a combination of the three mentioned methods of analytical, numerical and experimental approaches such as semi-analytical methods or other supplementary approaches. The mentioned methods (complementary methods) are a supplementary for the three mentioned methods (analytical, numerical and experimental methods).

A semi-analytical formulation has been presented for obtaining the viscosity of solids (such as metals) using the steady state creep model of the short fiber composites. For achieving this aim, fluid mechanics theory was used for determining the viscosity. Sometimes, obtaining the viscosity is experimentally difficult and intricate. Therefore, the present model may be beneficial to obtain the viscosity of the metals [50].

Another research work was presented for simple and fast estimation of the creep plastic behavior of the short fiber composites by a new approach based on neural network prediction. The method has been introduced to diminish the solution process. Furthermore, as an important application of the approach and the model mentioned in Ref. [51], spaceships and turbine blades are commonly under the stationary creep effects. Thus, analysis of the steady state creep is necessary and essential in various industries. Therefore, analysis of the second-stage creep behavior is mandatory for analysis of failure and fracture and creep resistance of the fibrous composites. One of the major applications of the research work [51] is the design of the fibrous composites with optical fibers and devices [51].

A new analytical approach was presented for analyzing the steady state creep in a short fiber composite SiC/Al6061 (Silicon Carbide/Aluminum 6061) under axial load based on high order displacement functions. The creep behavior of the matrix was described by an exponential law with elastic behavior of fibers. Predicting the creep behavior of the short fiber composite "Silicon Carbide/Aluminum 6061" using high order functions analytically was the novelty of the work. The research work was presented in order to prevent unwanted happenings, in addition to the control of the second-stage creep deformation rate. As a result, the creep behaviors are controllable, and the diagrams have smooth gradients and slopes [52]. Indentation creep behavior at room temperature and its mechanism of Ti-10 V-2Fe-3Al alloy with dual phase structure were investigated. Micro-indentation creep tests were performed under the maximum indenter load ranging from 1500 to 4500 mN and the holding time of 300 s. The effect of indenter load on creep behavior and creep parameters, like creep rate, creep strain rate, indentation stress and creep stress exponent, were analyzed at the steady state creep. The results revealed that creep parameters exhibited significantly indentation depth dependent. At the secondary stage of creep, creep strain rate and creep rate increased with the increase of maximum indenter load, whereas creep stress and creep stress exponent exhibited an opposite trend. Especially, creep stress exponent of 7.65 ± 1.25 in power-law creep behavior of Ti-10 V-2Fe-3Al alloy, which was consistent with dislocation process, indicated that the secondary stage of creep may be dominated by dislocation climb. Furthermore, experimental data and figures were used to evaluate and comprehend likely creep mechanisms during a micro-indentation scheme of Ti-10 V-2Fe-3Al alloy [53]. Briefly, micro-indentation creep behavior with various indenter loads was investigated; creep rate and indentation stress showed a linear relationship with indenter loads; high creep strain rate can be induced by increasing indenter load at holding time; indenter load effects on creep stress exponent are similar to that of hardness and the creep behavior at secondary stage may be dominated by dislocation climb [53].

Steady state creep was characterized for Ni-8YSZ solid oxide fuel/electrolysis cell (SOFC/SOEC) substrate material. Intrinsic and extrinsic factors affecting creep behaviors such as compositional ratio, porosity and mechanical loading configuration were assessed. Mechanical tests were supported by analytical and numerical calculations. The results indicated a diffusion-dominated creep mechanism under both compressive and tensile creep conditions. Creep appeared to be dominated by the ceramic phase. Porosity significantly reduced creep resistance. The activation energy was discussed based on loading configuration, temperature and porosity [54]. In brief, the creep of porous Ni-8YSZ anode substrates was investigated systematically; the porosity, composition ratio and loading configuration have effects on creep; the suitability of different testing methods was successfully analyzed and derived creep parameters can be used for materials' validation and modeling [54].

Creep deformation and fracture behaviors of Sanicro 25 alloy were obtained based on long-term creep strain tests. The multiscale precipitation behaviors were calculated thermodynamically and inspected by examination of the microstructure of the as-crept alloy [55]. Briefly, multiscale precipitation behaviors of Sanicro 25 during long-term creep were revealed; fine-distributed nanoscale precipitates were found as the main strengthening phases of Sanicro 25 at elevated temperatures and preferred grain orientation of Sanicro 25 before and after creep under different conditions were investigated by electron back-scattered diffraction [55]. *In situ* neutron diffraction measurements were performed on monocrystalline samples of the Ni-based superalloy CMSX-4 during N-type γ' raft formation under the tensile creep conditions of 1150°C/100 MPa and subsequently on a rafted sample under the low temperature/high stress creep conditions of 715°C/825 MPa.

During 1150°C/100 MPa creep, the γ' volume fraction decreased from ~70 to ~50%, the lattice parameter misfit was partly relieved and the load was transferred from the creeping γ matrix to the γ' precipitates. In the process of cooling back to room temperature, a fine distribution of γ' precipitates formed in the γ channels, and these precipitates were present in the 715°C/825 MPa creep regime. Under low temperature/high stress creep, the alloy with rafted γ'

microstructure exhibited superior creep strength to the cuboidal γ' microstructure produced following a standard heat treatment. A lengthy creep incubation period was observed, believed to be associated with {111} ⟨110⟩ dislocations hindering propagation of {111} ⟨112⟩ dislocations. Following the creep incubation period, extensive macroscopic creep strain accumulated during primary creep as the γ-phase yielded. Finally, the diffraction data suggest a loss of precipitate/matrix coherency in the (0 k0) interfaces as creep strain accumulated [56]. Bending creep deformation mechanism for nickel nanobeam was investigated using molecular dynamics simulation. Low temperature creep deformation ($T < 0.3T_m$) was found to be guided by jog formation and glide motion of grain boundary, whereas lattice diffusion, grain boundary migration and sliding are the controlling mechanism for high-temperature deformation ($T > 0.5T_m$). The occurrence of tertiary creep regime was observed only at high-temperature deformation due to creep instability caused by cavity formation. It was revealed through dislocation analysis that intrinsic Frank partial dislocations are the driving factor for cavity generation leading to intergranular fracture [57]. Briefly, it is for the first time, the study of bending creep deformation and the underlying mechanism at nanoscale level; jog formation helps the grain boundary movement during low temperature bending creep deformation; intrinsic Frank dislocations cause cavity formation at tertiary regime for high-temperature bending creep [57].

The as-received cast and forged (C&F) P91 steel was subjected to the creep test at temperature of 620–650°C for applied stress of 120 MPa. The room temperature tensile test was conducted after normalizing and tempering (N&T) treatment of the ruptured creep specimen. The standard tensile test specimen was prepared from the gauge section of creep-ruptured specimen. The N&T treatment was performed to restore the microstructure and mechanical properties of virgin P91 steel (N&T P91 steel). The microstructure of creep-fractured specimen in ruptured state and N&T condition were characterized by using field-emission scanning electron microscope (FE-SEM) with energy dispersive X-ray spectroscopy (EDS). The fracture surface morphology of crept specimen and the tensile tested specimen was also studied. The effect of prior creep deformation on the mechanical strength was more significant in the sample with longer creep rupture life [58].

Briefly, creep test was performed in temperature range of 620–650°C and at stress of 120 MPa; the effect of creep rupture life on precipitate size and fraction area of precipitates; the effect of creep rupture life on fracture surface morphology of creep tested cast-forged P91 steel; the tensile properties of N&T creep fractured material compared with N&T cast-forged P91 steel and the fracture surface morphology of tensile fractured surface by using FESEM were studied [58] (see **Figures 2, 3**).

The creep behavior and microstructural evolution of 8030 alloy at 90–150°C and 50–90 MPa of applied tensile stress were investigated by creep testing and transmission electron microscopy. The 8030 alloy possesses excellent creep resistance at low temperatures. The sizes of a small number of subgrains increase during the creep process due to subgrain merging. A creep activation energy of 123.2 kJ/mol is close to that of the lattice self-diffusion in aluminum, implying that a lattice self-diffusion mechanism is dominant at 150°C/90 MPa (**Figures 4** and **5**), [59].

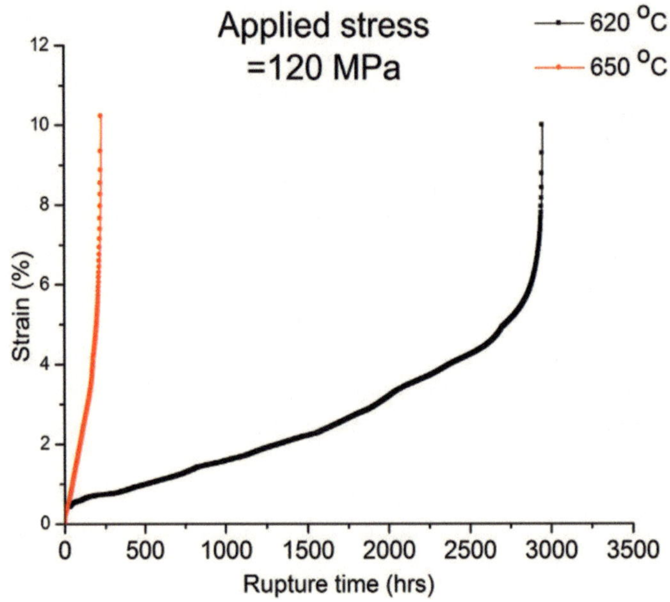

Figure 2. Creep rupture behavior of P91 steel [58].

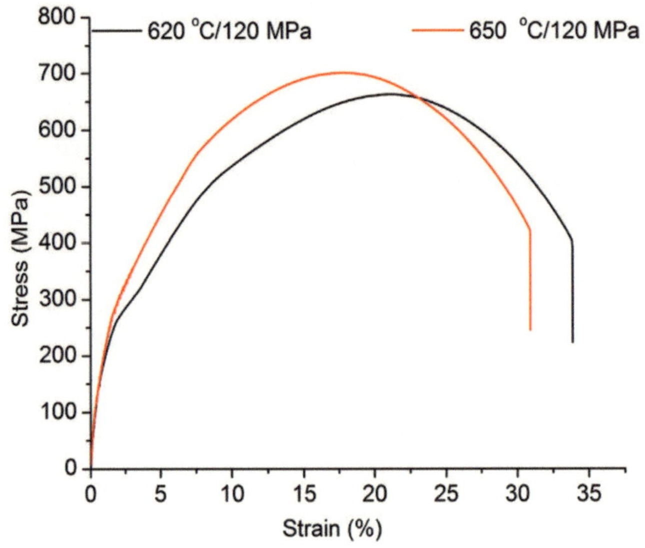

Figure 3. Engineering stress-strain curve [58].

Briefly, the 8030 alloy possesses excellent creep resistance at low temperatures; Al_3Fe-phase can improve the creep resistance of 8030 alloy; there are two kinds of creep mechanism at low temperature; the creep threshold stress is close to 0 MPa at 90 MPa/150°C [59].

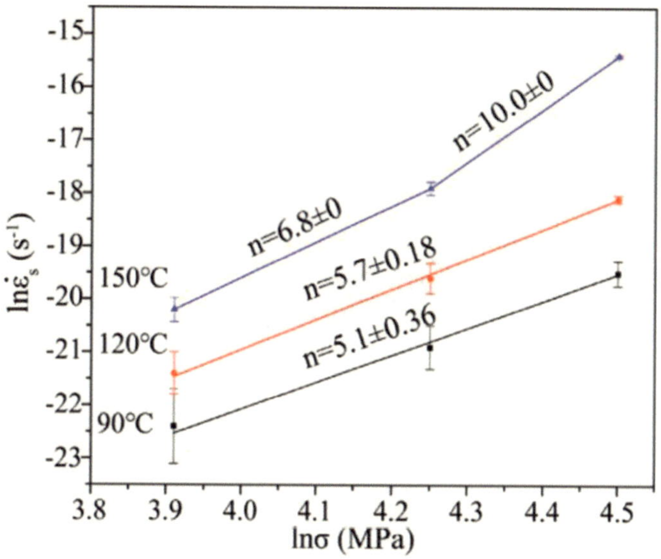

Figure 4. Relationship between ln $\dot{\varepsilon}_s$ and ln σ of 8030 alloys. (error bars: One standard deviation) [59].

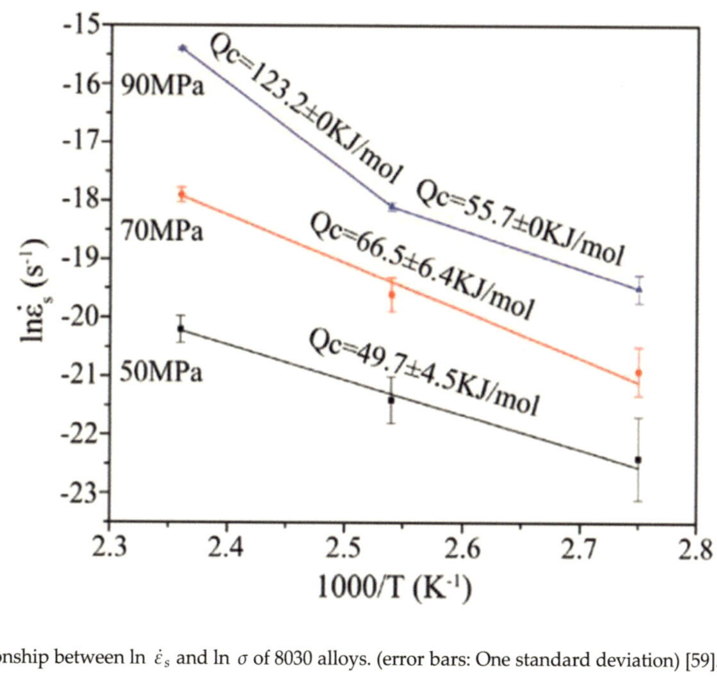

Figure 5. Relationship between ln $\dot{\varepsilon}_s$ and ln σ of 8030 alloys. (error bars: One standard deviation) [59].

Study of creep behavior of base metal (without weld) and welded specimens of P91B steel over a range of temperatures (600–650°C) and stresses (50–180 MPa) showed similar values of minimum creep rates for both specimens at higher stress regime (>100 MPa), whereas significantly higher creep rates in the case of welded specimens at lower stress regime. Considering that welded specimen is composed of two distinct structural regimes, that is, weld-affected zone and base metal, a method has been proposed for estimating the material parameters describing creep behavior of those regimes.

Stress-strain distribution across welded specimen predicted from finite element analysis based on material parameters revealed preferential accumulation of stress and creep strain at the interface between weld zone and base metal (**Figures 6** and **7**). This is in agreement with the experimental finding that creep rupture preferentially occurs at intercritical heat-affected zone in welded specimens owing to ferrite-martensite structure with coarse $Cr_{23}C_6$ particles [60]. Briefly, comparison of creep properties of welded and virgin specimens of P91B steel was performed; at lower stresses (<100 MPa), welded samples show higher minimum creep rate; creep rupture at intercritical heat-affected zone (IC-HAZ) in welded specimens; FEA showing accumulation of creep strain in weld/base metal interface and precipitate-free soft ferrite matrix accumulates strain and weakens IC-HAZ [60].

A group of segmented polyurethane copolymers with different hard segment (HS) contents were successfully synthesized. The microstructure of the PU copolymers was characterized via Fourier transform infrared spectroscopy (FTIR), differential scanning calorimetry (DSC), dynamic mechanical analysis (DMA) and small-angle neutron scattering (SANS) [61]. The creep behaviors of two alloys were studied under the temperature of 700°C and three applied stresses (560, 650 and 720 MPa) in order to investigate the creep mechanisms of Inconel718 and Allvac718Plus, using multiple microstructural analysis methods, including scanning electron microscope (SEM), field-emission scanning electron microscope (FE-SEM) and transmission electron microscopy (TEM).

Result showed that steady state region is not observed in these two alloys. Creep curves of two alloys were composed of primary region and tertiary region. Tertiary region occupies a dominant position. Ductile dimples were observed on creep fracture surface, indicating creep mechanisms of two alloys are the formation of creep voids. Also, the formation of creep voids is found to be correlated with three factors which are dislocation multiplication, dislocation motion and dislocation obstacles. Inconel718 has higher dislocation multiplication rate, larger dislocation motion rate and more δ-phases and fewer bands than Allvac718Plus. Therefore, Inconel718 has higher creep voids rate than Allvac718Plus, leading to the result that Inconel718 has shorter creep life than Allvac718Plus [62].

A spark plasma sintering (SPS) apparatus was used to perform uniaxial compressive creep tests on dense SPS-processed fine-grained alumina (**Figure 8**). Experiments were carried out in the 1125–1250°C temperature range under an applied stress of 80–120 MPa. Creep rates, stress exponent and apparent activation energy were determined. The microstructure of deformed samples was characterized by HRSEM. The creep rates, stress exponent (1.9–2.1) and apparent activation energy (454 kJ/mol) values obtained are in a good agreement with data reported in the study regarding creep of fine-grained polycrystalline alumina. These results, together with microstructural observations, suggest that the creep mechanism involved was grain boundary sliding (GBS), accommodated by dislocation climb and controlled by diffusion along the grain boundaries. It was thus demonstrated that an SPS apparatus can be employed as an accurate high-temperature creep testing system [63].

http://dx.doi.org/10.5772/intechopen.71184

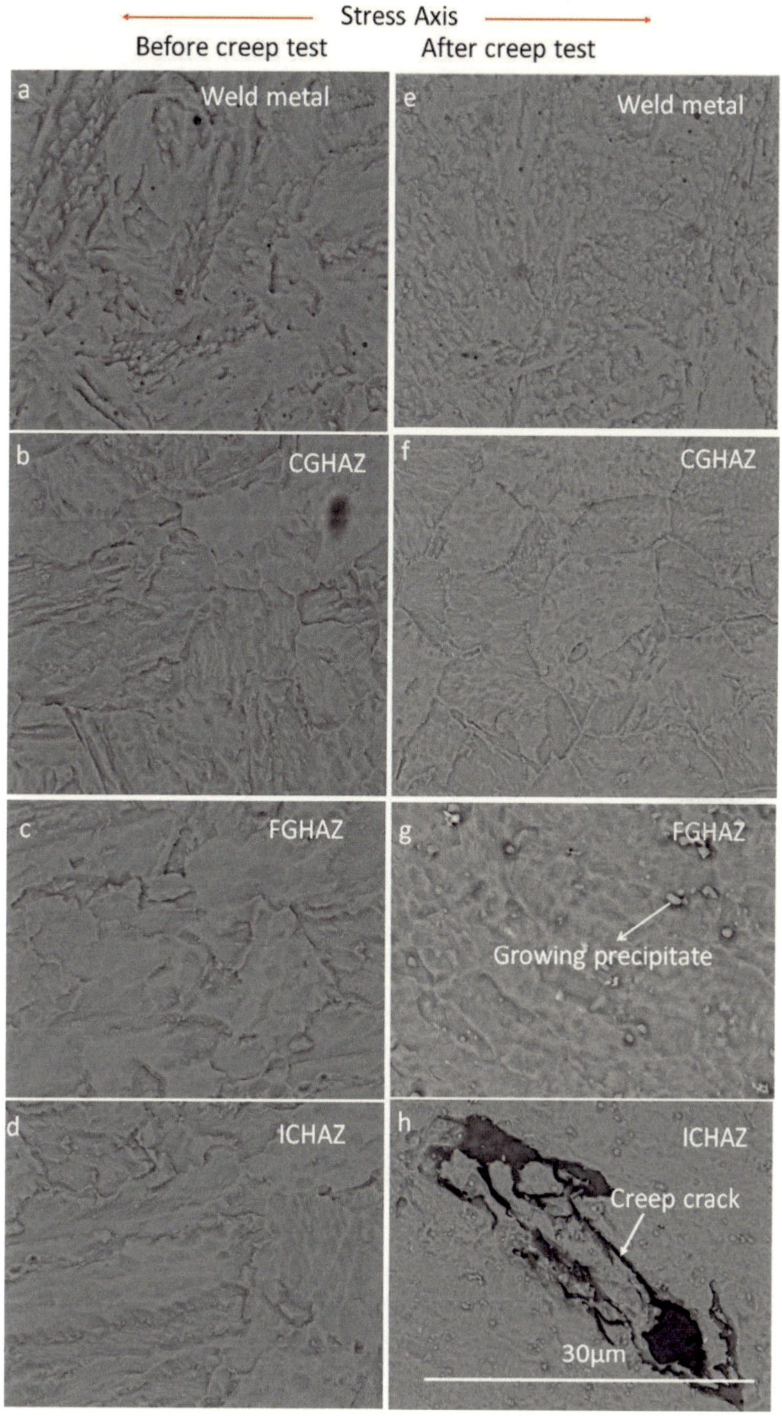

Figure 6. SEM-BSE micrographs showing the microstructures at different locations across the weldments: (a–d) before creep testing and (e–h) after creep testing (creep exposure at 650°C/50 MPa and rupture time is 3772 h) [60].

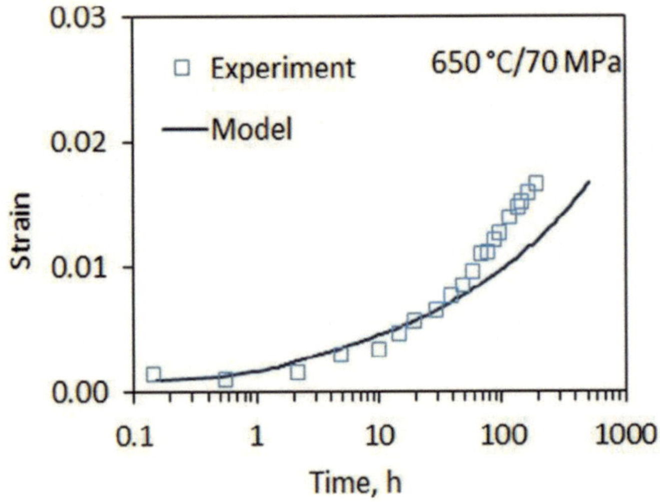

Figure 7. Comparison of creep strain time curves between experiment and FEA for composite structure at 650°C/70 MPa [60].

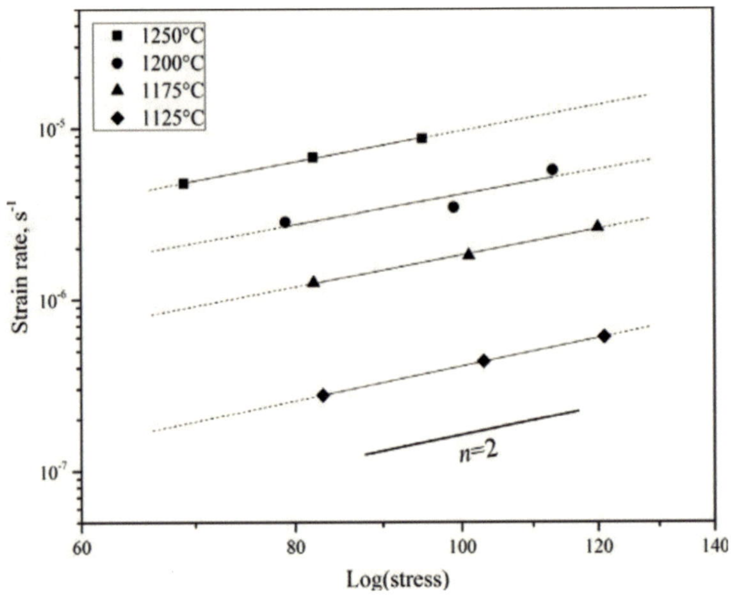

Figure 8. Creep rates of alumina as a function of true stress [63].

Copper gives creep strain versus time curves at 75°C, which look very similar to those recorded at much higher temperatures. Thus, an extended secondary stage where the strain rate is constant is observed. Considering the high creep exponent that can be up to 75, one would expect a creep curve with rapidly increasing strain rate but that is not found. The difference of creep of pure metals at high temperatures is so large that we can talk about an entirely new material class with respect to creep. To explain the observations, a recently developed dislocation model (Sandstrom, 2017) for cell structures is used.

A new creep model was presented where a back stress based on the dislocations in the cell walls is introduced. Unbalanced sets of dislocations without matching dislocations of opposite signs are formed in the cell walls. Since the unbalanced content is not exposed to static recovery, it forms a stable back stress. It was shown that the computed back stress can fully explain the observations and reproduce both creep curves and results for slow strain rate tensile tests [64]. Also, the creep behavior of alumina has been investigated [65–77]. There are numerous instances of the structural application of GFRP in the construction of buildings and bridges [78, 79]. Exceptional reviews of the widespread studies on the topic may be found in some important references [80–82]. Also, some studies were done regarding the proposition of quantum mechanical tunneling of dislocations at very low temperature [83–88].

Till now, numerous constitutive models have been presented to address viscoplastic, plasticity and the creep behaviors of materials [89–107]. Also, the time-dependent creep deformation of a metallic component under applied stress and temperature was studied by many researchers [92, 99, 104, 105].

Creep rupture and failure of the creeping metals are dedicated to the basic explanation of the creep which happens extensively in high-temperature deformation of the creeping metals. Particular concentration and attention are paid to the analysis of long-term strength, which characterizes the stress at which the creeping metal does not fail after a predetermined time. Lokoshchenko [106] details experimental and analytical results determined by Soviet and Russian scientists who are absent in presently accessible publications and demonstrates theoretical models and methods to attain long-term strength in metals [106].

As another interesting book, the third edition of fundamentals of creep in metals and alloys [105] stays generally up to date for the creeping metals, there are a wide range of improvements and updates that are either pleasing, or necessary, to make sure that the book continues to meet the needs of the investigators in the general area of steady state creep plasticity (time-dependent plasticity, viscoplastic and viscoelastic). As well, updating the areas presently covered in the second edition with new advances, the third edition will broaden its scope beyond metals and alloys to include ceramics, covalent solids, minerals and polymers, hence addressing the fundamentals of creep in all fundamental classes of the creeping materials [105].

6. Solved problems

Problem 6.1. Determine the viscosity of the creeping fibrous composites semi-theoretically.

Solution:
A semi-analytical formulation is presented for obtaining the viscosity of solids (such as metals) using the steady state creep model of the short fiber composites. For achieving this aim, fluid mechanics theory is used for determining the viscosity. Sometimes, obtaining the viscosity is

experimentally difficult and intricate. Therefore, the present model may be beneficial to obtain the viscosity of metals (see **Figure 9**).

Figure 9 graphically shows a simple simulation and analogy between the steady state creep behavior of the short fiber composite and viscosity of the viscous fluid. The shear stress is generally defined by the following equation in the creeping solids at any given arbitrary $r = r_0$ and $z = z_0$ mathematically,

$$\tau_{rz}^{matrix} = a_0 \ln z + a_1 \tag{10}$$

where the constants "a_0" and "a_1" are introduced in Eqs. (17a, b). On the other hand, the shear stress in the fluids is given by:

$$\tau_{rz}^{fluid} = \mu \dot{\gamma}_{rz}^{fluid} \tag{11}$$

where the parameters "μ" and "$\dot{\gamma}$" are the viscosity and shear strain rate, respectively. The shear strain rate "$\dot{\gamma}$" is defined as follows:

$$\dot{\gamma}_{rz} = \frac{\partial \dot{w}}{\partial r} + \frac{\partial \dot{u}}{\partial z} \tag{12}$$

Now, the average shear stress at the interface is obtained for determining the viscosity "μ" as follows:

$$\tau_{rz}^{matrix}\Big|_{average}^{r=a} = \frac{1}{l}\int_0^l (a_0 \ln z + a_1)\, dz \tag{13}$$

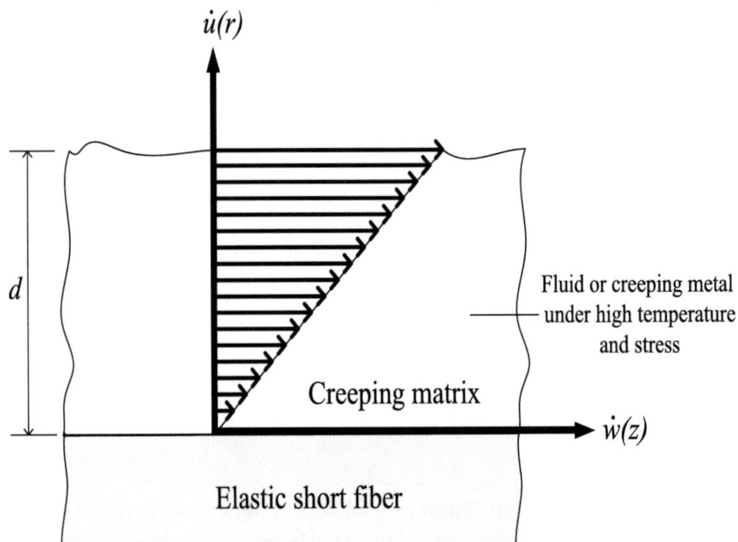

Figure 9. Simulation between creeping composite and viscous fluid.

Thus, the viscosity "μ" is determined by the combination of Eqs. (10)–(13), which yields

$$\mu = \frac{\tau_{rz}^{matrix}\Big|_{average}^{r=a}}{\dot{\gamma}_{rz}^{matrix}} \tag{14}$$

To summarize the calculations, previously published research results [13] are used for obtaining the viscosity "μ" (Eq. (14)). Therefore, the viscosity of the creeping matrix in the short fiber composite is determined using fluid mechanics and creep theories. Finally, the viscosity of the metals at the mentioned temperatures "μ" is semi-analytically obtained by the following relation:

$$\left[\tau_{matrix}\right]_{creep\ of\ composite}^{obtained\ using\ available\ results\ [13,22-24]} = \left[\tau_{viscous\ fluid}\right]_{fluid}^{viscosity\ "\mu"\ is\ unknown} = \mu\dot{\gamma}_{fluid} = \mu\left(\frac{\partial\dot{w}}{\partial r} + \frac{\partial\dot{u}}{\partial z}\right) \tag{15}$$

where $\dot{u}$ (radial displacement rate) and $\dot{w}$ (axial displacement rate) are obtained using the obtained results presented in [13, 22–24]. Moreover, the creeping shear stress "τ_{matrix}" is presented in Eqs. (17a, b). For the fibrous composite employed here "*SiC/Al6061*," the volume fraction of the fibers is approximately *15%* and the fibers have an aspect ratio of *7.4* and $k = 0.76$, which are according to the Ref. [36]. In addition, the second-stage creep constants of the creeping matrix material, *A* and *B*, in Eq. (16) at 300°C are considered as $A = exp.(-24.7)$ and $B = 6.47$, given by Morimoto et al. [36].

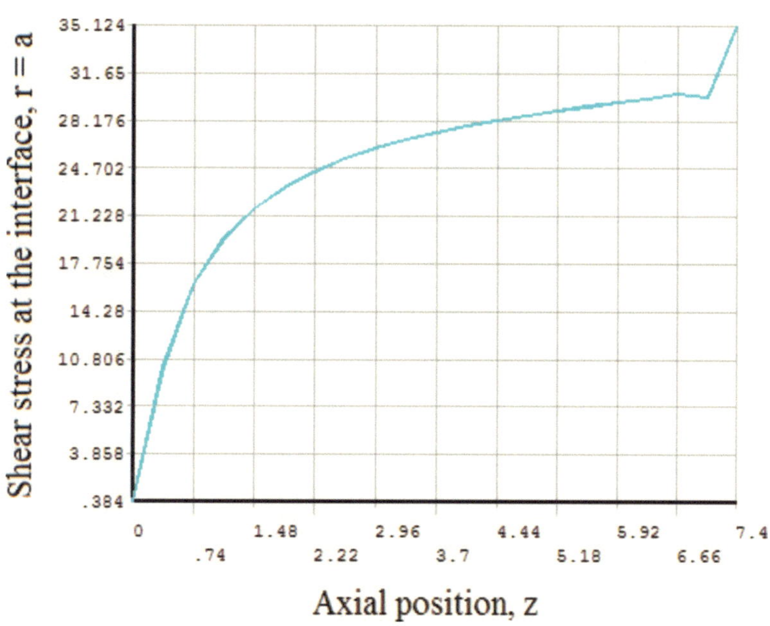

Figure 10. Shear stress behavior in the creeping metal matrix composite in the steady state creep (at interface, $r = a$, with the assumption of $a = 1$).

$$\dot{\varepsilon}_e = Aexp\left(\frac{\sigma_e}{B}\right) \tag{16}$$

where σ_e and $\dot{\varepsilon}_e$ are the equivalent stress and equivalent strain rate of the creeping matrix, respectively. A quantitative example is presented in this section. The interfacial shear stress is in the following form mathematically:

$$\tau_i = \tau_{rz}^{fluid}\Big|_{r=a} = \mu\dot{\gamma}_{rz}^{fluid}\Big|_{r=a} \tag{17}$$

$$\tau_i = 3.75\ln\left(\frac{z}{l}\right) + 27.2 \tag{18}$$

Shear stress presented in Eqs. (17a, b) is graphically shown in **Figure 10**.

The average shear stress, τ_{rz}^{matrix} (at interface, $r = a$), is obtained by Eq. (13). Also, the shear strain rate, $\dot{\gamma}_{rz}^{matrix}$, is determined by Eq. (12). By a simple calculation, the viscosity,μ, for the mentioned metal *Al6061* under the mentioned conditions is approximately equal to $4.8 \times 10^{15}\lambda$ Pa.s, in which the parameter of "λ" is a positive and small number. Also, FEM nodal solution of the

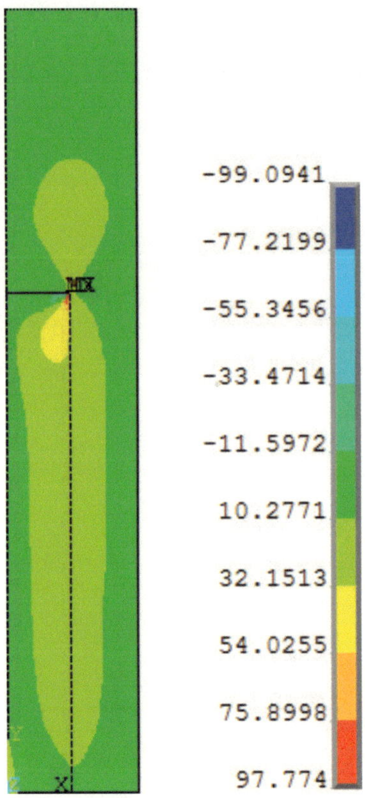

Figure 11. FEM nodal solution of the unit cell for predicting the shear stress behavior (the distribution of the shear stresses in the unit cell) [x = r(radial), y = z(axial) and z = θ(circumferential)].

unit cell for predicting the shear stress behavior is shown in **Figure 11** for better understanding the shear stress behavior in the steady state creep of the short fiber composites. **Figures 10** and **11** are presented with an assumption of $a = 1$.

For comparison purpose, the viscosity of pitch at $298\ K$ is equal to 2.3×10^8 Pa.s [34]. The results of the present method for obtaining the viscosity have a good agreement with the other available research results. The value of the mentioned viscosity is logical in comparison with the available results. In this problem, a novel approach was introduced to obtain the viscosity of the metals using the creeping property of the composites.

In addition, the behavior of the creeping metals under applied stress and temperature was simulated with the viscous fluids. The present method is useful and applicable for obtaining the viscosity, because the expensive and time-consuming experimental methods have many difficulties in determining the viscosity of the metals at high temperatures. Also, sometimes obtaining the viscosity of the metals at some temperatures is very intricate. Finally, this method can be used for obtaining the viscosity of the metals under different conditions using the steady state creeping property of the composites [50].

Problem 6.2. The Cauchy stress tensor components at one point of a creeping material (second-stage creep) considering Newtonian fluid, in which the bulk viscosity coefficient is zero, are given by:

$$\sigma_{ij} = \begin{bmatrix} -1 & 1 & -1 \\ 1 & -5 & 3 \\ -1 & 3 & -3 \end{bmatrix} Pa \tag{19}$$

Obtain the viscous stress tensor components in a creeping material (such as creeping matrix in composites).

Solution:

If the bulk viscosity coefficient is zero, Stokes' condition, we will have, $p = p_{ave} = p_0$, and moreover, we may determine

$$\sigma_{ij} = -p\delta_{ij} + \tau_{ij} \tag{20}$$

Also, have

$$\kappa = \lambda + \frac{2}{3}\mu = 0 \tag{21}$$

It yields

$$\sigma_{ii} = -3p \tag{22}$$

$$p = -\frac{\sigma_{ii}}{3} = -\frac{(-1-5-3)}{3} = 3 \tag{23}$$

Therefore, we have

$$\tau_{ij} = \sigma_{ij} + p\delta_{ij} = \begin{bmatrix} -1 & 1 & -1 \\ 1 & -5 & 3 \\ -1 & 3 & -3 \end{bmatrix} + \begin{bmatrix} 3 & 0 & 0 \\ 0 & 3 & 0 \\ 0 & 0 & 3 \end{bmatrix} = \begin{bmatrix} 2 & 1 & -1 \\ 1 & -2 & 3 \\ -1 & 3 & 0 \end{bmatrix} Pa \quad (24)$$

Problem 6.3. Assume that the Cauchy stress tensor components at the point P of a creeping material (steady state creep) are given by the following form:

$$\sigma_{ij} = \begin{bmatrix} 1 & 6 & 7 \\ 6 & 7 & 9 \\ 7 & 9 & 4 \end{bmatrix} GPa \quad (25)$$

a) Determine the hydrostatic stress (mean stress).

b) Obtain the deviatoric and spherical part of the stress tensor of "σ".

Solution:

a) The hydrostatic stress (mean stress) is given by the following equation:

$$\sigma_m = \sigma_{Hyd} = \frac{\sigma_{ii}}{3} = \frac{1+7+4}{3} = 4 \quad (26)$$

b) The spherical part of stress tensor "σ" is given by the following equation:

$$\sigma_{ij}^{sph} = \frac{I_\sigma}{3}\delta_{ij} = \sigma_{Hyd}\delta_{ij} = \begin{bmatrix} \sigma_{Hyd} & 0 & 0 \\ 0 & \sigma_{Hyd} & 0 \\ 0 & 0 & \sigma_{Hyd} \end{bmatrix} = \begin{bmatrix} 4 & 0 & 0 \\ 0 & 4 & 0 \\ 0 & 0 & 4 \end{bmatrix} \quad (27)$$

And the Deviatoric part becomes as follows:

$$\sigma_{ij} = \sigma_{ij}^{sph} + \sigma_{ij}^{dev} \quad (28)$$

Therefore,

$$\sigma_{ij}^{dev} = \sigma_{ij} - \sigma_{ij}^{sph} \quad (29)$$

Finally,

$$\sigma_{ij}^{dev} = \begin{bmatrix} -3 & 6 & 7 \\ 6 & 3 & 9 \\ 7 & 9 & 0 \end{bmatrix} GPa \quad (30)$$

http://dx.doi.org/10.5772/intechopen.71184

Problem 6.4. The below second-stage creep data have been taken on an aluminum alloy at 400°**C** and a constant stress of 25 MPa (see **Table 1**). Draw the following data as strain versus time, and after that obtain the steady state or minimum creep rate generally and approximately. Comment: The preliminary and immediate strain is not included (Based on University of California San Diego's (UCSD) exams).

Solution:

These creep data are plotted as shown in **Figure 12**.

Time (min)	Strain	Time (min)	Strain
0	0.000	16	0.135
2	0.025	18	0.153
4	0.043	20	0.172
6	0.065	22	0.193
8	0.078	24	0.218
10	0.092	26	0.255
12	0.109	28	0.307
14	0.120	30	0.368

Table 1. Aluminum alloy at 400°C and a constant stress of 25 *MPa*.

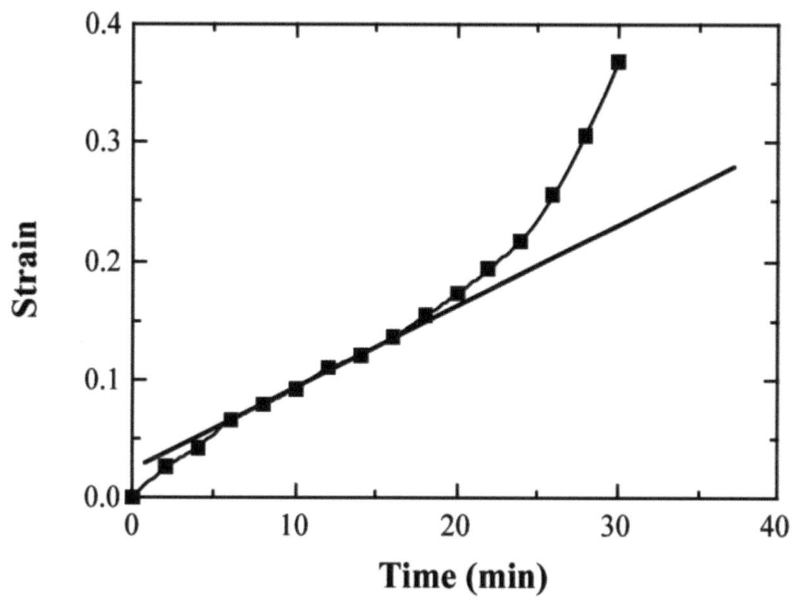

Figure 12. Plotted creep data.

The steady state creep rate (Δε/Δt) is the slope of the linear region. The straight line has been superimposed on the curve.

Problem 6.5. Steady state creep rate data for nickel at 1000°C (1273 K) are given in **Table 2**:

If it is known that the activation energy for creep is 272,000 J/mol, compute the steady state creep rate at a temperature of 850°C (1123 K) and a stress level of 25 MPa (3625 psi) (based on University of California San Diego's (UCSD) exams).

Solution:

Taking natural logarithms of both sides of the following equation yield,

$$\dot{\varepsilon}_s = K\sigma^n \exp\left(-\frac{Q_c}{RT}\right) \tag{31}$$

$$ln(\dot{\varepsilon}_s) = \ln(K) + n\ln(\sigma) - \frac{Q_c}{RT} \tag{32}$$

With the given data, there are two unknowns in this equation (namely K and n). Using the data provided in the problem statement, we can set up two independent equations as follows:

$$ln\left(1 \times 10^{-4}\right) = \ln(K) + n\ln(15) - \frac{272000}{8.31 \times 1273} \tag{33}$$

$$ln\left(1 \times 10^{-6}\right) = \ln(K) + n\ln(4.5) - \frac{272000}{8.31 \times 1273} \tag{34}$$

Therefore, we have

n = 3.825, K = 466 1/s.

Thus, it is now possible to solve at 25 *MPa* and *1123 K* using the mentioned equation.

$$\dot{\varepsilon}_s = K\sigma^n \exp\left(-\frac{Q_c}{RT}\right) = 466 \times 25^{3.825} \exp\left(-\frac{272000}{8.31 \times 1273}\right) = 2.28 \times 10^{-5}\ s^{-1} \tag{35}$$

$\dot{\varepsilon}_s\left(s^{-1}\right)$	**σ[MPa (psi)]**
10^{-4}	15 (2175)
10^{-6}	4.5 (650)

Table 2. Data of nickel at 1000°C (1273 K).

Author details

Vahid Monfared

Address all correspondence to: vahid_monfared_57@yahoo.com; vahid_monfared@alum.sharif.edu

Department of Mechanical Engineering, Zanjan Branch, Islamic Azad University, Zanjan, Iran

References

[1] Cox HL. The elasticity and strength of paper and other fibrous materials. British Journal of Applied Physics. 1952;**3**:72-79

[2] Mileiko ST. Steady state creep of a composite with short fibres. Journal of Materials Science. 1970;**5**:254-261

[3] McLean D. Viscous flow of aligned composites. Journal of Materials Science. 1972;**7**: 98-104

[4] Fukuda H, Chou TW. An advanced shear-lag model applicable to discontinuous fiber composites. Journal of Composite Materials. 1981;**15(1)**:79-91

[5] McLean M. Creep deformation of metal matrix composites. Composites Science and Technology. 1985;**23**:37-52

[6] Lee YS, Batt TJ, Liaw PK. Stress analysis of a composite material with short elastic fibre in power law creep matrix. International Journal of Mechanical Sciences. 1990;**32**(10): 801-815

[7] Pachalis JR, Kim J, Chou TW. Modeling of creep of aligned short-fiber reinforced ceramic composites. Composites Science and Technology. 1990;**37**:329-346

[8] Wang YR, Chou TW. Analytical modeling of creep of short fiber reinforced ceramic matrix composite. Journal of Composite Materials. 1992;**26**(9):1269-1286

[9] Durodola JF, Ruiz C, Derby B. Uniaxial creep of long fibre reinforced metal matrix composites. Composites Engineering. 1994;**4**(12):1241-1255

[10] Zhang J. Modeling of the influence of fibers on creep of fiber reinforced cementitious composite. Composites Science and Technology. 2003;**63**(13):1877-1884

[11] Gao XL, Li K. A shear-lag for carbon nanotube-reinforced polymer composites. International Journal of Solids and Structures. 2005;**42**:1649-1667

[12] Mondali M, Abedian A, Ghavami A. A new analytical shear-lag based model for prediction of the steady state creep deformations of some short fiber composites. Materials & Design. 2009;**30**:1075-1084

[13] Mondali M, Monfared V, Abedian A. Non-linear creep modeling of short fiber composites using Hermite polynomials, hyperbolic trigonometric functions and power series. Comptes rendus Mecanique. 2013;**341**(7):592-604

[14] Weng GJ, Sun CT. Effects of fiber length on the elastic moduli of randomly oriented chopped-fiber composites. In: Tsai SW, editor. ASTM-STP-674Composite Materials: Testing and Design. 1979. pp. 149-162

[15] Hsueh CH. A modified analysis for stress transfer in fiber-reinforced composites with bonded fiber ends. Journal of Materials Science. 1995;**30**:219-224

[16] Nayfeh AH, Abdelrahman WG. Micromechanical modeling of load transfer in fibrous composites. Mechanics of Materials. 1998;**30**:307-324

[17] Jiang Z, Liu X, Li G, Lian J. A new analytical model for three-dimensional elastic stress field distribution in short fibre composite. Materials Science and Engineering. 2004;**A366**:381-396

[18] Abedian A, Mondali M, Pahlavanpour M. Basic modifications in 3D micromechanical modeling of short fiber composites with bonded and debonded fiber end. Computational Materials Science. 2007;**40**:421-433

[19] Schjødt-Thomsen J, Pyrz R. Non-linear creep modelling of single-fibre model composites. Composites Science and Technology. 2000;**60**(9):1791-1800

[20] Kumar S, Singh RN. The creep response of uni-directional fiber-reinforced ceramic composites: A theoretical study. Composites Science and Technology. 2001;**61**(4):461-473

[21] Ohno N, Ando T, Miyake T, Biwa S. A variational method for unidirectional fiber-reinforced composites with matrix creep. International Journal of Solids and Structures. 2002;**39**(1):159-174

[22] Monfared V, Mondali M, Abedian A. Steady state creep behavior of short fiber composites by mapping, logarithmic functions (MF) and dimensionless parameter (DP) techniques. Archives of Civil and Mechanical Engineering. 2012;**12**(4):455-463

[23] Monfared V, Mondali M, Abedian A. Steady state creep analysis of polymer matrix composites using complex variable method. Proceedings of the Institution of Mechanical Engineers Part C-Journal of Mechanical Engineering Science. 2013;**227**(10):2182-2194

[24] Monfared V, Mondali M, Abedian A. Novel mathematical approaches for analyzing time dependent creep deformations in reinforced metals. Journal of Mechanical Science and Technology. 2013;**27**(11):3277-3285

[25] Dragon TL, Nix WD. Geometric factors affecting the internal stress distribution and high temperature creep rate of discontinuous fiber reinforced metals. Acta Metallurgica et Materialia. 1990;**38**(10):1941-1953

[26] Povirk GL, Needleman A, Nutt SR. An analysis of residual stress formation in whisker-reinforced al/SiC composites. Materials Science and Engineering. 1990;**A125**(2):129-140

[27] Levy A, Papazian JM. Elastoplastic finite element analysis of short-fiber-reinforced SiC/al composites: Effects of thermal treatment. Acta Metallurgica et Materialia. 1991;**39**(10):2255-2266

[28] Park YH, Holmes JW. Finite element modeling of creep deformation in fibre-reinforced ceramic composites. Journal of Materials Science. 1992;**27**(23):6341-6351

[29] Davis LC, Allison JE. Micromechanics effects in creep of metal-matrix composites. Metallurgical and Materials Transactions A. 1995;**26**(12):3081-3089

[30] Zhu-feng Y. Statistic modeling of the creep behavior of metal matrix composites based on finite element analysis. Applied Mathematics and Mechanics. 2002;**23**(4):421-434

[31] Mondali M, Abedian A, Adibnazari S. FEM study of the second stage creep behavior of Al6061/SiC metal matrix composite. Computational Materials Science. 2005;**34**:140-150

[32] Ghavami A, Abedian A, Mondali M. Finite difference solution of steady state creep deformations in a short fiber composite in presence of fiber/matrix debonding. Materials & Design. 2010;**31**:2616-2624

[33] Monfared V, Mondali M. Semi-analytically presenting the creep strain rate and quasi shear-lag model as well as FEM prediction of creep Debonding in short fiber composites. Materials & Design. 2014;**54**:368-374

[34] Edgeworth R, Dalton BJ, Parnell T. The pitch drop experiment. European Journal of Physics. 1984;**5**:198-200

[35] Nieh TG. Creep rupture of a silicon-carbide reinforced aluminum composite. Metallurgical and Materials Transactions A. 1984;**15**:139-146

[36] Morimoto T, Yamaoka T, Lilholt H, Taya M. Second stage creep of silicon carbide whisker/6061 aluminum composite at 573K. Journal of Engineering Materials and Technology. 1988;**110**:70-76

[37] Zhu SJ, YX L, Wang ZG, Bi J. Creep behavior of TiC-particulate-reinforced Ti alloy composite. Materials Letters. 1992;**13**(4-5):199-203

[38] Čadek J, Oikawa H, Šustek V. Threshold creep behaviour of discontinuous aluminium and aluminium alloy matrix composites: An overview. Materials Science and Engineering A. 1995;**190**(1-2):9-23

[39] Matsuda N, Akaike J, Hongo K, Matsuura K. The effect of second phase on the creep deformation of 6061Al matrix composites. Materials Science and Engineering A. 1997;**234-236**:751-754

[40] Cseh G, Bär J, Gudladt HJ, Lendvai J, Juhász A. Indentation creep in a short fibre-reinforced metal matrix composite. Materials Science and Engineering A. 1999;**272**(1):145-151

[41] Wang M, Zhao Y, Zhou L, Zhang D. Study on creep behavior of Ti–V–Cr burn resistant alloys. Materials Letters. 2004;**58**(26):3248-3252

[42] Kouadri-Boudjelthia A, Imad A, Bouabdallah A, Elmeguenni M. Analysis of the effect of temperature on the creep parameters of composite material. Materials & Design. 2009;**30**(5): 1569-1574

[43] Mahmudi R, Geranmayeh AR, Khanbareh H, Jahangiri N. Indentation creep of lead-free Sn–9Zn and Sn–8Zn–3Bi solder alloys. Materials & Design. 2009;**30**(3):574-580

[44] Olbricht J, Yawny A, Young ML, Eggeler G. Mechanical and microstructural observations during compression creep of a short fiber reinforced AlMg metal matrix composite. Materials Science and Engineering A. 2009;**510-511**:407-412

[45] Morscher GN, John R, Zawada L, Brewer D, Ojard G, Calomino A. Creep in vacuum of woven Sylramic-iBN melt-infiltrated composites. Composites Science and Technology. 2011;**71**(1):52-59

[46] Soubielle S, Diologent F, Salvo L, Mortensen MA. Creep of replicated microcellular aluminium. Acta Materialia. 2011;**59**(2):440-450

[47] Tang XG, Hou M, Zou J, Truss R, Zhu Z. The creep behaviour of poly(vinylidene fluoride)/"bud-branched" nanotubes nanocomposites. Composites Science and Technology. 2012;**72**(14):1656-1664

[48] Li LT, Lin YC, Zhou HM, Jiang YQ. Modeling the high-temperature creep behaviors of 7075 and 2124 aluminum alloys by continuum damage mechanics model. Computational Materials Science. 2013;**73**:72-78

[49] Tang LC, Wang X, Gong LX, Peng K, Zhao L, Chen Q, LB W, Jiang JX, Lai GQ. Creep and recovery of polystyrene composites filled with graphene additives. Composites Science and Technology. 2014;**91**:63-70

[50] Monfared V. Predicting the viscosity of solids using steady-state creep behavior of the fibrous composites semi-theoretically. Results in Physics. 2017;**7**:1433-1436

[51] Monfared V. Neural network based simulation of micro creeping fibrous composites SIC/AL6061 for plastic behavior. Journal of Theoretical and Applied Mechanics, Sofia. 2017;**47**(1):36-48

[52] Monfared V. Role of high order functions in analysis of the creep behavior of a short fiber composite "silicon carbide/aluminum 6061". Silicon. 2017;**9**(3):339-345

[53] Ma X, Li F, Zhao C, Zhu G, Li W, Sun Z, Yuan Z. Indenter load effects on creep deformation behavior for Ti-10V-2Fe-3Al alloy at room temperature. Journal of Alloys and Compounds. 2017;**709**:322-328

[54] Wei J, Malzbender J. Steady state creep of Ni-8YSZ substrates for application in solid oxide fuel and electrolysis cells. Journal of Power Sources. 2017;**360**:1-10

[55] Zhang Y, Jing H, Xu L, Zhao L, Han Y, Liang J. Microstructure and texture study on an advanced heat-resistant alloy during creep. Materials Characterization. 2017;**130**:156-172

[56] Coakley J, Ma D, Frost M, Dye D, Seidman DN, Dunand DC, Stone HJ. Lattice strain evolution and load partitioning during creep of a Ni-based superalloy single crystal with rafted γ' microstructure. Acta Materialia. 2017;**135**:77-87

[57] Reddy KV, Md M, Pal S. Mechanistic study of bending creep behaviour of bicrystal nanobeam. Computational Materials Science. 2017;**136**:36-43

[58] Pandey C, Mahapatra MM, Kumar P, Saini N. Effect of creep phenomena on room-temperature tensile properties of cast & forged P91 steel. Engineering Failure Analysis. 2017;**79**:385-396

[59] Jiang X, Zhang Y, Yi D, Wang H, Deng X, Wang B. Low-temperature creep behavior and microstructural evolution of 8030 aluminum cables. Materials Characterization. 2017; **130**:181-187

[60] Baral J, Swaminathan J, Chakrabarti D, Ghosh RN. Effect of welding on creep damage evolution in P91B steel. Journal of Nuclear Materials. 2017;**490**:333-343

[61] Lin C, Tian Q, Chen K, He G, Zhang J, Liu S, Almásy L. Polymer bonded explosives with highly tunable creep resistance based on segmented polyurethane copolymers with different hard segment contents. Composites Science and Technology. 2017;**146**:10-19

[62] Ni T, Dong J. Creep behaviors and mechanisms of Inconel718 and Allvac718plus. Materials Science and Engineering A. 2017;**700**:406-415

[63] Ratzker B, Sokol M, Kalabukhov S, Frage N. Using a spark plasma sintering apparatus as a tool in a compressive creep study of fine-grained alumina. Ceramics International. 2017;**43**(12):9369-9376

[64] Sandström R. Formation of a dislocation back stress during creep of copper at low temperatures. Materials Science and Engineering A. 2017;**700**:622-630

[65] Folweiler RC. Creep behavior of pore free polycrystalline aluminum oxide. Journal of Applied Physics. 1961;**32**(5):773-778

[66] Arshaw SIW, Norton FH. Deformation behavior of polycrystalline aluminum. Journal of the American Ceramic Society. 1962;**45**(10):479-486

[67] Coble RL, Guerard YH. Creep of polycrystalline aluminum oxide. Journal of the American Ceramic Society. 1963;**46**(7):353-354

[68] Paladino AE, Coble RL. Effect of grain boundaries on diffusion-controlled processes in aluminum oxide. Journal of the American Ceramic Society. 1963;**46**(3):133-136

[69] Cannon RM, Rhodes WH, Heuer AH. Plastic deformation of fine-grained alumina (Al2O3):I, interface-controlled diffusional creep. Journal of the American Ceramic Society. 1980;**63**(1–2):46-53

[70] Heuer AH, Tighe NJ, Cannon RM. Plastic deformation of fine-grained alumina (Al2O3): II, basal slip and non accommodated grain-boundary sliding. Journal of the American Ceramic Society. 1980;**63**(1-2):53-58,1980

[71] Xue LA, Chen I. Deformation and grain growth of low-temperature-sinteredhigh- purity alumina. Journal of the American Ceramic Society. 1990;**73**(11):3518-3521

[72] Wang J, Raj R. Interface effects in super plastic deformation of alumina containing zirconia, titania or hafnia as a second phase. Acta Metallurgica et Materialia. 1991;**39**(11):2909-2019

[73] Nutt SR, Lipetzky P. Creep deformation of whisker-reinforced alumina. Materials Science and Engineering A. 1993;**166**(1):199-209

[74] Flacher O, Blandin JJ, Plucknett KP. Effects of zirconia additions on the super plasticity of alumina-zirconia composites. Materials Science and Engineering A. 1996;**221**:102-112

[75] Chevalier J, Olagnon C, Fantozzi G, Gros H. Creep behavior of alumina, zirconia and zirconia-toughened alumina. Journal of the European Ceramic Society. 1997;**7**(6):859-864

[76] Ruano OA, Wadsworth J, Sherby OD. Deformation of fine-grained alumina by grain boundary sliding accommodated by slip. Acta Materialia. 2003;**51**(12):3617-3634

[77] Bernard-Granger G, Guizard C, Duclos R. Compressive creep behavior in air of a slightly porousas-sintered polycrystalline α-alumina material. Journal of Materials Science. 2007;**42**(8):2807-2819

[78] Scott DW, Lai JS, Zureick A-H. Creep behavior of FRP composites: Are view of technical literature. Journal of Reinforced Plastics and Composites. 1995;**14**:588-617

[79] Keller T. Use of fibre reinforced polymers in bridge construction. In: Structural Engineering Documents. Vol. 7. IABSE, Zurich, Switzerland: International Association for Bridge and Structural Engineering; 2003

[80] Lupinc V. In: Bressers J, editor. Creep and Fatigue in High Temperature Alloys. London: Applied Science Publishers; 1981. pp. 7-40

[81] Lagneborg R. In: Bressers J, editor. Creep and Fatigue in High Temperature Alloys. London: Applied Science Publishers; 1981. pp. 41-71

[82] Wilshire B, Burt H. Damage evolution during creep of steels. International Journal of Pressure Vessels and Piping. 2008;**85**:47-54

[83] Mott NF. Creep in metal crystals at very low temperatures. Philosophical Magazine. 1956;**1**(6):568-572

[84] Glen JW. The creep of cadmium crystals at liquid helium temperatures. Philosophical Magazine. 1956;**1**(5):400-408

[85] Arko AC, Weertman J. Creep deformation of Cd and Hg at liquid helium temperatures. Acta Metallurgica. 1969;**17**:687-699

[86] Tesh JR, Whitworth RW. Plastic deformation and creep if sodium chloride crystals at liquid helium temperatures. Physica Status Solidi. 1970;**39**:627-633

[87] Osetskii AI, Soldatov VP, Startsev VI, Natsik VD. Temperature dependence and activation parameters of creep in Zn in the temperature range 1.5 to 80 K. Physica Status Solidi A. 1974;**22**:739-748

[88] Startsev VI, Soldatov VP, Natsik VD, Abraimov VV. Role of quantum mechanisms and thermal heating in low-temperature creep of metals. Physica Status Solidi A. 1980;**59**:377-388

[89] Lubahn JD, Felgar RP. Plasticity and Creep of Metals. Wiley series on the science and technology of materials. Canada: Wiley; 1961

[90] Garofalo F. Fundamentals of Creep and Creep-Rupture in Metals, MacMillan Series in Materials. United States: MacMillan and Co; 1965

[91] Wilcox BA. Steady-State Creep of Dispersion-Strengthened Metals. United States: NASA; 1966

[92] Rabotnov YN. Creep Problems in Structural Memberse. NewYork, NY: American Elsevier Publishing Company. Inc.; 1969

[93] Greenfield P. Creep of Metals at High Temperature (Mechanics Engineering Monograph). Great Britain: Mills & Boon; 1972

[94] Frost HJ, Ashby MF. Deformation-Mechanism Maps: The Plasticity and Creep of Metals and Ceramics. United States (University of Michigan): Pergamon Press; 1982

[95] Boyle JT, Spence J. Stress Analysis for Creep. 1st ed. Southampton: Butterworth, UK: Butterworth-Heinemann; 1983

[96] Evans RW, Wilshire B. Creep of Metals and Alloys (Predictive and Quantitative Metallurgy Series). London (Institute of Metals): CRC Press; 1985

[97] Zyczkowski M. Creep in structures: 4th IUTAM Symposium. Cracow, Poland. 1990: 10–14 (IUTAM Symposia), Springer; Softcover reprint of the original 1st ed. 1991 edition

[98] Khan AS, Huang S. Continuum Theory of Plasticity. 1st ed. New York: John Wiley & Sons; 1995

[99] Nabarro FRN, Villiers HL de. The Physics of Creep.1995. Taylor & FrancisInc, Bristol, PA

[100] Mishra RS, Earthman JC. Creep Deformation: Fundamentals and Applications. Pittsburgh, PA, United States: Tms; 2002

[101] Betten J. Creep Mechanics. 2nd ed. Germany: Springer; 2005

[102] Abe F, Kern TU. Creep-Resistant Steels (Woodhead Publishing Series in Metals and Surface Engineering). 1st ed. England: Woodhead Publishing; 2008

[103] Rusinko A, Rusinko K. Plasticity and Creep of Metals. Chennai, India: Springer; 2011

[104] Hyde TH, Sun W, Hyde CJ. Applied Creep Mechanics. NewYork, NY: McGraw-Hill Education; 2014

[105] Kasnner ME. Fundamentals of Creep in Metals and Alloys. 3rd ed. Amsterdam, The Netherlands: Butterworth Heinemann is an imprint of Elsevier; 2015

[106] Lokoshchenko AM. Creep and Creep Rupture of Metals. 1st ed. England: CRC Press; 2017

[107] Charit I, Zhu YT. Mechanical and creep behavior of advanced materials: A SMD Symposium Honoring Professor K. Linga Murty. The Minerals, Metals & Materials Series, Springer; 1st ed. 2017

Chapter 11

Advanced Methods for Creep in Engineering Design

William Harrison, Mark Whittaker and
Veronica Gray

Additional information is available at the end of the chapter

http://dx.doi.org/10.5772/intechopen.72319

Abstract

There are many applications where the combination of stress and elevated temperature require creep to be considered during the design process. For some applications, an evaluation of rupture life for given conditions is sufficient, however, for components such as those in gas turbine aeroengines, the accumulation of creep strain over time and the effect this has on other phenomena, such as high-temperature fatigue must be considered. In this chapter, modern creep curve modelling methods are applied to alloys used in gas turbine applications over a wide range of test conditions. Also, different creep hardening modelling methods are discussed along with their application to transient creep showing the deficiencies of simplistic models. Models are related to micromechanical properties where possible, and creep damage models are evaluated and applied to different applications using finite element analysis (FEA).

Keywords: creep, creep damage, stress relaxation, finite element analysis (FEA)

1. Introduction

When designing components for high temperature applications, the effects of creep must be considered. For simple cases, prediction methods that evaluate creep rupture life, t_F, based on applied stress and temperature are sufficient. These 'single-point' prediction methods such as those proposed by Norton [1], Larson and Miller [2] and more recently, the hyperbolic tangent method [3] and Wilshire equations [4], vary in complexity and in their ability to predict creep properties of the full range of applied conditions [5, 6]. Similarly, minimum creep rates, $\dot{\varepsilon}_m$, can be related to applied stress and temperature using similar equations or by using the Monkman-Grant relationship [7],

$$M = t_F \dot{\varepsilon}_m \tag{1}$$

where M is the Monkman-Grant constant. The minimum creep rate, although useful for simple calculations, does not represent the full range of creep behaviour for an alloy at any given

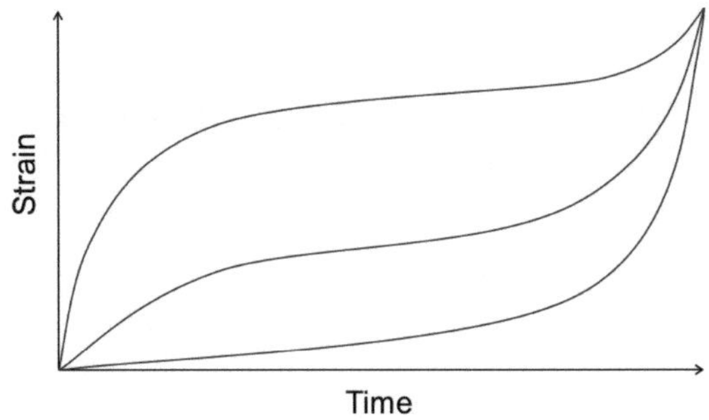

Figure 1. Schematic representation of different creep curves with equivalent rupture times and minimum creep rates.

applied conditions. On applying load to a material at high temperature the initial creep rate is high before decreasing to a minimum rate, a phenomenon known as primary creep. As deformation continues, the creep rate increases to failure during tertiary creep, giving the creep curve the characteristic shape shown in **Figure 1**. The shape of a creep curve can vary significantly depending on material and applied conditions [8]. This can have a considerable influence when calculating the lives of engineering components, especially in components where the effects of creep deformation and damage interact with other phenomenon such as fatigue during high-temperature fatigue or thermomechanical fatigue (TMF) [9].

Evaluating creep behaviour based on single point prediction methods alone cannot fully describe the shape of a creep curve since different curves can exhibit the same rupture life and minimum creep rate (**Figure 1**). Furthermore, to evaluate the creep behaviour of complex shaped components, simple calculations must be replaced by a more multifaceted approach such as finite element analysis [10]. To develop creep models suitable for finite element analysis, the full creep curve of a material must be obtainable over a wide range of conditions and an appropriate hardening model must be used. Furthermore, in order to predict rupture during variable creep conditions, a suitable damage model must be used.

2. Creep deformation

2.1. Full creep curve prediction methods

Many approaches exist for evaluating creep deformation behaviour from simple time-hardening models which only predict tertiary creep, to phenomenological models which aim to predict the micro-mechanisms of creep. Two main approaches for predicting full creep curves are investigated in this section. The first approach involves evaluating the times to given creep strain levels for a range of tests using a similar approach to stress rupture. Parameters are then derived to describe how this relationship varies with increasing creep strain. Examples of this method include an empirical relationship proposed by Gray and Whittaker [11] and a method based on the Wilshire equations [12, 13]. The former method relates times to strain, t_ε, to applied test conditions using

$$t(\varepsilon)\exp\left(\frac{-Q_c^*}{RT}\right) = M(\varepsilon)\left(1-\frac{\sigma}{\sigma_N}\right)^{P(\varepsilon)} \tag{2}$$

where the σ_N is the temperature dependent tensile strength, yield stress or proof stress. The parameters $M(\varepsilon)$ and $P(\varepsilon)$ are related to strain using a Frechet and Lognormal distribution with $\mu = 0$, respectively

$$M(\varepsilon) = A_1 \exp\left(-\left(\frac{\varepsilon}{A_2}\right)^{-A_3}\right) \tag{3}$$

$$P(\varepsilon) = \frac{A_4}{\varepsilon A_5 \sqrt{2\pi}} \exp\left(-\frac{[\ln(A_6\varepsilon)]^2}{2A_5^2}\right) + A_7 \tag{4}$$

where A_{1-7} are material constants, derived from experimental data. This method has been successfully applied to many creep resistant metals and alloys [11].

The Wilshire approach has been show to extrapolate stress rupture times and minimum creep rates well [4, 5, 12–15]. An extension of this method uses an equation of similar form to relate times to a given creep strain to applied stress and temperature,

$$\frac{\sigma}{\sigma_{\mathrm{TS}}} = \exp\left\{-k_3\left[t_\varepsilon \exp\left(\frac{-Q_c^*}{RT}\right)\right]^w\right\} \tag{5}$$

where σ and T are the applied test stress and temperature respectively, t_ε is the time to a given creep strain, σ_{TS} is the ultimate tensile strength at applied test temperature, and, Q_c^* is the activation energy for creep evaluated for $1/T$ against $\sigma/\sigma_{\mathrm{TS}}$. The material constants k_3 and w are obtained from the gradient and intercept from plotting $\ln(\ln(\sigma/\sigma_{\mathrm{TS}}))$ against $t_\varepsilon \exp\left(-Q_c^*/RT\right)$. Eq. (5) has been successfully used to represent creep data for alloys such as Ti834 [12] and Alloy720Li [13]. **Figure 2** shows creep strain data for Inconel100 with Eq. (5) representing creep strain data from 0.05 to 2% creep strain. Values for k_3 and w and in some cases Q_c^*, are not always constant across the full range of creep conditions. Studies [13, 14] have shown that in the case of creep rupture, there may be 2 or more sets of these parameters to account for different dominant creep mechanisms at different applied conditions. These changes in parameters usually occur at phenomena such as above and below the yield stress where dislocation networks change considerably [15]. Similar 'break points' have been observed for predicted times to strain for the nickel based superalloy, Alloy720Li [13].

Different relationships have been proposed to relate the parameters of Eq. (5) for different values of strain. For the nickel based superalloy Alloy720Li, w was found to vary minimally as strain increased and was assumed to be constant. A power law relationship was used to relate k_3 to creep strain, ε:

$$k_3 = k_{3,0} + k_{3,1}\varepsilon^{-k_{3,2}} \tag{6}$$

where $k_{3,0}$, $k_{3,1}$ and $k_{3,2}$ are material constants obtained from experimental data [13]. Care must be taken when evaluating the trends in w and k_3 because negative creep rates are possible if they deviate significantly with test data.

An alternative method to predicting full creep curves involves using equations to represent time verses creep strain behaviour and relating the parameters of these equations to applied

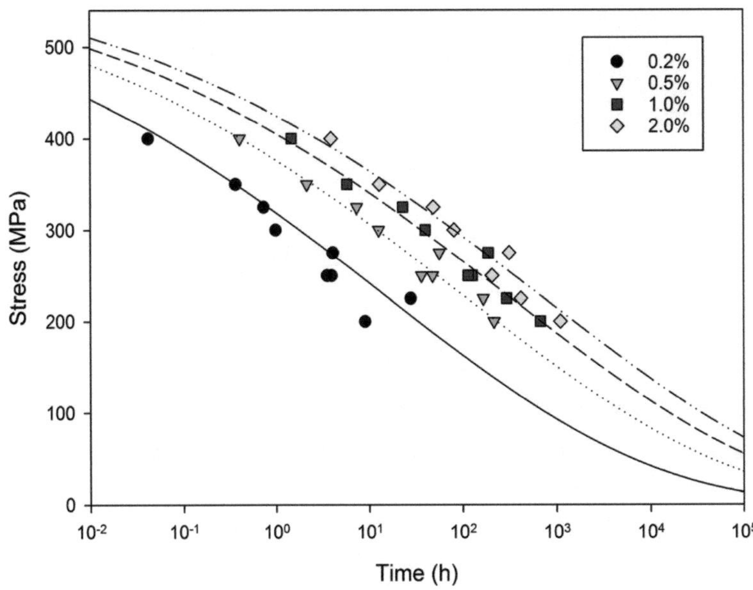

Figure 2. Times to 0.2%, 0.5%, 1% and 2% strain for Inconel100 with fits obtained using Eq. (5).

stress and temperature. Examples of this method include the theta-projection method [16, 17], a model by Dyson and McLean [18], and, a true stress method proposed by Wu et al. [19]. Both methods split the creep curve in to primary and tertiary regions, using equations of similar form to characterise each region. In the case of the theta method, creep strain is related to time using:

$$\varepsilon = \theta_1\left(1 - e^{-\theta_2 t}\right) + \theta_3\left(e^{\theta_4 t} - 1\right) \tag{7}$$

where θ_{1-4} are the 4-θ coefficients obtained from the experimental test data. These parameters are determined by minimising ϕ in the expression:

$$\phi^{n-1} = \sum_{i=1}^{n-1}\left[\varepsilon_i - \theta_1\left(1 - e^{-\theta_2 t_i}\right) + \theta_3\left(e^{\theta_4 t_i} - 1\right)\right]^2 \tag{8}$$

for n data points where ε_i and t_i are strain and time values for each data point. The expression $\theta_1\left(1 - e^{-\theta_2 t}\right)$ represents the decreasing rate of primary creep and $\theta_3\left(e^{\theta_4 t} - 1\right)$ represents tertiary creep. The parameters θ_1 and θ_3 represent the magnitude of primary and tertiary creep respectively and are referred to as scale parameter, whereas θ_2 and θ_4 characterise the curvature of the primary and tertiary phases and are termed rate parameters (**Figure 3**). **Figure 4** shows how this method has been used to accurately represent the curve shapes of many different metals and alloys over a range of applied conditions.

Various methods have been proposed to relate θ_{1-4} to applied test conditions. Evans [17] used a multi-linear approach whereby:

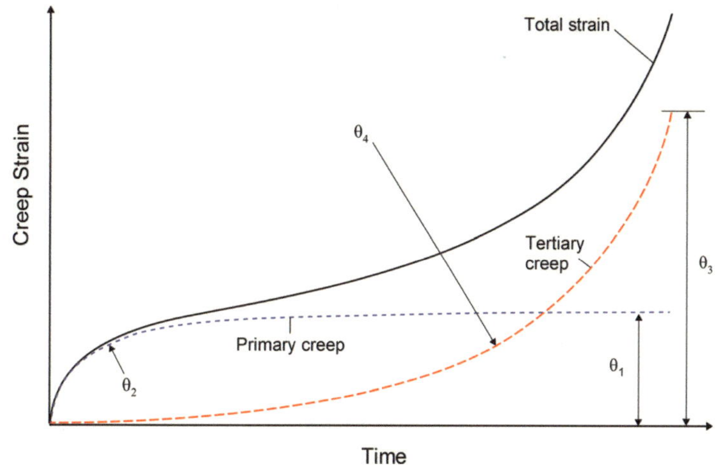

Figure 3. Schematic representation of a creep curve represented by Eq. (7), showing primary and tertiary regions.

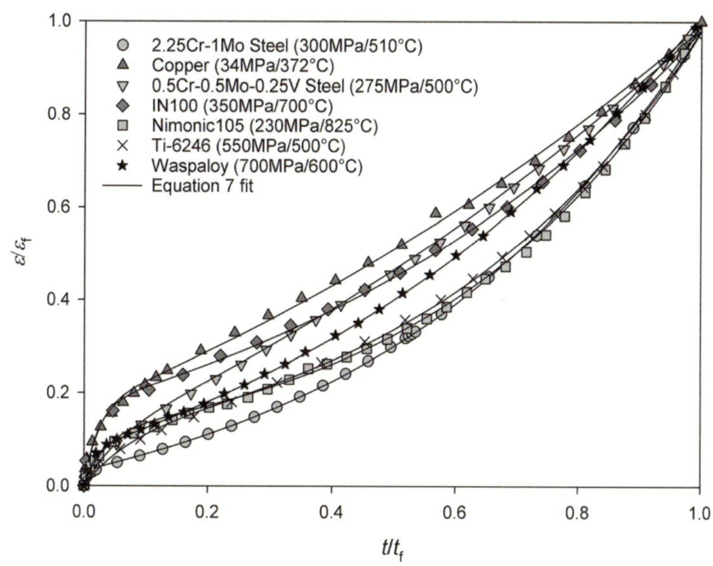

Figure 4. Examples of creep curves represented by the theta-projection method (Eq. (7)).

$$\ln(\theta_k)_h = a_k + b_k\sigma_h + c_kT_h + d_k\sigma_hT_h, \qquad k = 1\text{–}4 \tag{9}$$

where a_k, b_k, c_k and d_k (k = 1–4) are obtained by linear regression. This empirical approach interpolates the θ-coefficients well, however produces unrealistic values when extrapolating to applied conditions beyond existing experimental data. Poor extrapolation is problematic when evaluating the creep behaviour of complex engineering components using finite element analysis (FEA) since some regions of the components may have values of σ and T that do not correspond to those tested through uniaxial creep tests [10]. An alternative approach is to use a power law expression to relate θ_{1-4} to σ and an Arrhenius expression to account for temperature effects. For the titanium aluminide alloy Ti-45Al-2Mn-2Nb the stress was normalised against temperature compensated tensile strength, σ_{TS}, which was sufficient to account for temperature effects for θ_1 and θ_3 [20]. The equations used to relate θ_{1-4}. To applied test conditions are

$$\theta_1 = A_1\left(\frac{\sigma}{\sigma_{TS}}\right)^{n_1} \tag{10}$$

$$\theta_2 = A_2\left(\frac{\sigma}{\sigma_{TS}}\right)^{n_2} \exp\left(\frac{-Q_2^*}{RT}\right) \tag{11}$$

$$\theta_3 = A_3\left(\frac{\sigma}{\sigma_{TS}}\right)^{n_3} \tag{12}$$

$$\theta_4 = A_4\left(\frac{\sigma}{\sigma_{TS}}\right)^{n_4} \exp\left(\frac{-Q_4^*}{RT}\right) \tag{13}$$

where Q_k^*, A_k and n_k (k = 1, 4) are determined from θ-coefficents obtained using Eq. (8). The parameters of Eqs. (10) and (11) are determined from the gradients and intercepts of linear lines of best fit from plots of $\ln(\theta_1)$ and $\ln(\theta_3)$ against $\ln(\sigma/\sigma_{TS})$ respectively. θ_2 and θ_4 display more dependence on temperature and are evaluated from plots of $\ln\left[\theta_2 \exp\left(Q_2^*/RT\right)\right]$ and $\ln\left[\theta_4 \exp\left(Q_4^*/RT\right)\right]$ against $\ln(\sigma/\sigma_{TS})$ respectively. Since, stress is normalised using temperature dependent tensile strength Q_2^* and Q_4^* are lower than values for activation energy evaluated against σ alone. A common activation energy of 330 KJmol^{-1} was found to apply to both Q_2^* and Q_4^*, a value similar to the measured interdiffusion coefficient for γ-TiAl [20] indicating that this mechanism plays an important role in creep of these types of alloy. The relationship of the θ-coefficients with respect to stress and temperature for Ti-45Al-2Mn-2Nb can be seen in **Figure 5**.

An alternative approach to the θ-projection method is a true stress model proposed by Wu et al. [19]. This method relates total strain, ε to time, t, by

$$\varepsilon = \varepsilon_0 + \varepsilon_p\left[1 - \exp\left(-\frac{t}{t_{tr}}\right)\right] + \frac{1}{M'}\left[\exp(M'kt) - 1\right] \tag{14}$$

where ε_0 is the instantaneous strain, ε_p represents the magnitude of primary creep and t_{tr} is the transient time. M' and k are related to the calculated rates of dislocation glide, climb and grain boundary sliding. This equation has a similar form to Eq. (7) where $\theta_1 = \varepsilon_p$, $\theta_2 = 1/t_{tr}$, $\theta_3 = 1/M'$ and $\theta_4 = M'k$. M' and k are related to calculated rates of dislocation glide, climb and grain boundary sliding obtained from experimental creep curves.

2.2. Hardening methods

Various methods to predict creep curves and hence creep rates during uniaxial creep have been proposed, however, to predict creep behaviour in engineering components using FEA, a

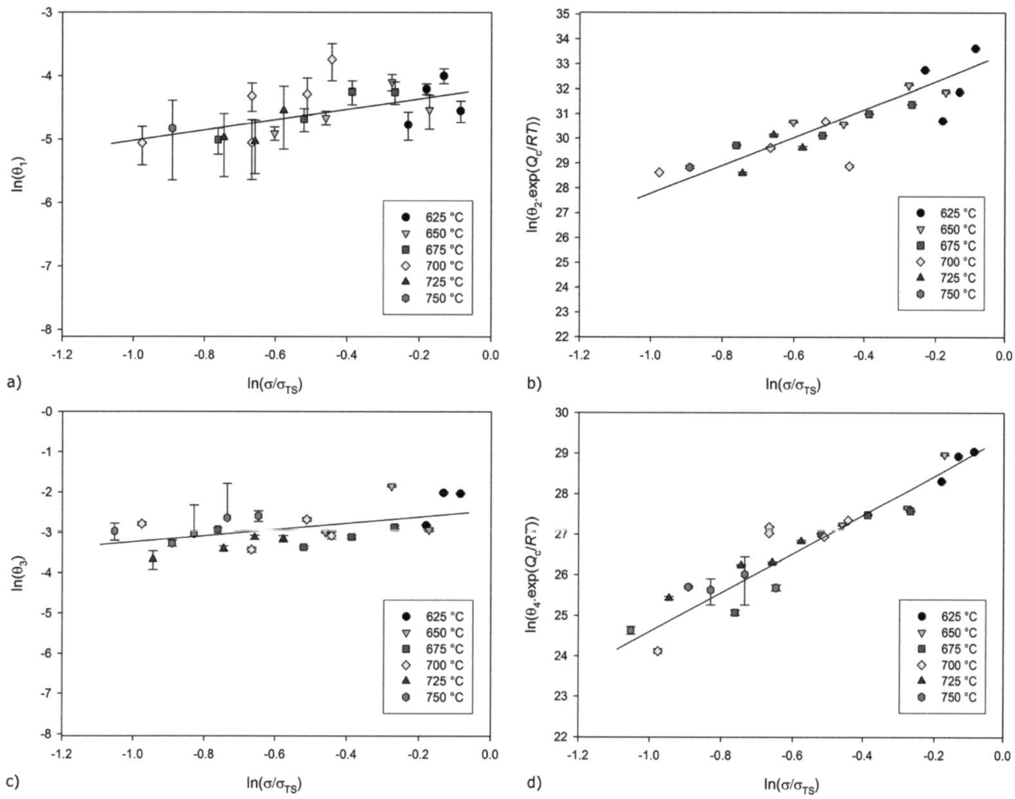

Figure 5. The stress and temperature dependence of: (a) θ_1; (b) θ_2; (c) θ_3 and (d) θ_4 for the titanium aluminide alloy Ti-45Al-2Mn-2Nb [20].

suitable hardening model is required. During uniaxial constant stress creep, creep rate, $\dot{\varepsilon}_c$, is dependent on applied stress, temperature and the time and/or strain for which the conditions have been applied.

$$\dot{\varepsilon}_c = f(\sigma, T, t) \tag{15}$$

However, in real engineering situations, the stress and/or temperature may change, resulting in the calculation of a different creep curve. Therefore, creep behaviour is now dependent on the position on the creep curve at new applied conditions and therefore more information is required to calculate creep rate. The simplest case is to consider that the new creep rate relates to the current creep time on the new curve (time-based hardening). This method, although easy to implement, fails to accurately account for strain rate changes for all but minor changes in applied conditions. For large changes in creep conditions, the shape of the creep curve changes significantly resulting in poor predictions. An alternative method is to relate creep rate to the applied conditions and creep strain history (strain-based creep hardening). For this case, the strain rate is calculated at the point on a creep curve which relates to the current total creep strain. This method, although more accurate that time-based hardening, produces inaccuracies when creep conditions change from primary creep dominated to tertiary creep dominated.

To address the shortcomings of these methods, a life-fraction hardening method is often used. This method uses an effective time based on t/t_F where t_F is the creep rupture life for the applied stress and temperature. A graphical representation of these hardening methods is given in **Figure 6**.

An alternative method of creep hardening is to base creep rate on the applied conditions and various material state variables

$$\dot{\varepsilon} = \Phi(\sigma, T, \xi_1, \xi_2, \dots, \xi_\alpha, \dots, \xi_n) \tag{16}$$

Where ξ_α (α = 1, n) are internal material state variables, each dependent on the loading history of the material. Evans [17], proposed that creep rate is dependent internal state variables representing dislocation hardening (*H*), recovery (*R*) and damage (*W*):

$$\dot{\varepsilon} = \dot{\bar{\varepsilon}}_0(1 + H + R + W) \tag{17}$$

where $\dot{\bar{\varepsilon}}_0$ is the initial effective creep rate of the virgin material. Each of these state variables represents the many mechanisms which control creep rate. These variables are related to both the strain and time history by the proportionality constants $\widehat{H}$, $\widehat{R}$ and $\widehat{W}$, such that:

$$\dot{H} = -\widehat{H}\dot{\varepsilon} \tag{18}$$

$$\dot{R} = \widehat{R} \tag{19}$$

$$\dot{W} = \widehat{W}\dot{\varepsilon} \tag{20}$$

The initial effective creep rate and proportionality constants can be related to applied condition using the θ coefficients from Eq. (7),

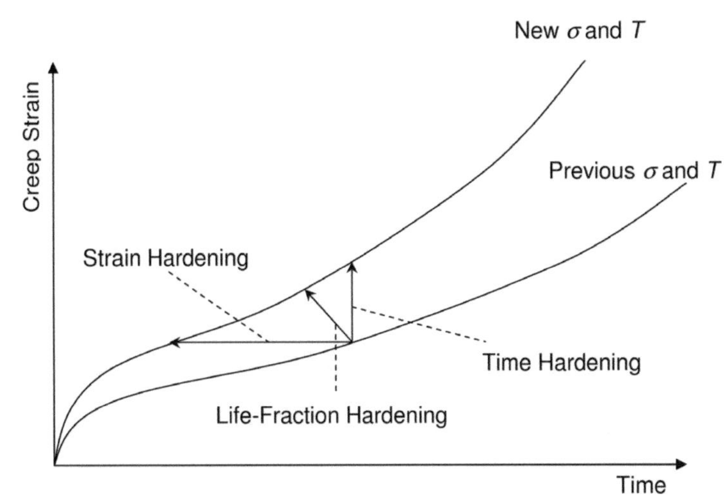

Figure 6. Schematic representation of time, strain and life-fraction hardening.

http://dx.doi.org/10.5772/intechopen.72319

$$\dot{\bar{\varepsilon}}_0 = \theta_1\theta_2 + \theta_3\theta_4 \tag{21}$$

$$\hat{H} = \frac{\theta_2}{\theta_1\theta_2 + \theta_3\theta_4} \tag{22}$$

$$\hat{R} = \frac{\theta_2\theta_3\theta_4}{\theta_1\theta_2 + \theta_3\theta_4} \tag{23}$$

$$\widehat{W} = \frac{1}{\theta_3} \tag{24}$$

If damage is assumed to be purely a tertiary process, the effective creep rate, $\dot{\bar{\varepsilon}}$, can be calculated from Eqs. (21)–(24) using [17]:

$$\dot{\bar{\varepsilon}} = \dot{\bar{\varepsilon}}_0\left(\frac{1 + H + R + \hat{R}}{\hat{H}\dot{\bar{\varepsilon}}_0}W\right) \tag{25}$$

For virgin material, *H*, *R* and *W* are equal to zero, however as creep continues their values are updated for each time increment, ∂t, using [17]:

$$H_{i+1} = H_i - \hat{H}\dot{\bar{\varepsilon}}_0(1 + H_i + R_i)\partial t \tag{26}$$

$$R_{i+1} = R_i + \hat{R}\partial t \tag{27}$$

$$W_{i+1} = W_i + \frac{\widehat{W}\hat{R}}{\hat{H}}(1 + W_i)\partial t \tag{28}$$

Transient uniaxial creep tests have been used to test the ability of different hardening methods to predict creep at non-constant stress and temperature for the nickel based superalloy, Waspaloy [8]. These creep tests applied temperature and stress which changed between two sets of conditions, one at high stress and low temperature, the other at low stress and high temperature, with both sets of conditions predicted to give similar lives approximately 10 days. The applied conditions were changed every 24 hours with a 1 hour transition period to allow sufficient time for the temperature to change, during which the stress was held at the lower level of the two conditions. **Figure 7** shows the creep rates observed from a transient creep test in which the temperature was changed from 600 to 750°C with applied stresses of 880 MPa and 390 MPa respectively, along with the test data for the equivalent isothermal constant stress creep tests. Immediately after each applied condition change, it can be seen that the creep rates are initially high before decreasing in what appears to be regions of 'pseudo' primary creep. In general, the creep rates observed during transient creep are higher than those observed for isothermal constant stress creep. The creep rates for the test at 600°C with an applied load of 880 MPa display a large proportion of primary creep, shown by the initially high creep rate followed by a gradual decrease, whereas the test at 750°C and 390 MPa is dominated by tertiary creep, with a minimum creep rate achieved after only 10% of life.

Eq. (8) was used to numerically describe the isothermal constant stress creep data to allow different hardening methods to be evaluated against transient creep rates (**Figure 8**). Time

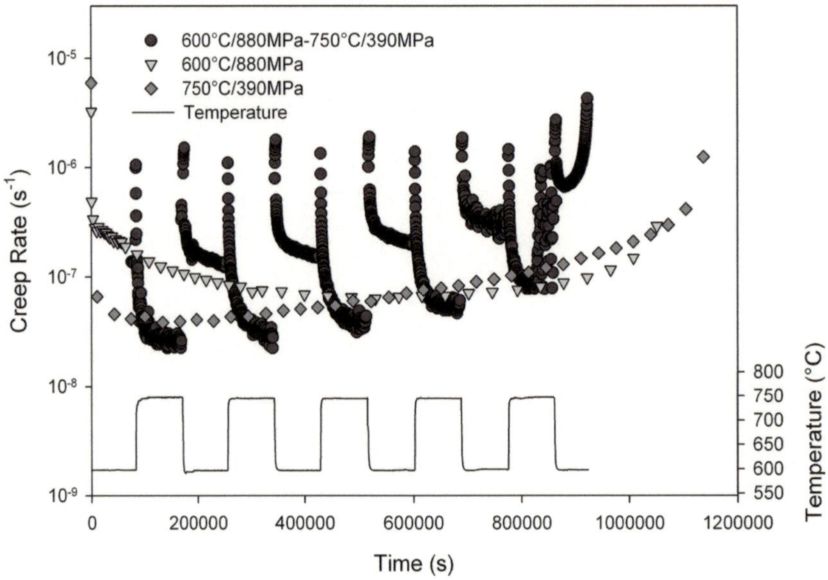

Figure 7. Creep rates obtained from transient and isothermal constant stress creep tests for Waspaloy at 600°C/880 MPa and 750°C/390 MPa [8].

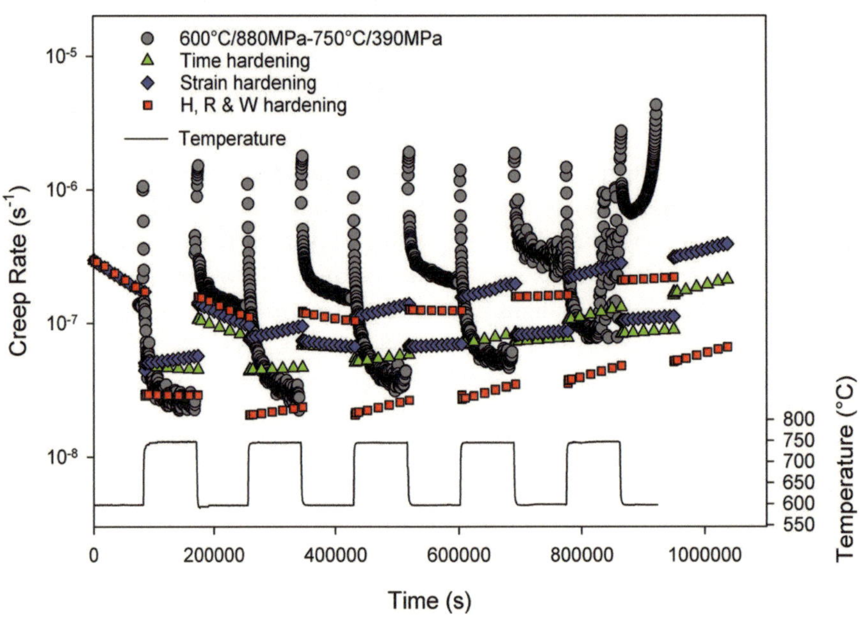

Figure 8. Creep transient creep behaviour for Waspaloy with predicted creep rates using time-hardening, strain hardening and a hardening model based on internal state variables.

hardening strain hardening and the hardening method based on internal state variables all predict the first cycle well, however for subsequent cycles the predictions using time and strain based hardening fail to predict the overall trend in creep rate, as well as the local peaks in rate after each load change. In this case, since the predicted rupture lives at each set of conditions is similar, life-fraction hardening would display a similar trend to time-based hardening. The creep rates predicted for a hardening method based on internal state variables describes the trend in rate more accurately but also fails to predict the 'pseudo' primary creep at the beginning of each cycle.

At both sets of creep conditions, creep occurs by diffusion controlled movement of dislocations, however, the dominant mechanism by which this occurs is varies between the high-stress/low-temperature and low-stress/high-temperature cases. At the high stress condition, the stress exceeds the yield point resulting in the formation of new dislocations. This leads to higher dislocations in the material at high stress than in the low stress material which only contains dislocations present after forming. Furthermore, at the higher temperature state, more thermal energy is available for diffusion controlled creep mechanisms such as climb. Whereas at low temperatures but higher stresses, precipitation cutting becomes more dominant. Therefore, when changing between two sets of creep conditions, the creep rate is dependent on the loading history of the material.

3. Creep damage and rupture

For simple uniaxial models failure can be creep rupture times may be predicted using simple equations such as those proposed to Norton [1], Larson and Miller [2]. However, for cases where the stress and temperature evolve over time a different approach is required. Creep damage models allow the accumulation of damage to be predicted regardless of applied conditions. Kachanov [21] proposed that the increase in creep rate during tertiary creep could be related to an increase in stress caused by a decrease effective cross sectional area due to the nucleation and growth of grain boundary cavities and triple point cracking. Later, Rabotnov [22] extended this idea by introducing a continuum damage parameter, ω, to represent this decrease in effective cross section. Leckie and Hayhurst [23] generalised this approach and Othman and Hayhurst [24] extended it to include the effects of primary creep. The constitutive model based on the θ-projection method includes internal state variables for dislocation hardening, H, recovery, R, and creep damage, W. Of particular interest to the case of creep rupture is W which represents the sum of all damage processes that occur during tertiary creep. W is zero for virgin material and increases as creep continues. Failure is predicted when this parameter exceeds a critical value, W_F. The value of W_F can be obtained from experimentally obtained creep tests using

$$W_F = \frac{1}{\theta_3}\left[\varepsilon_F - \theta_1\left(1 - e^{-\theta_2 t_F}\right)\right] \tag{29}$$

where ε_F is the creep strain at failure and θ_{1-3} are obtain using Eq. (8). W_F must then be related to applied creep conditions. For the γ-TiAl alloy, Ti-45Al-2Mn-2Nb, the following expression was found to represent the stress rupture data well

$$W_F = c + ln\left(\frac{\sigma}{\sigma_{TS}}\right)K \tag{30}$$

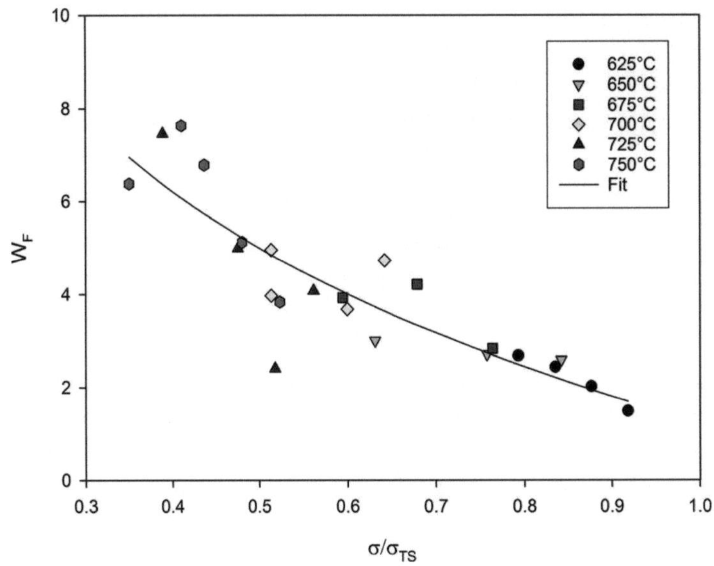

Figure 9. The dependence of W_F on applied creep conditions for Ti-45Al-2Mn-2Nb.

where σ_{TS} is the temperature dependent ultimate tensile strength and c and K are derived from experimental data by plotting W_F against $\ln(\sigma/\sigma_{TS})$. A plot of critical damage against applied stress is given in **Figure 9**. Since this data is derived from uniaxial test data, the applied stress, σ, may represent the von-Mises stress, $\overline{\sigma}$, or the maximum principle stress, σ_1.

Where multiaxial creep data is available, the dependence of W and W_F on σ_1 can be evaluated. For Waspaloy, W_F, can be related to applied stress using:

$$W_F = 237.3 \exp\left[-5.828\left(\frac{\sigma_1}{\sigma_{TS}}\right)\right] \tag{31}$$

where σ_{TS} is the temperature dependent tensile strength [17]. For other alloys such as the titanium alloy Ti6246, less dependence on loading direction was found and W_F can be related to $\overline{\sigma}$ [10].

4. Finite element analysis

Once suitable methods to interpolate creep curves for any given condition and suitable hardening models have been derived, it is possible to compile models for use in finite element analyses (FEA). The constitutive model based on the θ-project method [17] and the Wilshire creep curve method [13] have both been incorporated into FEA as user defined subroutines for the commercially available software Abaqus [10]. In order to evaluate creep deformation and damage, the user must define the magnitude of creep strain accumulated at each time step, based on stress, temperature and any user defined state variables. Furthermore, if implicit integration is used, the derivative of the creep strain increment with respect to stress must also be defined. User defined state variables such as those used to represent internal material damage (W) must also be evaluated. These models have been used to predict creep behaviour in various engineering applications [10, 25].

4.1. Stress relaxation

An application to consider is stress relaxation due to creep. This can be predicted using a simple single element finite element model with a prescribed displacement boundary condition. **Figure 10** shows a good correlation of the predicted and experimentally obtained results of a two stage stress relaxation experiment in the nickel based superalloy Alloy720Li using a model based on the strain hardening formulation of the Wilshire creep curve extrapolation method.

4.2. Notched bars

The value of FEA comes from its ability to predict mechanical behaviour in engineering components with complex geometries. Components such as those found in gas turbine aeroengines operate at high temperatures for extended periods of time and any geometric feature of the component that concentrates stress will exhibit higher creep rates than surrounding regions resulting in a redistribution of stress. Accurate predictions of this stress redistribution are particularly important for cases where component lives are influenced by fatigue as the stress field affects subsequent fatigue calculations. Furthermore, creep failure can initiate in subsurface locations due to a combination of stress redistribution and subsurface damage [10].

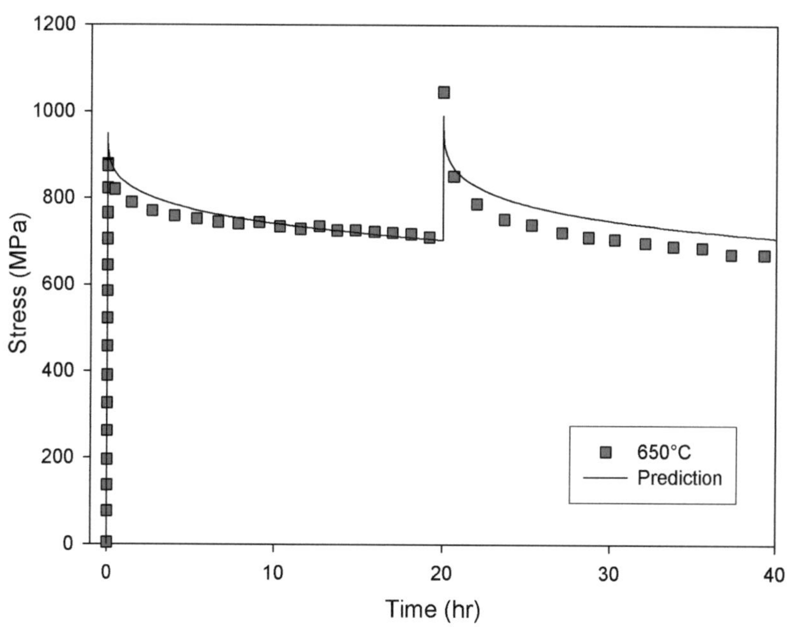

Figure 10. Uniaxial stress relaxation in Alloy720Li at 650°C with prediction using the Wilshire creep curve extrapolation method (Eqs. (5) and (6)) with strain-based hardening.

Predictions of stress redistribution in a Waspaloy round circumferentially notched bar after 1000 s of creep deformation using the θ-projection method can be seen in **Figure 11**.

The constitutive model based on the θ-projection technique contains material state variables for dislocation hardening, H, recovery, R, and damage, W. The latter allows for predictions of material failure as W exceeds a critical value W_F. In this model, material failure is simulated by reducing element stiffness when $W/W_F > 1$. In Abaqus, this is achieved by using a field variable dependent elastic modulus which decreases by an order of magnitude as the failure criteria is exceeded. Using this method, predictions of creep rupture can be made for cases when the stress and/or temperature is not constant. Predictions of the rupture times of both round and v-shaped circumferentially notched Waspaloy creep specimens correlate well with experimental data and predicted fracture paths are consistent with those observed experimentally [17]. Due to the dependence of W_F on stress triaxiality, different paths of failure are observed for different shaped notched with more diffuse damage occurring with lower stress intensities (K_T). Predicted creep rupture paths in round and v-shaped circumferentially notched Waspaloy specimens are shown in **Figure 12**.

4.3. Small punch creep

The small punch creep test is often used to characterise the mechanical properties of materials where only small quantities exist, such as during novel alloy development or for remnant life assessment [25–27]. The test consists of applying load to the centre of a small disc of material, typically 0.5 mm thick with a diameter of 9.5 mm, using a hemispherical punch. Punch load and displacement are then related to uniaxial stress and strain using conversion factor, k_{SP}. During small punch creep, the stress state evolves as punch displacement increases and therefore finite element analyses can provide useful information about the accumulation of creep strain and damage. Furthermore, by comparing the fracture path to the predicted stress, a greater understanding of the role of the stress state on creep rupture can be obtained.

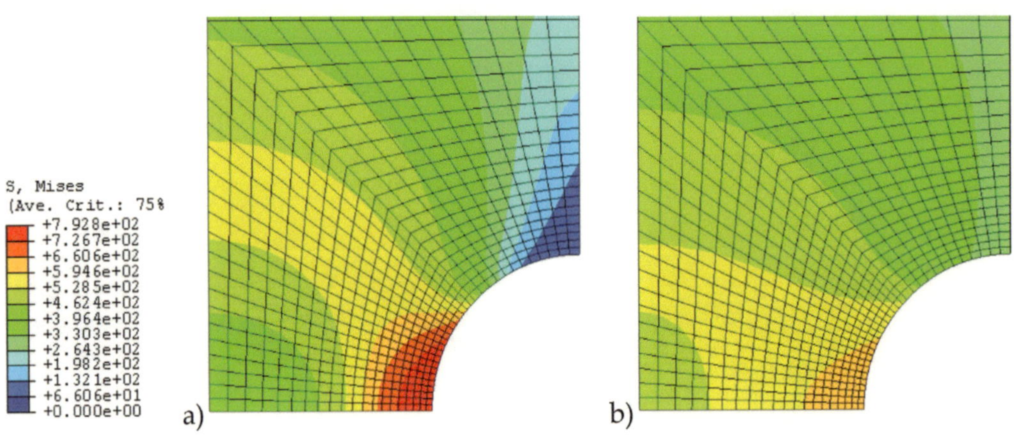

Figure 11. Stress distribution in a Waspaloy round circumferentially notched bar at (a) 0 s and (b) 1000 s at 650°C with an applied stress equating to 800 MPa across the specimen neck.

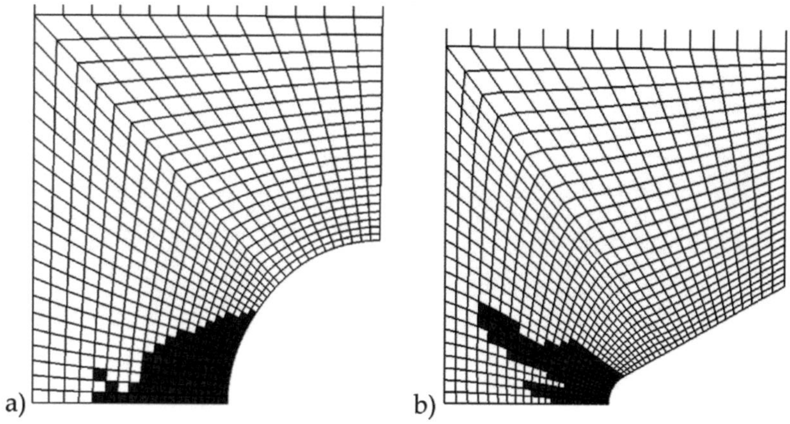

Figure 12. Predicted damage in a Waspaloy (a) round notched (K_T = 1.4) and (b) v-shaped notched (K_T = 2.234) test specimens.

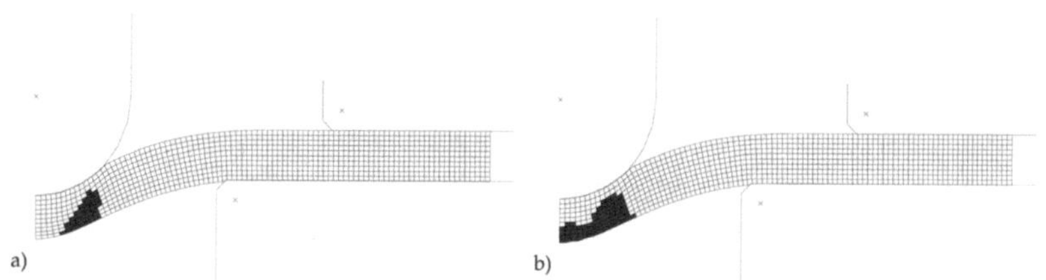

Figure 13. Predicted damage in a TiAl small punch test with damage based on (a) von-Mises stress and (b) maximum principle stress [25].

Predictions of the small punch creep behaviour of Ti-45Al-2Mn-2Nb have been made using the constitutive model based on the θ-projection method (**Figure 13**).

4.4. High-temperature fatigue

The effects of creep in engineering components that are exposed to cyclic loading at high temperatures cannot be ignored. A fatigue crack propagating through material exhibiting creep damage will advance more rapidly as creep cavities and microvoids provide a preferential path for growth. In this case fatigue lives are negatively affected by creep damage and cracks propagate along grain boundaries (intergranular). However, redistribution of stress due to creep can reduce stress at stress concentration features such as cracks and notches, reducing the driving force for crack propagation, hence increasing fatigue lives. This has been clearly shown in the titanium alloy Ti-6Al-4V whereby increasing the temperature from 450 to 500°C increases the fatigue lives of notched specimens, whereas a further increase to 550°C has a negative effect on life [28]. Therefore, during high temperature fatigue, component life is a dependent on time-dependent (creep) and time-independent (fatigue) damage mechanisms. FEA can be used to predict stress relaxation around notches and cracks, as well as to evaluate damage ahead of an advancing crack. Using a quasi-static model, the accumulation of creep deformation and damage ahead of an advancing fatigue crack in the titanium alloy, Ti-6246 has been predicted (**Figure 14**) [10].

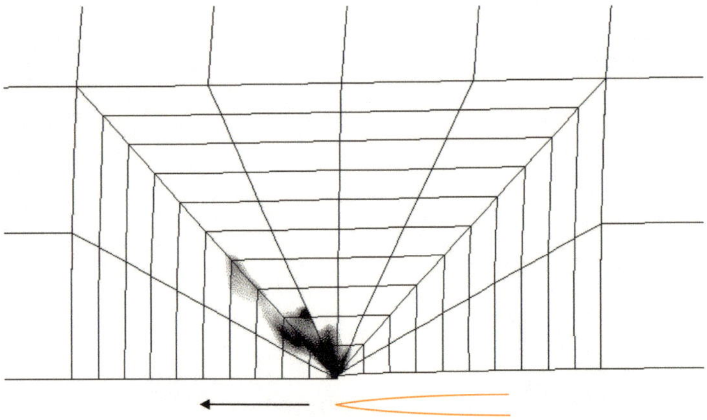

Figure 14. Predicted creep damage around the crack tip after 6×10^4 s at 773 K (R = 0.1, ΔK = 22) in Ti-6246 [10].

5. Conclusions

Although simple single point prediction methods can be useful for certain applications, accurate predictions of the mechanical response of some engineering components that operate at elevated temperatures require creep models that predict the full shape of the creep curve. These models, implemented in FEA have been used to predict creep behaviour in cases where stress and/or temperature are not constant, such as during stress relaxation. Furthermore, attempts have been made to relate model parameters to observed micromechanical behaviour. Creep damage models have been used to provide useful predictions of creep life which can be used to evaluate time dependent damage during other load cases, such as during high temperature fatigue and thermo-mechanical fatigue.

Author details

William Harrison*, Mark Whittaker and Veronica Gray

*Address all correspondence to: w.harrison@swansea.ac.uk

Swansea University, Swansea, Wales, UK

References

[1] Norton FH. The Creep of Steels at High Temperatures. New York: McGraw-Hill; 1929. 112 p

[2] Larson FR, Miller J. A time-temperature relationship for rupture and creep stresses. Trans. ASME. 1952;**74**:765-775

[3] Williams SJ, Bache MR, Wilshire B. 25 Year Perspective Recent developments in analysis of high temperature creep and creep fracture behaviour. Materials Science and Technology. 2010;**26**(11):1332-1337

[4] Wilshire B, Battenbough AJ. Creep and fracture of polycrystalline copper. Materials Science and Engineering A. 2007;**443**:156-166

[5] Whittaker MT, Harrison WJ, Lancaster RJ, Williams S. An analysis of modern creep lifing methodologies in the titanium alloy Ti6-4. Materials Science and Engineering: A. 2013; **577**:114-119

[6] Abdallah Z, Gray V, Whittaker M, Perkins K. A critical analysis of the conventionally employed creep lifing methods. Materials. 2014;**7**(5):3371-3398

[7] Monkman FC, Grant NJ. An empirical relationship between rupture life and minimum creep rate. In: Grant NJ, Mullendore AW, editors. Deformation and Fracture at Elevated Temperatures. Boston: MIT Press; 1965

[8] Harrison WJ, Whittaker MT, Deen C. Creep behaviour of Waspaloy under non-constant stress and temperature. Materials Research Innovations. 2013;**17**(5):323-326

[9] Whittaker M, Lancaster R, Harrison W, Pretty C, Williams S. An empirical approach to correlating thermo-mechanical fatigue behaviour of a polycrystalline Ni-base superalloy. Materials. 2013;**6**(11):5275-5290

[10] Harrison W. Creep modelling of Ti6246 and Waspaloy using ABAQUS [thesis]. UK: University of Wales Swansea; 2007

[11] Gray V, Whittaker M. Development and assessment of a new empirical model for predicting full creep curves. Materials. 2015;**8**(7):4582-4592

[12] Abdallah Z, Perkins K, Williams S. Advances in the Wilshire extrapolation technique—Full creep curve representation for the aerospace alloy Titanium 834. Materials Science and Engineering: A. 2012;**550**:176-182

[13] Harrison W, Whittaker M, Williams S. Recent advances in creep modelling of the nickel base superalloy, alloy 720Li. Materials. 2013;**6**(3):1118-1137

[14] Whittaker MT, Wilshire B. Creep and creep fracture of 2.25Cr-1W steels (Grade 23). Materials Science and Engineering A. 2010;**527**(18):4932-4938

[15] Whittaker MT, Harrison WJ, Deen C, Rae C, Williams S. Creep deformation by dislocation movement in Waspaloy. Materials. 2017;**10**(61):1-14

[16] Evans RW, Parker JD, Wilshire B. An extrapolation procedure for long-term creep strain and creep life prediction, with special reference to 0.5Cr0.5Mo0.25V ferritic steels. In: Wilshire B, Owen DRJ, editors. Recent Advances in Creep and Fracture of Engineering Materials and Structures. Swansea: Pineridge Press; 1982. p. 135

[17] Evans RW. A constitutive model for the high-temperature creep of particle-hardened alloys based on the θ projection method. Proceedings of the Royal Society of London A: Mathematical, Physical and Engineering Sciences. 2000;**456**(1996):835-868

[18] Dyson BE, McLean M. Microstructural evolution and its effects on the creep performance of high temperature alloys. In: Microstructural Stability of Creep Resistant Alloys for High Temperature Plant Applications. Institution of Metals, London; 1998, pp. 371-393

[19] Wu X, Williams S, Gong D. A true-stress creep model based on deformation mechanisms for polycrystalline materials. Journal of Materials Engineering and Performance. 2012;**21** (11):2255-2262

[20] Harrison W, Abdallah Z, Whittaker M. A model for creep and creep damage in the γ-titanium aluminide Ti-45Al-2Mn-2Nb. Materials. 2014 Mar 14;**7**(3):2194-2209

[21] Kachanov LM. Rupture time under creep conditions. International journal of fracture. 1999 Apr 1;**97**(1-4):11-18

[22] Rabotnov YN. Creep Problems in Structural Members. Amsterdam: North-Holland; 1969

[23] Leckie FA, Hayhurst DR. Constitutive equations for creep rupture. Acta Metallurgica. 1977;**25**(9):1059-1070

[24] Othman AM, Hayhurst DR. Multi-axial creep rupture of a model structure using a two parameter material model. International Journal of Mechanical Sciences. 1990;**32**(1):35-48

[25] Lancaster RJ, Harrison WJ, Norton G. An analysis of small punch creep behaviour in the γ titanium aluminide Ti–45Al–2Mn–2Nb. Materials Science and Engineering: A. 2015;**626**: 263-274

[26] Kobayashi KI, Kajihara I, Koyama H, Stratford GC. Deformation and fracture mode during small punch creep tests. Journal of Solid Mechanics and Materials Engineering. 2010;**4**(1):75-86

[27] Blagoeva D, Li YZ, Hurst RC. Qualification of P91 welds through small punch creep testing. Journal of Nuclear Materials. 2011;**409**(2):124-130

[28] Whittaker MT, Harrison W, Hurley PJ, Williams S. Modelling the behaviour of titanium alloys at high temperature for gas turbine applications. Materials Science and Engineering: A. 2010;**527**(16):4365-4372

Chapter 12

Fundamental Models for the Creep of Metals

Rolf Sandström

Additional information is available at the end of the chapter

http://dx.doi.org/10.5772/intechopen.70726

Abstract

Analysis of creep properties has traditionally been made with empirical methods involving a number of adjustable parameters. This makes it quite difficult to make predictions outside the range of the original data. In recent years, the author has formulated basic models for prediction of creep properties, covering dislocation, particle and solid solution hardening. These models do not use adjustable parameters. In the present chapter, these models are further developed and utilised. The dislocation mobilities play an important role. The high-temperature climb mobility is extended to low temperatures by taking vacancies generated by plastic deformation into account. This new expression verifies the validity of the combined climb and glide mobility that has been used so far. By assuming that the glide rate is controlled by the climb of the jogs, a dislocation glide mobility is formulated. The role of the mobilities is analysed, and various creep properties are derived. For example, secondary creep rates and strain versus time curves are computed and show good agreement with experimental data.

Keywords: creep, dislocation, mobility, model, creep strain

1. Introduction

If metals are exposed to load at high temperatures, a slow deformation called creep takes place. A characteristic feature of creep is that it occurs even when the load is kept constant. This should be contrasted to plastic deformation at ambient temperatures where an increase in the load is needed to generate further plastic strain. Creep in metals has been studied for many decades, and a number of excellent textbooks exist on the subject, see for example [1–4]. A common way of measuring the creep deformation is to apply a fixed load to a specimen and then record its elongation as a function of time. The relative increase in the elongation is referred to as the (creep) strain. Its time derivative is called the creep strain rate. Another important quantity is the stress, i.e. the load divided by the specimen cross section. After sufficient test time, the specimen ruptures. The rupture time and the specimen elongation at rupture are recorded.

A creep test typically shows three stages: primary, secondary and tertiary. During the primary stage, the creep rate is initially high but shows a gradual decrease until it reaches a constant value, which corresponds to the secondary stage. This stage usually takes the longest time. Finally, the creep rate starts to increase. This is the tertiary stage that eventually leads to rupture. The dependence of the creep strain as a function of time is referred to as a creep (strain) curve. Although the described shape of the creep curve is the most common one, there are many variants.

Creep-exposed materials can be found in many types of high-temperature plants, for example, fossil-fired power plants. The desire is to make the materials as strong as possible, and the creep rate low to ensure a long lifetime. The deformation rate is controlled by the movement of the dislocations, i.e. the line defects that are present in large numbers. If the movement of the dislocations is fast, the creep strength is low. To increase the creep strength, the movement of the dislocations must be reduced. Typically, the most efficient way to hinder this movement is to have a high content of other (forest) dislocations. There is a strong interaction between different dislocations. The second most used way of increasing the strength is to introduce particles in the microstructure. Particles or precipitates are a very potent way of raising the strength. A third way is to have elements in solid solution. The difference in size between the solute and the matrix atoms makes it more difficult for the dislocations to move.

Most creep investigations concern metals above half the melting point. The stress dependence of the strain rate in the secondary stage has always generated much interest in creep research because it has been assumed to reflect the operating dislocation mechanism. In **Figure 1**, such dependence is illustrated for 0.5Cr0.5Mo0.25V steel at 565°C over a wide range of stresses.

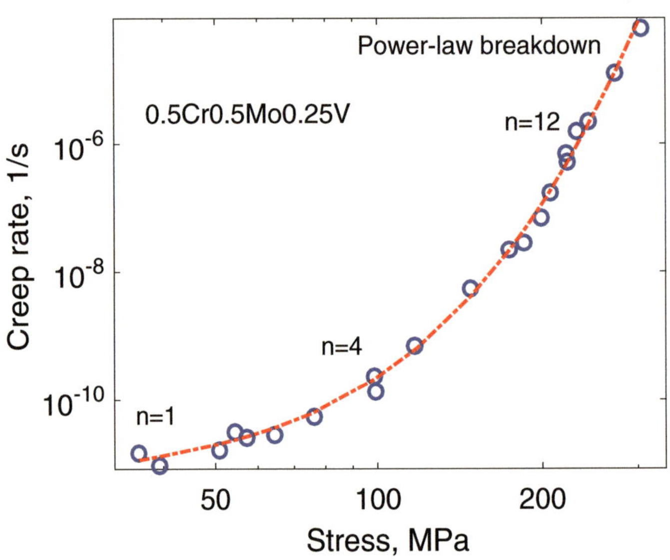

Figure 1. Creep rate versus stress for 0.5Cr0.5Mo0.25V steel at 565°C. The *n* values indicate the exponent in the power-law creep law. At large stresses, the creep rate increases exponentially with the stress. This is referred to as power-law breakdown. Some of the data points are extrapolated. (After Wilshire [5].).

The slope of the curve is the creep exponent n. At intermediate stresses (and temperatures), the creep exponent is typically in the range 3–8. In the figure, it is 4. At high stresses (and at low temperatures), the creep exponent is much higher, in the figure illustrated with $n = 12$. At still higher stresses, the creep rate varies exponentially with stress, which is called power-law breakdown. At very low stresses, the n value is small, sometimes approaching unity. 0.5Cr0.5Mo0.25V steel is clearly a particle-strengthened material. Other particle-strengthened alloys can show much higher creep exponents than in **Figure 1**.

In the past, the creep exponent has often been used to identify the operating dislocation mechanisms. For intermediate exponents (3–8), climb of dislocations is in general considered as the operating mechanisms although glide has also been assumed for certain alloy types. This will be discussed below. At high stresses, glide has been suggested as the dominating mechanism. At low creep exponents approaching unity, diffusion creep has been assumed as the main mechanism. The consistent change of operating mechanism with stress has been challenged, see for example [5].

In recent years, basic creep models have been formulated. With the help of these models, the assumptions mentioned above will be reanalysed in the present chapter.

2. Dislocation model

To understand the creep process, the key quantity is the dislocation density and its variation with time. Models for the dislocation density development during creep have been available for a long time [6]. Recently, these models have been expanded and derived more precisely. We will use the following formulation [7]:

$$\frac{d\rho}{d\varepsilon} = \frac{m}{bc_L}\rho^{1/2} - \omega\rho - 2\tau_L M\rho^2/\dot{\varepsilon} \tag{1}$$

where ρ is the dislocation density, ε the strain, m the Taylor factor, b the Burgers vector, τ_L the dislocation line tension, M the dislocation mobility and $\dot{\varepsilon}$ the strain rate. The value of the dislocation mobility M will be discussed below. The dynamic recovery parameter ω for pure metals and the work-hardening constant c_L are given by the following expressions [8, 9]:

$$\omega = \frac{m}{b}d_{\text{int}}\left(2 - \frac{1}{n_{\text{slip}}}\right) \tag{2}$$

$$c_L = \frac{m^2\alpha G}{\omega(R_m - \sigma_y)} \tag{3}$$

where d_{int} is the interaction distance between dislocations where dislocations of opposite sign get close enough to annihilate each other and thereby reduce the dislocation content. This distance is taken as the core diameter of the dislocations. n_{slip} is the number of active slip systems, which is 12 for fcc alloys. α is a constant in the Taylor equation, G is the shear modulus and R_m and σ_y are the true tensile strength and yield strength at ambient temperatures.

According to Eqs. (2) and (3), ω and c_L are temperature-independent constants. This is also in close agreement with experiments [7, 9]. In some papers, ω has been found to be temperature dependent [10, 11]. However, in these cases the final term in Eq. (1), the static recovery term, has not been considered, which should cover at least part of the temperature dependence. Eq. (1) can be considered as a basic equation for the development of the dislocation density. A detailed derivation of Eq. (1) can be found in [8].

Eq. (1) describes the development of the dislocation density with strain. The first term on the right-hand side is responsible for the generation of dislocations, i.e. work hardening. The two other terms take into account the annihilation of dislocations. The second term gives a strain-controlled recovery, which is called dynamic recovery. The final term is referred to as static recovery. It is time controlled since the strain rate appears in the denominator. The terminology of recovery is not consistent in the literature, which one has to be aware of.

3. Dislocation mobilities

3.1. Climb mobility

The dislocation mobility M in Eq. (1) is an important quantity. It describes the speed v of the dislocations:

$$v = Mb\sigma \tag{4}$$

where σ is the applied stress. Dislocations can propagate by glide along their slip system or by climb perpendicular to the slip system. The latter process requires emission or absorption of atoms by diffusion, and it is slower than glide. The climb mobility at high temperatures (>0.4 T_m where T_m is the melting temperature) was derived by Hirth and Lothe [12]:

$$M_{\text{climb}} = \frac{D_{s0}b}{k_B T} e^{\frac{\sigma b^3}{k_B T}} e^{-\frac{Q}{R_G T}} \tag{5}$$

where T is the absolute temperature, σ the applied stress, D_{s0} the pre-exponential coefficient for self-diffusion, Q the activation energy for self-diffusion, k_B Boltzmann's constant and R_G the gas constant.

At lower temperatures, the climb rate is influenced by the generation of vacancies due to plastic deformation. A brief derivation of this effect will be given here since it cannot be found in the literature. When a climbing dislocation is forced to move, it will emit or absorb vacancies. When gliding dislocations cut each other, jogs in the form of steps of the length of a Burgers vector are formed on them. The jogs are often sessile and must then climb when they move and hence emit or absorb vacancies.

Mecking and Estrin [13] have estimated the number of vacancies produced mechanically in a unit volume per unit time as:

$$P = 0.5\frac{\sigma\dot{\varepsilon}}{Gb^3} \tag{6}$$

The quantities in this equation have been defined above. In [13] the constant in Eq. (6) was estimated to 0.1. A detailed derivation shows that it is 0.5. The annihilation rate A for the excess vacancies was found to be

$$A = \frac{D_{vac}}{\lambda^2}(c - c_0) \tag{7}$$

where c_0 is the equilibrium vacancy concentration and $\Delta c = c - c_0$ is the excess concentration. D_{vac} is the diffusion constant for the vacancies. λ is the spacing between vacancy sinks. Assuming the presence of a substructure, λ can be related to the cell or subgrain size d_{sub} [13], which in turn can be found from the applied stress:

$$\lambda = d_{sub} = \frac{K_{sub}Gb}{\sigma} \tag{8}$$

where K_{sub} is a constant that typically takes values from 10 to 20. From Eqs. (6)–(8), we find the following expression for the excess vacancy concentration:

$$\frac{\Delta c}{c_0} = 0.5\frac{\sqrt{2}K_{sub}^2\dot{\varepsilon}b^2}{D_{self}}\frac{G}{\sigma} \tag{9}$$

In deriving Eq. (9), we have also made use of a relation for the self-diffusion coefficient:

$$D_{self} = c_0\Omega D_{vac} \tag{10}$$

where Ω is the atomic volume. It is now assumed that the climb rate is proportional to the total vacancy concentration. This is the same assumption as was made in [13]. The increase in the climb rate g_{climb} due to the presence of excess vacancy concentration from Eq. (10) is then

$$g_{\text{climb}} = 1 + \frac{\Delta c}{c_0} \tag{11}$$

To find the total climb mobility, the expression in Eq. (5) should be multiplied by g_{climb}:

$$M_{\text{climb enh}} = M_{\text{climb}}g_{\text{climb}} \tag{12}$$

3.2. The glide mobility

In a dislocation-free crystal, the glide mobility is very high. Edington measured a mobility of $M_0 = 1 \times 10^4$ 1/Pa/s for a copper single crystal [14]. In an alloy where a forest of dislocations is present, the mobility is much lower. During deformation as described above, jogs will be formed on the dislocation. Many times the jogs have to move perpendicular to their glide planes. This means that they are sessile, i.e. they have to move by climb [12], which is a slow process. It is natural to assume that it is the motion the jogs that controls the glide rate and that is what we will do. This is also what Hirth and Lothe have assumed [12].

The basis for the glide mobility is Eq. (12), since the jogs move by climb. However, there is an additional factor. Only a small fraction of each dislocation consists of jogs. Since the jogs move slowly, the forces on the dislocations are concentrated to the jogs. The average distance between jogs can be determined from the dislocation density ρ as $l_{jog} = 1/\sqrt{\rho}$. The force F on a dislocation is given by the Peach-Koehler formula $F = b\,\sigma\,l$ where l is the length of the dislocation. If l is chosen as l_{jog}, F will be the force on each jog. Consequently, the stress on the jogs is increased by

$$g_{glide} = \frac{l_{jog}}{b} = \frac{1}{b\sqrt{\rho}} \tag{13}$$

where the length of a jog is taken as the length of the Burgers vector. With the help of Taylor's equation,

$$\sigma = \sigma_y + \alpha m G b\sqrt{\rho} \tag{14}$$

where σ_y being the yield strength, Eq. (13) can be rewritten as

$$g_{glide} = \frac{\alpha m G}{\sigma - \sigma_y} \tag{15}$$

The glide mobility is obtained by multiplying the climb mobility by g_{glide}:

$$M_{glide} = M_{climb} g_{climb} g_{glide} \tag{16}$$

Eq. (16) applies to both edge and screw dislocations. With the assumptions considered, the climb and glide mobility are closely related. g_{glide} is roughly equal to the ratio between the shear modulus G and the applied stress σ. Since G is much larger than σ, g_{glide} is always significantly larger than unity. Thus, the glide mobility is larger than the climb mobility, which is a common assumption when modelling creep.

3.3. Cross slip mobility

Screw dislocations can change glide plane with the help of cross slip. This can simplify the annihilation of dislocations with opposite signs and thereby contribute to the recovery. Cross slip requires an additional activation energy E_{cs}. Püschl gave the following estimate of E_{cs} [15]:

$$E_{cs} = 0.012 G b^3 \frac{d_{SFE}}{b} \ln\left(\frac{2 d_{SFE}}{b}\right) \tag{17}$$

where d_{SFE} is the width of a stacking fault [12]:

$$d_{SFE} = \frac{Gb^2}{8\pi\gamma_{SFE}} \frac{(2-\nu_P)}{(1-\nu_P)} \tag{18}$$

where ν_P is Poisson's ratio and γ_{SFE} is the stacking-fault energy. With stacking-fault energies of 45 mJ/m^2 for copper and 166 mJ/m^2 for aluminium [16], the resulting values for E_{cs} become 560 and 40 kJ/mol, respectively. Thus, a pronounced temperature dependence is obtained. Eq. (17) is based on elasticity theory models. However, recently ab initio calculations have been performed providing similar results to those of Eq. (17) [17, 18]. Now, the influence of cross slip on the mobility can be introduced:

$$g_{\text{cross-slip}} = \exp\left(-\frac{E_{cs}}{R_G T}\right) \tag{19}$$

$$M_{\text{cross-slip}} = M_{\text{climb}} g_{\text{climb}} g_{\text{glide}} g_{\text{cross-slip}} \tag{20}$$

The consequences of the strong temperature dependence in the model for cross slip mobility will be discussed later in Section 6.

3.4. The climb-glide mobility

The results in Sections 3.1–3.3 are new. It has been recognised that the climb mobility in Eq. (5) predicted far too low creep rates at low temperatures and high stresses. It was thought that glide could be the controlling mechanism during these conditions. For this purpose, a combined climb and glide mobility was introduced [19]:

$$M_{\text{clglide}} = M_{\text{climb}} f_{\text{clglide}} \tag{21}$$

where f_{clglide} is given by

$$f_{\text{clglide}} = \exp\left(\frac{Q}{R_G T}\left(\frac{\sigma}{R_{\text{max}}}\right)^2\right) \tag{22}$$

where R_{max} is the true tensile strength at ambient temperatures. Eq. (22) has several important consequences at low temperatures. First, it reduces the activation energy for creep. Second, it increases the creep rate by a large factor. Third, it raises the value of the creep exponent dramatically. These results are in excellent agreement with experiments [9, 20]. Examples will be given below in Sections 4 and 5.

The derivation of Eq. (21) was inspired by the work of Kocks et al. [21]. They gave an empirical expression for the glide mobility. Unfortunately, it involved five unknown parameters and was therefore of little use directly. However, some of the parameters could be fixed by following a procedure due to Nes where an integrated climb and glide mobility was formulated [22]. The remaining parameters could be set with the help of work by Chandler [23].

Ideally, to describe creep, the fundamental models for the mobilities derived in Sections 3.1–3.3 should be used. However, they are difficult to use directly since g_{climb} involves the strain rate.

Instead, the equations in Sections 3.1–3.3 will be used to verify Eq. (22), which can then be applied to derive the creep rate. Direct comparison between g_{climb} and $f_{clglide}$ is shown in **Figures 2** and **3** for aluminium.

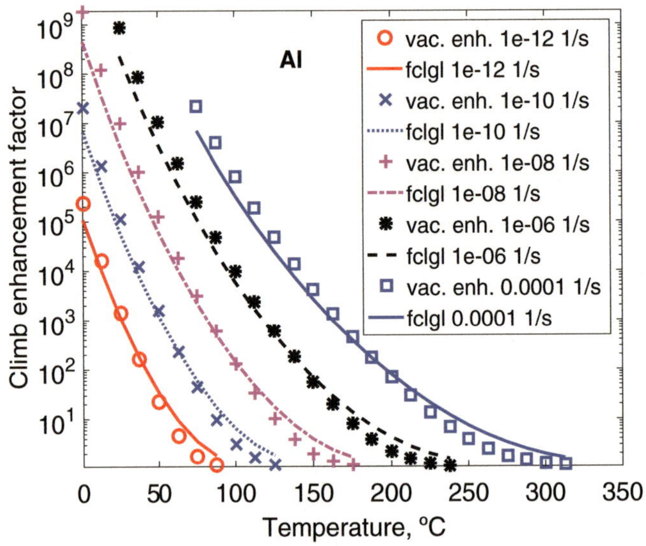

Figure 2. Climb enhancement factor versus temperature at five strain rates for aluminium. Eq. (11), for the increase in vacancy concentration due to plastic deformation, is compared with Eq. (22) for the climb-glide enhancement.

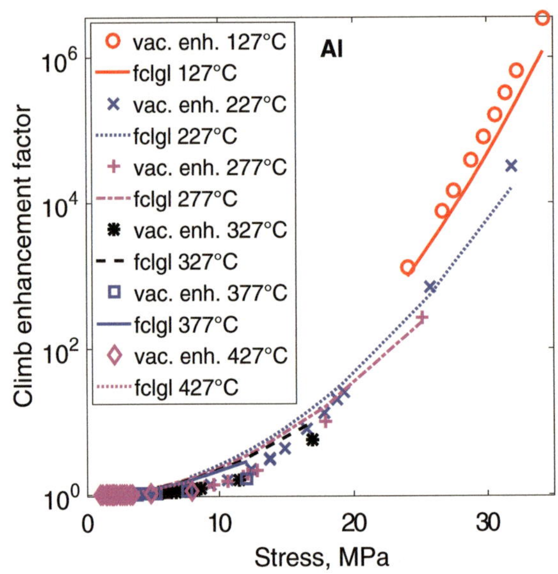

Figure 3. Climb enhancement factor versus stress at six temperatures for aluminium. Eq. (11) is compared with Eq. (22). The values of stresses and strain rates are taken from experimental creep data [24].

In **Figure 2**, a continuous set of parameters for temperature and strain rate are used, whereas in **Figure 3** experimental values are applied. It can be seen that the enhancement in vacancy concentration due to plastic deformation can fully explain the increase in creep rate in relation to the high-temperature climb mobility.

A second example of the comparison is given in **Figure 4** for copper.

Again the agreement between the two sets of models is quite good. The temperature, stress and strain rate dependences are well covered in **Figures 2–4**.

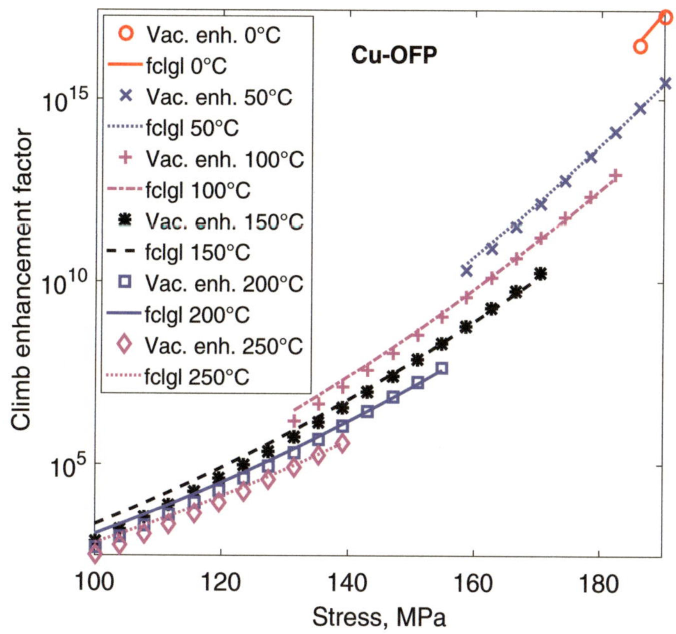

Figure 4. Climb enhancement factor versus stress at six temperatures for Cu-OFP. Eq. (11) for the increase in vacancy concentration due to plastic deformation is compared with Eq. (22) for the climb-glide enhancement.

4. Secondary creep

4.1. Pure elements

Our understanding of the creep process is largely based on the creep recovery theory [6]. The key feature of this theory is that the recovery rate is sufficiently rapid that the dislocation density can be kept constant during secondary creep. If the dislocation density is continuously rising, the creep deformation will slow down and eventually stop, which is contrary to observations. Thus, there is a balance between the generation and the annihilation of dislocations during creep. If we assume stationary conditions, the strain derivative in Eq. (1) vanishes. The resulting expression for secondary strain rate is

$$\dot{\varepsilon}_{\text{sec}} = 2\tau_{\text{L}} M \rho^{3/2} / \left(\frac{m}{bc_L} - \omega \rho^{1/2} \right) \tag{23}$$

Using Taylor's Eq. (14), Eq. (23) can be expressed in terms of stresses:

$$\dot{\varepsilon}_{sec} = h(\sigma - \sigma_i) \text{ with } h(\sigma) = 2\tau_L M(T,\sigma)\frac{\sigma^3}{(\alpha m G b)^3} / \left(\frac{m}{bc_L} - \omega\frac{\sigma}{\alpha m G b}\right) \quad (24)$$

$$\sigma_{disl} = \alpha m G b \rho^{1/2} = \sigma - \sigma_i \quad (25)$$

where σ_{disl} is the dislocation stress. σ_i is an internal stress that can have contributions from the yield strength, solid solution hardening and particle hardening. At low stresses, the mobility M is given by Eq. (5) that is independent of stress. If no internal stress is present, Eq. (24) gives an approximate power-law expression with a stress exponent of 3. Such a stress exponent is often observed at high temperatures for austenitic stainless steels [25].

With our present knowledge, the natural assumption is that static recovery is controlled by climb. This is analysed in Section 6. This means that it is the climb mobility in Eq. (12) that should be used in Eq. (24). In addition, we saw in Section 3.4 that the expression for the enhancement factor for the climb mobility g_{climb} due to the raised vacancy concentration in Eq. (11) agreed with the climb-glide enhancement factor $f_{clglide}$ in Eq. (22). Since the implication of $f_{clglide}$ is known to successfully have described experimental data, this gives further support to the use of Eq. (12).

Eq. (24) will now be applied to pure aluminium. For σ_i, the Peierls stress will be used. Although a Peierls stress is not usually considered to be of importance for fcc alloys, recent studies suggest that this conclusion is not true for Al. With ab initio methods, Shin and Carter

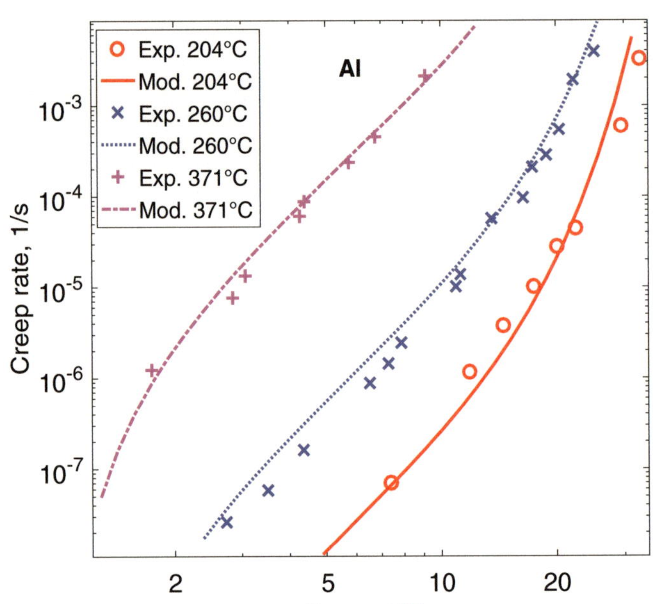

Figure 5. Secondary creep rate versus stress for pure aluminium. Eq. (24) is compared to experimental data from [27].

found the following value for the Peierls stress of edge dislocation σ_{pe} [26]. The value for screw dislocations was much smaller:

$$\sigma_{pe} = 4.9 \times 10^{-5} G \tag{26}$$

The application of Eq. (24) is illustrated in **Figure 5**.

At intermediate stresses in **Figure 5**, the slope of the curves is about 4.5, which is the value of the stress exponent. At higher stresses, the slope increases that is referred to as power-law breakdown. At low stresses, there is also an increase of the stress exponent. This is due to the presence of the internal stress in Eq. (26). It can be seen that the model in Eq. (24) can handle these three regions of the creep rate versus stress curves quite well.

4.2. Solid solution

The presence of elements in solid solution has two effects on the secondary creep rate. It gives rice to a drag stress or a break stress, and it increases the activation energy for creep. The increase in the activation energy is $U_j^{\max}$, which is the maximum interaction energy between a dislocation and a solute j [28]:

$$U_j^{\max} = \frac{\beta}{b} = \frac{1}{\pi} \frac{(1+\nu_P)}{(1-\nu_P)} G \Omega_0 \varepsilon_j \tag{27}$$

where Ω_0 is the atomic volume and ε_j the linear lattice misfit of solute j. The additional contribution to the activation energy is taken into account by multiplying the dislocation mobility by the factor f_Q:

$$f_Q = e^{-U_j^{\max}/RT} \tag{28}$$

For slowly diffusing solutes, the contribution to the internal stress, cf. Eq. (24), is the drag stress [28]:

$$\sigma_j^{drag} = \frac{v c_{j0} \beta^2}{b D_j k_B T} I(z_0) \tag{29}$$

where v is the dislocation speed, cf. Eq. (4), c_{j0} is the concentration of solute j and D_j the diffusion constant for solute j. $I(z_0)$ is an integral of $z_0 = b/r_0 k_B T$ where r_0 is the dislocation core radius. $I(z_0)$ often takes values of around 3.

The use of Eq. (29) is illustrated for Al-Mg alloys in **Figure 6**. The drag stress is added to the internal stress in Eq. (24). The factor in Eq. (28) is also taken into account which raises the activation energy for creep by the amount $U_j^{\max}$, where j refers to Mg.

In the same way as for pure aluminium, there are three stages of stress dependence. In the middle range of stresses, power-law behaviour is obeyed. At low stresses, there is a slight increase in the creep exponent due to the presence of the Peierls stress that is the same as for pure aluminium. At high stresses, power-law breakdown takes place with an increase in the stress exponent.

The modelling in **Figure 6** is based on the climb mobility, so climb is assumed to be the controlling mechanism over the full-stress range. In the literature, it has frequently been assumed that glide is controlling in the middle stress range, see for example [31]. This should be due to a larger effect of solid solution hardening on gliding than on climbing dislocations. This is difficult to understand, since solid solution hardening has about the same effect for both mechanisms [32]. In addition, why there should be transitions in mechanism at low and high stresses is not obvious. Sometimes, it is assumed that the solutes break away from the dislocations at high stresses, but that is predicted to take place at much higher stresses than where the transition takes place [12]. Considering glide as a controlling mechanism is not consistent with the glide mobility in Eq. (16). If that is applied, the experimental results in **Figure 6** cannot be reproduced.

For fast diffusion elements such as interstitials, the drag stress according to Eq. (29) is usually negligibly small. Instead, the solutes are locked to the dislocations, and they have to break away to become mobile. The size of the break stress that should be added to the internal stress σ_i in Eq. (24) is given by [28, 32]

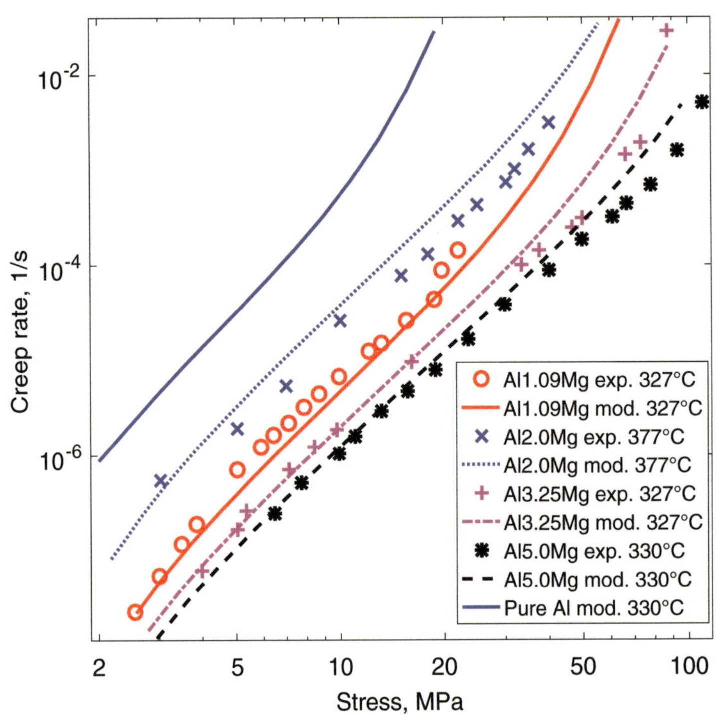

Figure 6. Secondary creep rate versus stress for Al-Mg alloys. Eq. (24) with the stress contribution to the internal stress from Eq. (29) and the increase in activation energy from Eq. (28) is compared to experimental data [29–31].

$$\sigma_{j_{\text{break}}} = \frac{U_j^{\max} c_{j0}}{b^3} \int_{y_L}^{y_R} e^{-\frac{W_j(y)}{kT}} dy \tag{30}$$

The break stress is proportional to $U_j^{\max}$. $W_j(y)$ is the interaction energy between a solute j and a dislocation at a distance y. The integration is performed over a distance of about $\pm 20\ b$. The application of Eq. (30) is illustrated for copper with 100 at. ppm phosphorus (Cu-OFP). At low temperatures, phosphorus gives a strong improvement of the creep strength. The solute phosphorus has two effects. It increases the creep activation energy by $U_j^{\max}$, and a break stress is added to the internal stress in Eq. (24). The result is shown in **Figure 7** for oxygen-free copper (Cu-OF) with and without phosphorus.

From **Figure 7**, it is obvious that the model can describe the influence of phosphorus on the creep rate quantitatively. No adjustable parameters are used in this figure (or in any other figure in the chapter). For Cu-OFP the creep exponent is about 65 in **Figure 7**. This demonstrates the validity of Eq. (24) also deep down in the power-law breakdown regime.

4.3. Stress-strain relations

Handling of stress-strain curves is not the subject of this chapter. The only message is that Eq. (1) can also be used to predict stress-strain curves. For further details, the reader is referred to [7, 33].

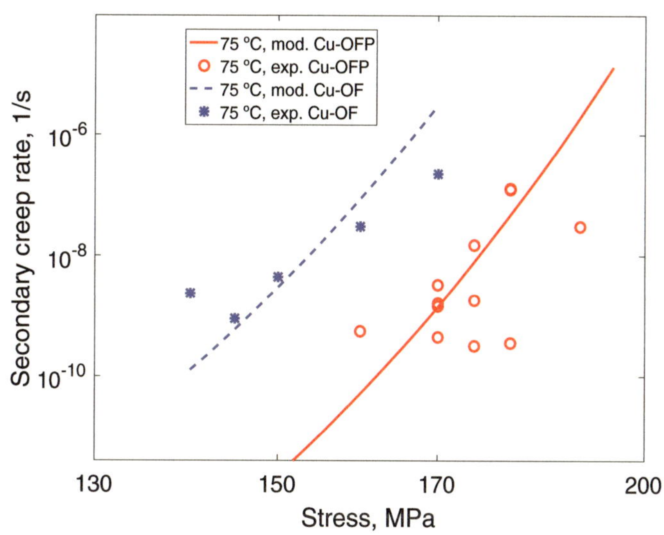

Figure 7. Secondary creep rate versus stress for oxygen-free copper with and without phosphorus at 75°C. Eq. (24) with the stress contribution to the internal stress from Eq. (30) and the increase in activation energy from Eq. (28) is compared to experimental data. After [28].

5. Creep strain curves

In the primary stage, the creep rate is in most cases larger than in the secondary stage. This is assumed to be due to a lower dislocation density and thereby a lower dislocation stress. The additional driving stress during primary creep becomes [34]

$$\sigma_{\text{primadd}} = \sigma_{\text{dislstat}} - \sigma_{\text{disl}} \tag{31}$$

where σ_{dislstat} is the dislocation density during stationary conditions, which is given by the difference between the applied stress σ and the internal stress σ_i:

$$\sigma_{\text{dislstat}} = \sigma - \sigma_i \tag{32}$$

The total stress during primary creep is given by

$$\sigma_{\text{prim}} = \sigma + \sigma_{\text{primadd}} = 2\sigma - \sigma_{\text{disl}} - \sigma_i \tag{33}$$

where Eqs. (31) and (32) have been inserted. Eq. (33) gives the stress that should be used in the expression for the secondary creep rate (Eq. (24)) to obtain the general expression for the creep rate:

$$\dot{\varepsilon}(\sigma) = \dot{\varepsilon}_{\text{sec}}(2\sigma - \sigma_{disl} - \sigma_i) \tag{34}$$

When the secondary stage is reached, $\sigma = \sigma_i + \sigma_{\text{disl}}$ and $\dot{\varepsilon}(\sigma)$ are equal to the secondary creep rate, according to Eq. (24) as it should. When σ_{disl} is smaller than its stationary value, the creep rate is higher, which is characteristic for primary creep. The use of Eq. (34) is illustrated in **Figure 8**. In **Figure 8a**, a creep strain versus time curve is shown for Cu-OFP. The creep rate versus time for the same case is given in **Figure 8b**. In the double logarithmic diagram, a straight line is obtained, which is in close agreement with the experimental data. This type of relation that is referred to as the φ model is quite frequently observed. It has several different

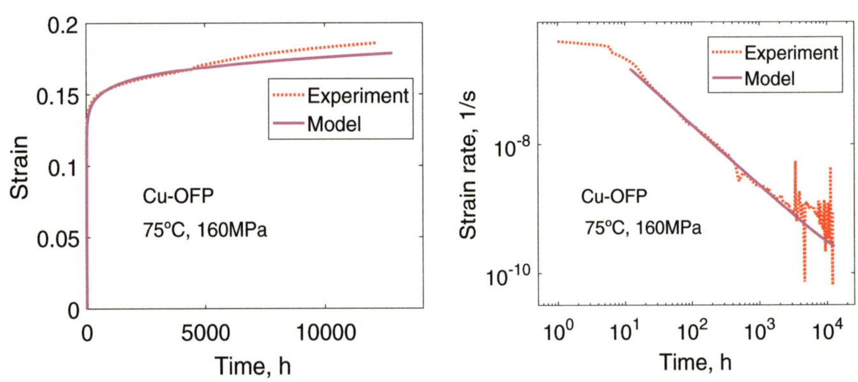

Figure 8. Creep strain results according to Eq. (34) are compared to experimental data for Cu-OFP at 75°C and 160 MPa: (a) creep strain versus time and (b) creep rate versus time. After [34]. The cusps on this curve are due to the necessity of reloading the specimen after a certain strain.

names [9]. It is most well known for 9% Cr steels, see for example [35]. It is evident that Eq. (34) can describe primary creep quite well.

6. Creep recovery theory

At ambient temperatures, for example, in steels, a gradually decreasing creep rate is observed at constant load until the deformation in practice stops completely. This is known as logarithmic creep. In constant strain rate tests, the deformation increases as long as the load is increasing. This behaviour takes work hardening and dynamic recovery into account but not static recovery. The derivation of the expression for the dynamic recovery constant is based on pure glide [36]. It has been suggested many times in the literature that dynamic recovery is controlled by cross slip, see for example [15, 37]. Qualitatively, this might seem logically. Cross slip allows the screw dislocations to annihilate each other, leading to the partial recovery that is characteristic for deformation processes at ambient temperatures in steels. The problem is that the available models for cross slip that were summarised in Section 3.3 give temperature dependencies that are orders of magnitude larger than the observed ones. Unfortunately, we must conclude that we do not understand the role of cross slip at present. In fact, we can model observed dynamic recovery considering just glide.

Most measured creep curves are characterised by a primary, a secondary and a tertiary stage. Work hardening and recovery are continuously taking place. According to the creep recovery theory, there is balance between work hardening and recovery in the secondary stage. Otherwise, there would be either a raise or a decrease in the dislocation density, giving a reduction or increase in the creep rate, respectively. Consequently, the dislocation density must remain approximately unchanged in the secondary stage.

At high temperatures ($T > 0.5\ T_m$), there is consensus that static recovery is controlled by climb in so-called class II alloys, which include most creep-exposed alloys of technical interest. This mechanism allows dislocations of opposite sign that attract each other to move towards each other and finally annihilate. On the other hand, in class I alloys, glide has been assumed to be controlling in a certain stress range with a creep exponent of 3. The most well-known type of alloy in this class is AlMg. The assumption is problematic for several reasons. (i) The reason for glide control is considered to be strong solid solution hardening. However, the effect of solid solution hardening is almost equally strong for climb [28, 32]. In addition, the glide rate is always faster than the climb rates according to Eqs. (12) and (16). (ii) At the upper end of the stress range with a creep exponent of 3, break-away of solutes from the dislocations is assumed to take place. The models for this effect suggest a much higher stress than the one observed. (iii) The most problematic issue is that the static recovery must be based on glide. From observations at ambient temperatures, it is unlikely to be possible. An alternative approach was presented in Section 4.2 fully based on a climb model. It was demonstrated that all the mentioned issues could be solved. In addition, a good fit to the data was found.

At or close to ambient temperatures, several metals give creep curves that have the same general appearance as at high temperatures ($T > 0.5\ T_m$) with primary, secondary and tertiary creep. For example, this is the case for Cu-OFP, see **Figure 8**. For a long time, it was assumed that creep was controlled by glide and cross slip at low temperatures [19], because the estimated climb rate was much too low to be of importance. However, this is not considered to be the case anymore. When taking the enhanced vacancy concentration into account, the observed creep rates for copper and aluminium can be fully explained by climb. This is evident from **Figure 7**. In addition, it clarifies why static recovery can take place at low temperatures, which is very difficult assuming the presence of only glide and cross slip.

7. Conclusions

- To model creep of alloys, the development of the dislocation density must be known. In recent years, a basic model for the dislocation density has been formulated that fulfils this requirement. Together with models for solid solution and particle hardening, the creep behaviour of many alloys can be described without the use of adjustable parameters.
- A new expression for the dislocation climb mobility has been derived. It extends the well-established formula of Hirth and Lothe to lower temperatures by taking the enhanced concentration of vacancies due to plastic deformation into account. The new expression can explain observed creep rates down to near-ambient temperatures.
- Assuming that glide of dislocations is controlled by the climb rate of their jogs, an expression for the dislocation glide mobility is formulated. It turns out that the glide rate is always higher than the climb rate. This suggests that climb is rate controlling in dislocation creep. In the chapter, it is illustrated for AlMg alloys that this might apply to class I alloys as well.
- It is suggested in the literature that cross slip is the controlling mechanism for dynamic recovery. An expression for the cross slip mobility is set up based on published models for the activation energy of cross slip. However, this expression cannot explain the observed rate of dynamic recovery. This is still only possible by assuming glide control.
- It is demonstrated that the dislocation model can describe a range of properties without the use of adjustable parameters. These properties include secondary creep and stress-strain curves.

Author details

Rolf Sandström

Address all correspondence to: rsand@kth.se

Materials Science and Engineering, KTH Royal Institute of Technology, Stockholm, Sweden

References

[1] Evans RW, Wilshire B. Creep of Metals and Alloys. Institute of Metals, London; 1985

[2] Penny RK, Marriott DL. Design for Creep. Chapman & Hall, London; 1995

[3] Abe F, Kern T-U, Viswanathan Eds R. Creep-Resistant Steels. Woodhead, Cambridge; 2008

[4] Kassner ME. Fundamentals of Creep in Metals and Alloys. Butterworth-Heinemann, Oxford; 2015

[5] Wilshire B. Observations, theories, and predictions of high-temperature creep behavior. Metallurgical and Materials Transactions A: Physical Metallurgy and Materials Science. 2002;**33**:241-248

[6] Lagneborg R. Dislocation mechanisms in creep. International Metallurgical Reviews. 1972;**17**:130-146

[7] Sandström R, Hallgren J. The role of creep in stress strain curves for copper. Journal of Nuclear Materials. 2012;**422**:51-57

[8] Sandström R. The role of microstructure in the prediction of creep rupture of austenitic stainless steels. In: ECCC Creep & Fracture Conference; Düsseldorf; 2017

[9] Sandstrom R. Basic model for primary and secondary creep in copper. Acta Materialia. 2012;**60**:314-322

[10] Kocks UF. Laws for work-hardening and low-temperature creep. Journal of Engineering Materials and Technology, Transactions of the ASME. 1976;**98**:76-85

[11] Hallén H. A theory of dynamic recovery in F.C.C. metals. Materials Science and Engineering. 1985;**72**:119-123

[12] Hirth JP, Lothe J. Theory of Dislocations. Malabar, Florida: Krieger; 1982

[13] Mecking H, Estrin Y. The effect of vacancy generation on plastic deformation. Scripta Metallurgica. 1980;**14**:815-819

[14] Edington JW. The influence of strain rate on the mechanical properties and dislocation substructure in deformed copper single crystals. Philosophical Magazine. 1969;**19**:1189-1206

[15] Püschl W. Models for dislocation cross-slip in close-packed crystal structures: A critical review. Progress in Materials Science. 2002;**47**:415-461

[16] Wu X-Z, Wang R, Wang S-F, Wei Q-Y. Ab initio calculations of generalized-stacking-fault energy surfaces and surface energies for FCC metals. Applied Surface Science. 2010;**256**: 6345-6349

[17] Du J-P, Wang C-Y, Yu T. Cross-slip process in model Ni(Al) solid solution: An embedded-atom method study. Computational Materials Science. 2014;**91**:192-199

[18] Nöhring WG, Curtin WA. Dislocation cross-slip in fcc solid solution alloys. Acta Materialia. 2017;**128**:135-148

[19] Sandstrom R, Andersson HCM. Creep in phosphorus alloyed copper during power-law breakdown. Journal of Nuclear Materials. 2008;**372**:76-88

[20] Sandström R. Fundamental models for creep properties of steels and copper. Transactions of the Indian Institute of Metals. 2016;**69**:197-202

[21] Kocks UF, Argon AS, Ashby MF. Thermodynamics and kinetics of slip. Progress in Materials Science. 1975;**19**:1

[22] Nes E, Marthinsen K. Modeling the evolution in microstructure and properties during plastic deformation of f.c.c.-metals and alloys—An approach towards a unified model. Materials Science and Engineering: A. 2002;**322**:176-193

[23] Chandler HD. Effect of unloading time on interrupted creep in copper. Acta Metallurgica et Materialia. 1994;**42**:2083-2087

[24] Mecking H, Styczynski A, Estrin Y. Steady state and transient plastic flow of aluminium and aluminium alloys. In: . Strength of Metals and Alloys (ICSMA 8). Oxford: Pergamon; 1989. p. 989-994

[25] Vujic S, Sandstrom R, Sommitsch C. Precipitation evolution and creep strength modelling of 25Cr20NiNbN austenitic steel. Materials at High Temperatures. 2015;**32**:607-618

[26] Shin I, Carter EA. Possible origin of the discrepancy in Peierls stresses of fcc metals: First-principles simulations of dislocation mobility in aluminum. Physical Review B: Condensed Matter and Materials Physics. 2013;**88**:064106:1-10

[27] Servi IS, Grant NJ. Creep and stress rupture behaviour of aluminium as a function of purity. Transactions of AIME. 1951;**191**:909-916

[28] Korzhavyi PA, Sandström R. First-principles evaluation of the effect of alloying elements on the lattice parameter of a 23Cr25NiWCuCo austenitic stainless steel to model solid solution hardening contribution to the creep strength. Materials Science and Engineering A. 2015;**626**:213-219

[29] Oikawa H, Honda K, Ito S. Experimental study on the stress range of class I behaviour in the creep of Al-Mg alloys. Materials Science and Engineering. 1984;**64**:237-245

[30] Sato H, Maruyama K, Oikawa H. Effects of the third element on creep behavior of Al-Mg and αFe-Be solid solution alloys. Materials Science and Engineering A. 1997;**234–236**: 1067-1070

[31] Horita Z, Langdon TG. High temperature creep of Al-Mg alloys. In: . Strength of Metals and Alloys (ICSMA 7). Pergamon; 1985. p. 797-802

[32] Sandstrom R, Andersson HCM. The effect of phosphorus on creep in copper. Journal of Nuclear Materials. 2008;**372**:66-75

[33] Sandström R. Influence of phosphorus on the tensile stress strain curves in copper. Journal of Nuclear Materials. 2016;**470**:290-296

[34] Sandström R. The role of cell structure during creep of cold worked copper. Materials Science and Engineering: A. 2016;**674**:318-327

[35] Abe F. Analysis of creep rates of tempered martensitic 9%Cr steel based on microstructure evolution. Materials Science and Engineering A. 2009;**510-511**:64-69

[36] Roters F, Raabe D, Gottstein G. Work hardening in heterogeneous alloys—A microstructural approach based on three internal state variables. Acta Materialia. 2000;**48**:4181-4189

[37] Jackson PJ. Dislocation modelling of shear in f.c.c. crystals. Progress in Materials Science. 1985;**29**:139-175